Oligopoly Dynamics

Springer
Berlin
Heidelberg
New York
Barcelona
Hong Kong
London
Milan
Paris
Tokyo

Tönu Puu · Irina Sushko
Editors

Oligopoly Dynamics

Models and Tools

With 98 Figures

Springer

Professor Dr. Tönu Puu
Umeå University
Centre for Regional Science
90187 Umeå, Sweden

Dr. Irina Sushko
National Academy of Sciences
Institute of Mathematics
3 Tereshchenkivska
01601 Kiev – 4, Ukraine

ISBN 3-540-43186-1 Springer-Verlag Berlin Heidelberg New York

Library of Congress Cataloging-in-Publication Data applied for
Die Deutsche Bibliothek – CIP-Einheitsaufnahme
Oligopoly Dynamics: Models and Tools / Tönu PUU; Irina Sushko Ed. –
Berlin; Heidelberg; New York; Barcelona; Hong Kong; London; Milan; Paris; Tokyo:
Springer, 2002
ISBN 3-540-43186-1

Springer-Verlag Berlin Heidelberg New York
a member of BertelsmannSpringer Science + Business Media GmbH

http://www.springer.de

Printed in Germany

Cover design: Erich Kirchner, Heidelberg

SPIN 10865915 42/2202-5 4 3 2 1 0 – Printed on acid-free paper

Preface

These proceedings are from a conference held at the *Centre for Regional Science* (CERUM) at Umeå University, Sweden, 17-18 June 2001. Unlike many conference proceedings, this volume contains only invited contributions on specified topics so as to make the book coherent and self-contained. The authors and editors hope that this coherence will make the volume useful also as a text for courses in industrial organisation. To this end two chapters on the history of oligopoly theory, from the beginnings with Cournot 1838, to the present day, and one chapter on modern methods for analysing iterated discrete time maps, have been inserted at the beginning of the book.

Unlike most current literature on games and oligopoly, this book is not focused on the usual topics of game theory: optimal strategies, dominance, and equilibrium. Rather it is the evolutionary dynamics, often of a complex type, including deterministic chaos, which are in focus. The contributions, after the historical and the methodological introductions, represent various segments of the research frontier in this area, though pains have been taken to tie some of the models to a number of most promising contributions from the frugal period 1929-1941, which have suffered from unjust neglect in the following industrial organisation literature.

As the reader easily realises, most of the contributions are multi-authored, with the authors, mathematicians and economists, appearing in different combinations, so this volume represents truly collaborative and interdisciplinary work. Indeed, such collaboration is necessity in order to cope with the dynamics of oligopoly models.

Current textbooks in industrial organisation use to display some of the classical models, by for instance Cournot and Hotelling, as they appeared very long ago, as historical relicts, almost without any further development, as if such development were not possible or interesting. For this reason we feel a fresh start is needed, especially focusing the recursive dynamics, which attracts little interest in current game theory.

In past years the study of complex dynamical systems has relied on computer simulations and on the local study of stability loss for Nash equilibria, whereas relatively little space has been devoted to global bifurcation studies. The present study is a bit different in this respect. The particular model studies have focus on global behaviour, using recently accumulating new methods. These methods for noninvertible maps in terms of critical lines and absorbing areas are described at some length in the introductory methodological chapter.

The editors wish to thank several institutions for their help with this project, first of all CERUM at Umeå University, in particular its director Professor Lars Westin, for organizing the workshop, as well as its sponsors: The *Bank of Sweden Tercentenary Foundation*, and the *Gösta Skoglund Foundation*.

We are also grateful to CERUM for their hospitality over a prolonged period, during which the editors were able to work together, the *Swedish Institute*, who financed the stay of one of us (Sushko), and the *Communications and Transport Research Board*, who financed the research of the other (Puu) over several years.

We would further like to thank the Econometrics Group at the *University of Southern Denmark* at Odense and its director Professor Gustav Kristensen, as well as the Department of Economics at the *University of Urbino*, Italy, for their hospitality in various phases of the collaboration. The last thanks are in particular extended to Professor Laura Gardini personally and to her family, as work intruded their home while the department was on the move.

Finally, we are most grateful to the Department of Geoinformatics of the *University of Economics and Business Administration* in Wien, Austria, and its director, Professor Manfred M. Fischer, for their hospitality during the final editing of this volume.

Wien, 30 November 2001

Tönu Puu Irina Sushko

Contents

HISTORY

1 A Century of Oligopoly Theory 1838 - 1941 1
Tönu Puu

2 The Development of Complex Oligopoly Dynamics Theory 15
J. Barkley Rosser, Jr

METHODS

3 Some Methods for the Global Analysis of Dynamic Games Represented by Iterated Noninvertible Maps 31
Anna Agliari, Gian-Italo Bischi, and Laura Gardini

COURNOT MODELS

4 The Role of Competition, Expectations and Harvesting Costs in Commercial Fishing 85
Gian-Italo Bischi and Michael Kopel

5 Cournot Duopoly with Kinked Demand According to Palander and Wald 111
Tönu Puu, Laura Gardini, and Irina Sushko

6 Duopoly with Piecewise Linear Discontinuous Reaction Functions 147
Irina Sushko, Laura Gardini, and Tönu Puu

7 A Cournot Duopoly with Bounded Inverse Demand Function 171
Anna Agliari and Tönu Puu

8 Chaos Synchronization and Intermittency in a Duopoly Game with Spillover Effects 195
Gian-Italo Bischi and Fabio Lamantia

9 Intrinsic Stability, Multi Stability and Instability 219
Shahriar Yousefi

10 The Interaction of Uncertainty and Information Lags in the Cournot Oligopoly Model 233
Carl Chiarella, Ferenc Szidarovszky, and Peiyan Zhu

HOTELLING MODELS

11 Hotelling Type Duopoly and Oligopoly 265
Tönu Puu and Laura Gardini

Contributors 311

1 A Century of Oligopoly Theory 1838-1941

Tönu Puu[1]

[1] Centre for Regional Science, University of Umeå, Sweden

1 The Beginnings: Cournot

Elementary economics textbooks recognise two opposite market forms: competition and monopoly. In the competitive case the firms are very numerous and thus small in relation to the total size of the market. In consequence they consider market price as being fixed independently of any action they can take on their own with regard to the variation range of their supply.

In the case of monopoly one single firm dominates the whole market. Its supply influences market price appreciably, and it takes advantage of this to increase its profits by deliberately limiting the supply, thus establishing a monopoly price. It can also apply various discriminatory policies provided it can segregate segments of consumers' demand and charge different prices from different categories of consumers.

Duopoly and oligopoly, though contextually the first step from monopoly towards perfect competition, is analytically not a case of intermediate complexity, but more complicated than any of the extremes. This is so as the duopolists have to take in consideration, not only the behaviour of all the consumers in terms of the entire market demand curve for the commodity. The duopolist must also take account of the behaviour of the competitor, inclusive possible retaliation on his own actions.

There is no place for doubt about the origins of duopoly and oligopoly theory. Nobody would deny that Augustin Cournot in 1838 in his *Recherces sur les Principes Mathématiques de la Théorie des Richesses* both formulated the problem, and presented a fairly extensive analysis of it. Further, it was one of the earliest economics treatises to be formulated as a mathematical model. Cournot was a professional of mathematics, but we should not exaggerate our astonishment about this fact, because excessive specialisation in science is a phenomenon of the 20th Century. After all, those days it was common for students to study all the disciplines of a faculty for a degree.

Cournot stated the problem as follows: Consider two firms selling a homogenous product in quantities q_1, q_2. Their total sales will be $q = q_1 + q_2$. Assume a demand function $q = D(p)$, with its inverse: $p = f(q) = D^{-1}(q)$. The profits of the two firms, or rather revenues, as Cournot disregarded production costs, are:

$$\Pi_1 = f(q_1 + q_2)q_1 \tag{1}$$

$$\Pi_2 = f(q_1 + q_2)q_2 \tag{2}$$

If there were production costs, say $g_1(q_1), g_2(q_2)$, we would just subtract the costs from the respective revenue functions, and this would make little difference for the analysis. This fact was realised by Cournot, so he suggested the production of mineral water direct from the sources as his case, obviously disregarding bottling costs.

The first firm would put $\partial\Pi_1 / \partial q_1 = 0$, the second $\partial\Pi_2 / \partial q_2 = 0$, which gives two equations:

$$f(q_1 + q_2) + f'(q_1 + q_2)q_1 = 0 \tag{3}$$

$$f(q_1 + q_2) + f'(q_1 + q_2)q_2 = 0 \tag{4}$$

These are implicit equations between the two variables, the supply quantities. Suppose we solve them explicitly, the first for q_1, the second for q_2, something which often is impossible in practice:

$$q_1 = \phi_1(q_2) \tag{5}$$

$$q_2 = \phi_2(q_1) \tag{6}$$

Of course, even if we can get the explicit solutions, from (3)-(4), we still have to make sure that they are indeed maxima and not minima, i.e. fulfil the second order conditions $2f' + f''q_1 < 0$ and $2f' + f''q_2 < 0$. Note that, even if we get closed form solutions, (5)-(6) they may still be multiple valued.

The equations (5)-(6) are called the reaction functions. They can be regarded in two different ways. We can regard them as a simultaneous system, and consider the Cournot equilibrium point $\overline{q}_1, \overline{q}_2$ which satisfies (5)-(6). If we see (5) and (6) as curves, this Cournot equilibrium point is the intersection point of the curves.

There are obvious questions pertinent to the equilibrium: *i*) does it exist, and *ii*) is it unique? We can by no means take an affirmative answer for granted. As we will see below, Palander (1936, 1939) and Wald (1936) actually constructed examples with no, or several coexistent Cournot equilibria.

Another way to see the reaction functions is to consider them in a dynamic process, as

$$q_1' = \phi_1(q_2) \tag{7}$$

$$q_2' = \phi_2(q_1) \tag{8}$$

where the dash means advancing the map one period, from period *t* to period (*t*+1). There was a lot of discussion about different adjustment schemes, "simultaneous" or "alternative" adjustments, but the significance of the differences was exaggerated which we see if we note that the composite maps, advanced two periods from *t* to (*t*+2), $q_1'' = \phi_1 \circ \phi_2(q_1)$ and $q_2'' = \phi_2 \circ \phi_1(q_2)$ are independent maps producing the same sequences of events no matter whether the competitors react simultaneously or take turns.

In the context of the reaction functions as a dynamical system a new question arises: *iii*) is the Cournot equilibrium stable?

Cournot actually regarded the reaction functions both ways, in terms of simultaneous equilibrium, and in terms of a recursive dynamical system. He left, of course, many loose ends and obvious exposures for objections, concerning the optimization variable, the homogeneity of the commodity, and the implied strategy in the adjustment process. In fact, one of the great merits of Cournot's work was keeping scientists busy for more than 150 years.

2 The First Critics: Bertrand

There has been some recent discussion in the journal *History of Political Economy* about whether there were any comments at all to Cournot's work over the very long period of 45 years, before Joseph Bertrand 1883 in *Jour-*

nal des Savants launched some extremely critical comments on Cournot's work.

It is claimed, by de Bornier (1992) and Dimand (1995), that there has been a widespread misunderstanding that there was a complete silence about Cournot's work during these 45 intervening years, and implicitly Schumpeter is blamed for his comments in his monumental *History of Economic Analysis*.

In fact Schumpeter says that, though "*Bertrand was ... the first to make an attack ... it was so inadequate...*" that it is in doubt whether it "*would have made much impression if ... others had not ... repudiated Cournot's solution*". Schumpeter also says that the criticism was "*eagerly seized upon, as an authoritative condemnation*" (of the prospects of mathematical economics) "*by people who understood neither mathematics nor economics*".

The counterexamples produced in the recent debate, however, are of the normal polite review character, some rather enthusiastic, and it seems that Schumpeter never said anything about these, only about Bertrand's aggressive criticism and about its inadequacy.

The point of controversy was the following: Cournot assumed the supply *quantities* of some homogeneous good to be the choice variables for the competitors, and put up the conditions for equilibrium, as well as the fundamentals of the dynamics in terms of the reaction functions, representing the optimal reactions with respect to supply. According to the criticism, given the good was homogeneous, one firm could, by undercutting the competitor's price, however slightly, always recover the entire market as its share. In the end this warfare would lead to collusive monopoly, or to competitive pricing at the level of marginal costs.

Unlike Cournot, Bertrand has not left any imprint upon economics, except his first critical argument which had to be elaborated by others. Like Cournot he was a mathematician, apparently appreciated by his contemporaries, as he was a professor at the prestigious École Polytéchnique (to which Évariste Galois was not even admitted as a student), and a member of academies such as Collège de France. Bertrand does not seem to have left any imprint on mathematics either. Therefore it might be doing him too much honour to continue to call the type of oligopoly where prices replace quantities "Bertrand oligopoly", but this is done by convention and habit.

The argument was elaborated in particular by Edgeworth (1897), whose criticism, unlike Bertrand's, according to Schumpeter, was not open to accusations of misinterpretation. Out of this argument the idea of product differentiation was born. Buyers would regard the commodity as heterogeneous, they would have preference for the product of one seller, and so only gradually desert their favourite brand when price differences became too pronounced. The now classical solution is due to Chamberlin (1932).

Another brand of development was due to Hotelling (1929) who assumed that, though the commodity itself was homogeneous, the competitors could establish bounded local monopolies, guarded by transportation costs. This ingenious contribution dealt with competitive price duopoly, just as in the general literature on monopolistic competition, and in addition some intriguing location paradoxes, such as clustering in space. There has been a tendency in subsequent literature to single out these location paradoxes, "two ice cream sellers on a beach" in vulgar text book terms, whereas the oligopoly aspects were left for the general discussions of product differentiation with no explicit reference to space.

3 On Strategy: von Stackelberg and Frisch

Another problematic matter with the Cournot model was the rather simplistic assumptions concerning the formation of expectations and of strategy including retaliation. The dynamic process implied that each firm observed the latest decisions taken by the competitor and assumed this as a basis for its own optimal decision in the next run. Of course, if the system actually was in a stable Cournot equilibrium, then there would not be much to object. However, if not, then both firms might in fact observe a more or less irregular cyclic change in market price and the competitor's supply, and then an assumption that the competitor would not revise the previous decision would look fairly untenable.

Two important attempts were addressed to this issue. One is due to Heinrich von Stackelberg (1934, 1938). It is suggested that the first firm may learn the reaction function of the second firm, and *vice versa*. We can substitute from (6) into (1) and from (5) into (2), thus obtaining:

$$\Pi_1 = f(q_1 + \phi_2(q_1))q_1 \tag{9}$$

$$\Pi_2 = f(\phi_1(q_2) + q_2)q_2 \tag{10}$$

Unlike the case of (1)-(2), (9)-(10) are functions of only one variable each. We can hence use total derivatives, putting $d\Pi_1 / dq_1 = 0$ and $d\Pi_2 / dq_2 = 0$, which yields:

$$f(q_1+\phi_2(q_1))+f'(q_1+\phi_2(q_1))q_1+\phi_2'(q_1)f'(q_1+\phi_2(q_1))q_1=0 \qquad (11)$$

$$f(\phi_1(q_2)+q_2)+f'(\phi_1(q_2)+q_2)q_2+\phi_1'(q_2)f'(\phi_1(q_2)+q_2)q_2=0 \qquad (12)$$

The derivatives of the reaction functions can be easily evaluated by implicit differentiation of (5)-(6), so

$$\phi_1'(q_2)=-\frac{f'(\phi_1(q_2)+q_2)+f''(\phi_1(q_2)+q_2)}{2f'(\phi_1(q_2)+q_2)+f''(\phi_1(q_2)+q_2)} \qquad (13)$$

$$\phi_2'(q_1)=-\frac{f'(q_1+\phi_2(q_1))+f''(q_1+\phi_2(q_1))}{2f'(q_1+\phi_2(q_1))+f''(q_1+\phi_2(q_1))} \qquad (14)$$

Note that equations (11)-(12) are only slightly more complicated than (3)-(4), as they have the additional terms involving the derivatives of the reaction functions (13)-(14).

The important issue, however, is that (11)-(12) are functions of one variable alone, and can, in principle, be solved each separately. So suppose we write the solutions:

$$q_1=\hat{q}_1 \qquad (15)$$

$$q_2=\hat{q}_2 \qquad (16)$$

These can be compared to (5)-(6). While von Stackelberg calls adhering to (5) or (6) followership, he calls adhering to (15)-(16) leadership. The leader takes the behaviour of the competitor in explicit consideration while calculating maximum profit.

We now get four strategic types: *i)* $q_1=\phi_1(\bar{q}_2)$, $q_2=\phi_2(\bar{q}_1)$, which is the Cournot case, *ii)* $q_1=\hat{q}_1$, $q_2=\phi_2(\hat{q}_1)$, where the first firm is a leader, the second a follower, *iii)* $q_1=\phi_1(\hat{q}_2)$, $q_2=\hat{q}_2$, which is the same with the roles reversed, and *iv)* $q_1=\hat{q}_1$, $q_2=\hat{q}_2$, both competitors wanting to be leaders.

All the three first cases are compatible, each of the firms does exactly what the competitor expects. The fourth is not, both expect the other to behave as a follower, but none of them does.

Ragnar Frisch (1933) proposed a very similar analysis, and we need not even set up the formulas anew. Only, (5)-(6) $q_1 = \phi_1(q_2)$ and $q_2 = \phi_2(q_1)$ are now not necessarily the actual reaction functions, they represent the retaliations each firm *expects* from the competitor, true or not. Otherwise, the analysis follows exactly as in the von Stackelberg case.

It is a bit surprising that not much progress has been made along these lines, though Frisch's contribution is mentioned by the Swedish Nobel Prize committee as the particular reason why he was given the prestigious award.

The next step would naturally have tried to make more explicit how expectations are formed about the behaviour of competitors. We can only report some attempts made by Schneider (1932, 1942), which look rather simplistic, with expectations formed as averages with constant weights from the previous decisions of both competitors, and later of, course, game theory.

4 Multiplicity and Nonexistence of Equilibria: Palander and Wald

In a short printed abstract of contributions to a Cowles Commission conference at Colorado Springs in 1936, and an extensive follow up, in Swedish, dating from 1939, Tord Palander focused some interesting dynamics problems in Cournot duopoly when the demand curve was kinked linear and the marginal revenue curve hence jumped up, producing two different local profit maximising intersections with the curve of marginal cost, which was assumed constant or even zero. The Cournot reaction functions, as well, then became piecewise linear, including a jump, and they could produce several, coexistent and locally stable equilibria. Assume the inverse demand function:

$$p = f(q_1+q_2) = \begin{cases} \alpha_1 - \beta_1(q_1+q_2) & q_1+q_2 \le \overline{q} \\ \alpha_2 - \beta_2(q_1+q_2) & q_1+q_2 > \overline{q} \end{cases} \tag{17}$$

where

$$\overline{q} = \frac{\alpha_1 - \alpha_2}{\beta_1 - \beta_2} \tag{18}$$

Note that to have the right kind of kink we must have $\alpha_1 > \alpha_2, \quad \beta_1 > \beta_2$.

The "kinked demand case" due to Sweezy (1939) for duopoly is something quite different. There demand was assumed to become less, not more, elastic as price went down. The explanation for this was not in terms of the behaviour of the consumers, but of the competitors, who were supposed to retaliate to increased demand, aimed at bringing price down, but not to reduced demand to bring about the reverse. This type of model was used to the purpose of explaining an allegedly observed extreme stability in duopoly pricing, and so had a strong flavour of ad hoc explanation. In this latter case the jump in marginal revenue was down, not up, but this is not the issue Palander dealt with, which no doubt is the more interesting.

The "kink" in the present case arises from the fact that any commodity at a very high price becomes an article of luxury, consumed only by a small group of wealthy consumers. As price goes down a wider group of consumers can afford it, and the demand elasticity increases considerably. The argument for this was given by Joan Robinson in her *Economics of Imperfect Competition* 1933, and subsequently such cases where these facts were simplified to a pair of straight lines with different slopes have for decades been used in microeconomics textbooks to analyse price discrimination.

Palander gave his argument in terms of two numerical cases, one with identical firms, another with a slight asymmetry between them. In the first case the reaction functions had two coexistent intersections, i.e. two Cournot points. Palander specified the basins for these, though nor completely, and also the basin for initial conditions from which the system in stead went to an attractive 2-periodic oscillation.

For another case, where the reaction functions did not intersect at all, and there hence did not exist any Cournot equilibrium, Palander recognised the existence of an attractive 3-periodic cycle. Given the numerical tools at that time, the accuracy of his calculations is in fact amazing, though he missed the coexistence of a 6-period cycle and again the complete characterisation of the basins.

It is noteworthy that, also in 1936, Abraham Wald considered the same type of problems, in an article which later became celebrated as the first rigorous statement of existence problems for multi-market equilibria, left open by Wálras and later elaborated by Arrow and Debreu.

Like Palander, Wald gave his argument in terms of two numerical examples, and also assumed the demand curve to start and end with straight line segments, but, unlike Palander, he did not assume the curves to meet under a sharp angle. Rather the two line segments were smoothly joined by either a circular segment, or by a demand curve where price was reciprocal to supply squared. The lines and curves were hence joined at tangency, so the demand function was not only continuous but differentiable. The first example, with a circular segment, seems miscalculated, but the second, hyperbola like, is most interesting. It produced reaction functions each composed by two falling and one rising segment. As both rising segments coincided, the case resulted in a nondenumerable infinity of coexistent Cournot equilibria.

Coexistence of equilibria, or oscillations, are the most we can get out of the cases of linear demand functions, at least when production costs are zero or constant as assumed by Palander and Wald. This is due to the fact that all falling pieces of any reaction function, derived from a linear demand function, have a constant slope of -0.5. Accordingly, all Cournot equilibria become locally stable, and this rules out more spectacular dynamics phenomena such as chaos.

5 Chamberlin and Hotelling's Location Paradox

We already mentioned the two final solutions to Bertrand's objection: Edward Chamberlin's, in *The Theory of Monopolistic Competition* of 1933, which just replaced quantity by price in Cournot's original model, and assumed the consumers to have preferences for certain goods even if they were close substitutes. Demand for any one good would depend on its own price and on the price charged by the competitor supplying some close substitute, say: $q_1 = D_1(p_1, p_2)$, $q_2 = D_2(p_1, p_2)$. Profits (revenues) were then quite trivially:

$$\Pi_1 = p_1 D_1(p_1, p_2) \tag{19}$$

$$\Pi_2 = p_2 D_2(p_1, p_2) \tag{20}$$

and we could just repeat every step quite as above in the discussion of the Cournot model, or the von Stackelberg or Frisch variants of it, just replacing quantity by price in the argument.

This is all very reasonable, as reasonable as it is non-challenging for further development, so it remained something discussed in all textbooks, but left more or less as it was. In textbooks it is this setup which is called Bertrand duopoly.

Much more intriguing was Hotelling's solution, because it was much more specific, and had certain implications, so, like the Cournot model, it has been both rejected and vindicated, if not for three half Centuries, for at least half a Century.

Harold Hotelling in his 1929 article *Stability in Competition*, printed in *The Economic Journal*, but, in fact a speech to the *American Mathematical Society* in 1928, considers the case where two vendors supply the same commodity. Like Cournot, Hotelling assumed the commodity to be perceived as perfectly homogenous by the consumers, but, the vendors being separated in geographical space, so that the consumers had to carry the costs of transportation in addition to the mill prices charged by the vendors, there arose separate monopoly areas belonging to each vendor. These were determined so that in each area the sum of the mill price charged by that vendor and the transport costs were lower than if the commodity had been bought from the competitor. At certain points the prices plus transportation costs from both suppliers broke even, and these become the boundary points.

Though there was monopoly, protected by transportation costs, within the market areas, there was also competition at the fringes, because, if one competitor lowered its price he would encroach upon the other's market area. It is obvious that Hotelling, quite as Bertrand and Edgeworth, preferred to make price, not quantity, core of the analysis. Prices determine the extents of the market areas, because the consumers buy the good where it (including transportation) is cheapest. It is surprising that Hotelling at the same time assumed demand to be *totally inelastic*. Each consumer, or rather the consumers living in one unit of market area, would buy one unit of the commodity, no matter what its prices was. In this case total demand became identical with the size of the market area.

It is obvious that Hotelling regarded this as a simplification in order to make the problem tractable. It is also obvious that Hotelling regarded the elastic demand case as the more reasonable, but confined only a short, though insightful, verbal discussion to it. The bulk of the article, including the whole formal analysis, concerns the case of inelastic demand. It is in fact doubtful whether the simplifying assumption did simplify anything at all. There is an inherent contradiction as to the role of prices, which has been the cause of considerable discussion.

Along with other simplifications, Hotelling also introduced one with respect to space. He considered a *one dimensional space*, exemplified by two vendors along an intercontinental railway, or in the single street of a Wild West movie type of American Town. In textbooks this has become two ice cream vendors on a beach. There is another important fact about space too: it is *bounded*. The case turns out completely different if the two vendors are not located on a fixed line segment, but in stead on a circle.

Anyhow, the setup becomes very clear with these simplifications. Each firm has one market endpoint fixed, the other one depends on the locations chosen, and on the prices charged by both competitors. The aim is obvious: to maximise profit, i.e. revenue, because, like Cournot, Hotelling disregarded production costs. Each firm had two parameters of choice: price and location. Location only affected the quantity sold, so an optimum choice of location would maximise sales, i.e. market area. It is here that the paradox comes. Given the location of the other vendor, each one would maximise market area by choosing a location next to the competitor (on the longer side of the total market). A process of relocation would eventually lead to both being located together in the midpoint of the whole fixed interval.

Hotelling was preoccupied with the welfare losses due to this, as much lower total transportation costs would accrue if the vendors located at the quartiles of the interval. Pricing in this setup, would imply either collusion, or competitive pricing at the level of marginal production cost. As we would be back at Bertrand's original objection, each firm could cut out the competitor and take the whole market by just an infinitesimally lower price. Once the vendors locate in the same point there are no transportation costs to grant differential prices for a homogenous good.

The next important step in the analysis was due to Lerner and Singer (1937). They pointed out the factual absurdities of inelastic demand, and proposed a case where each consumer (again rather market area unit) would still buy one unit of the commodity independently of price, but only up to a certain maximum reservation price. If the actual price (including transportation of course) was higher, they would buy nothing at all. This in fact, was a demand function, a step function. Lerner and Singer presented a truly ingenious analysis of the Hotelling case with such a step function, just using one single type of graph, and also considered several interesting cases of dynamics.

Finally, Smithies (1941) suggested to replace the step function by a simple linear demand function. Smithies's essentially verbal argument was very accurate, but he considered a full formal analysis impossible. That was a pity, the case is by no means intractable, and it resolves most of the problems pertinent to the Hotelling model. With linear demand, integration over space makes the profit functions quadratic in location and cubic in prices, so nice inner location optima can actually be found for considerable parameter ranges. Maybe it was Smithies's fear of the integrals, maybe it was the d'Aspremont's *et. al.* note (1979), which once again pinpointed the formal problems in Hotelling's original setup, that was the cause for nobody trying hands formally on the Hotelling problem with elastic demand.

6 Conclusion

In 1944 John von Neumann and Oscar Morgenstern launched game theory. As a new paradigm it was tried on every topic, and completely took over oligopoly problems. The actions of the competitors were cast in terms of probabilistic strategies, and a new kind of equilibria were sought for. In the present context, where we focus recursive dynamics, this tradition is of little interest. As for traditional oligopoly, it is sufficient to check out a relatively late textbook, such as Friedman's of 1983, to see that not much happened in this field since 1941 which we chose as closing date for this chapter.

As we have seen the extended decade 1929-1941 was particularly fertile in terms of important contributions. It would be interesting to speculate about why this period, including the Wall Street crash, the great depression and the beginning of the Second World War, contributed so much to the area which now is designated by the neutral and almost disparaging term "industrial organisation".

References

d'Aspremont, C., Gabszewicz, J. J.and Thisse, J. F., 1979, "On Hotelling's 'stability in competition'", *Econometrica* 47:1145-1150

Beckmann, M. J., 1968, *Location Theory* (Random House, New York)

Beckmann, M. J., 1976, "Spatial Price Policies Revisited", *Bell Journal of Economics* 7:619-630

Beckmann, M. J. and Thisse, J. F., 1986, "The Location of Production Activities", *Handbook of Regional and Urban Economics* 1:21-95 (Elsevier Science)

Bertrand, J., 1883, "Théorie Mathématique de la Richesse Sociale", *Journal des Savants* 48:499-508

de Bornier, J. M., 1992, "The 'Cournot-Bertrand debate': a historical perspective, *History of Political Economy* 24:623-656

Chamberlin, E. H., 1932, *The Theory of Monopolistic Competition* (Harvard University Press)

Cournot, A., 1838, *Recherches sur les Principes Mathématiques de la Théorie des Richesses* (Hachette, Paris)

Dimand, R. W., 1995, "Cournot, Bertrand, and Cherriman", *History of Political Economy* 27:563-578

Edgeworth, F. Y., 1897, "La Teoria Pura del Monopolio", *Giornale degli Economisti* 15:13-31

Fellner, W., 1949, *Competition Among the Few - Oligopoly and Similar Market Structures* (Alfred A. Knopf, New York)

Friedman, J. 1968, "Reaction functions and the theory of duopoly", *Review of Economic Studies* 35:257-271

Friedman, J., 1983, *Oligopoly Theory* (Cambridge University Press)

Frisch, R., 1933, "Monopole - Polypole - La notion de force dans l'économie, *Nationaloekonomisk Tidsskrift* 71 Tillaegshaefte

Greenhut, M. L. and Ohta, H., 1972, "Monopoly Output under Alternative Spatial Pricing Techniques", *American Economic Review* 62:705-713

Hotelling, H., 1929, "Stability in Competition", *Economic Journal* 41: 41-57

Lerner, A. P. and Singer, H. W., 1937, "Some notes on duopoly and spatial competition, *Journal of Political Economy* :145-186

Leontief, W., 1936, "Stackelberg on monopolistic competition", *Journal of Political Economy*,

von Neumann, J. and Morgenstern, O., 1944, *Theory of Games and Economic Behaviour* (Princeton University Press)

Palander, T. F., 1935, *Beiträge zur Standortstheorie* (Almqvist & Wiksell, Uppsala)

Palander, T. F., 1936, "Instability in competition between two sellers", in Abstracts of Papers Presented at the Research Conference on Economics and Statistics held by the Cowles Commission at Colorado College, *Colorado College Publications*, General Seies No. 208, Studies Series No. 21.

Palander, T. F., 1939, "Konkurrens och marknadsjämvikt vid duopol och oligopol" ("Competition and market equilibrium in duopoly and oligopoly"), *The Scandinavian Journal of Economics* (then Ekonomisk Tidskrift) 41:124-145, and 222-250

Ponsard, C., 1954, *Économie et Espace - essai d'integration du facteur spatial dans l'analyse économique* (Sedes, Paris)

Ponsard, C. (Ed.), 1988, *Analyse économique spatiale* (Presses Universitaires de France)

Puu, T., 1991, Chaos in duopoly Pricing, *Chaos, Solitons, & Fractals* 1:573-581

Puu, T., 2001, "Bertrand oligopoly revisited", *Discrete Dynamics in Nature and Society* 6:1-10

Puu, T., 2001, "Hotelling's 'ice-cream dealers with elastic demand'", *Annals of Regional Science* (in print)

Rand, D., 1978, "Exotic phenomena in games and duopoly models, *Journal of Mathematical Economics* 5:173-184

Robinson, J., 1933, *The Economics of Imperfect Competition*, Macmillan

Schneider, E., 1932, *Reine Theorie monopolistischer Wirtschaftsformen* (Tübingen)

Schneider, E., 1942, "En dynamisk teori for prisdannelsesprocessen under tilbudsduopol", Nationaloekonomisk Tidsskrift 80:210-232

Smithies, A. and Salvage, L. I., 1940, "A Dynamic Theory of Duopoly, *Econometrica* 8:130-

Smithies, A., 1941, "Monopolistic Price Policy in a Spatial Market", *Econometrica* 9:63-73

Smithies, A., 1941, "Optimum Location in Spatial Competition", *Journal of Political Economy* 49:423-439

Sraffa, P., 1926, "The Laws of Returns Under Competitive Conditions", *The Economic Journal* 36:535-550

von Stackelberg, H., 1934, *Marktform und Gleichgewicht*, (Wien)

von Stackelberg, H., 1938, "Probleme der unvollkommenen Konkurrenz," *Weltwirtschaftliches Archiv* 48:95-114

Stiegler, 1940, G.J., "Notes on the theory of duopoly", *Journal of Political Economy*

Sweezy, P. M., 1939, "Demand under conditions of oligopoly", *Journal of Political Economy* 47:568-573

Triffin, R., 1940, Monopolistic Competition and General Equilibrium Theory, (Cambridge)

Wald, A., 1936, "Über einige Gleichungssysteme der mathematischen Ökonomie", *Zeitschrift für Nationalökonomie* (Dezember) 637-670

Zeuthen, F., 1930, *Problems of Monopoly and Economic Warfare*, London

2 The Development of Complex Oligopoly Dynamics Theory

J. Barkley Rosser, Jr. [1]

[1] Program in Economics, James Madison University; VA 22807 Harrisonburg, USA

1 Introduction

This chapter will review the development of the theory of complex oligopoly dynamics from the 1970s to the year 2001 in its main strands. It will also provide certain speculations regarding possible future developments. This will serve as a link between the first chapter's discussion of the broader history of oligoply theory up to the 1940s and the more specific chapters that present current models in the rest of this book. But, to tell our story we first need to remind ourselves of certain ideas from the first chapter of this book. One is the very founding document of oligopoly theory, Cournot's seminal work of 1838. This is both because the specific model that he presented has been much studied for its ability to generate complex dynamics and also because of its more general foreshadowing of game theory. It has often been noted that the Cournot equilibrium is but a special case of the Nash (1951) equilibrium, the more general formulation used by modern industrial organization economists in studying oligopoly theory. Indeed, it is sometimes even called the Cournot-Nash equilibrium. Although many of the models of complex oligopoly dynamics use the specific Cournot model, many use more general game theoretic formulations. We note simply as an aside here that Cournot's work was the first to apply calculus to solving an economic optimization problem and also was the first to introduce supply and demand curves, albeit in the "Walrasian" form with price on the horizontal axis.

The other is the idea of multiple oligoply equilibria. For competitive markets Walras and Marshall both noted the possibility of multiple equilibria, but von Mangoldt (1863) was apparently the first to observe this case. Joan Robinson (1933, pp. 57-58) first suggested this possibility for the pure monopoly case because of the possibility of upwardly sloping marginal revenue curves in a world of segmented markets. Wald (1936) also showed this possibility for the Cournot case when marginal revenue slopes upward.

Puu (1995) would eventually demonstrate for the Robinson multiple equilibria monopoly case that if the monopolist follows a Newtonian algorithm search procedure chaotic dynamics can result. Puu (2000, Chap. 6) further shows for this case a wide variety of other complex dynamics including coexisting attractors, and through the analysis of critical lines, the appearance of tongues and riddled basins of attraction. Earlier, Bonanno (1987) examined the Robinson type case in a catastrophe theoretic context with smoothly shifting demand curves and fixed cost curves.

A prescient case is that of Palander's (1939) study of the possible existence of a three-period cycle in a case of no equilibrium in a Cournot model with kinked demand curves, although it is clear that Palander had no idea that such an example might imply what we now know to be chaotic dynamics. In this regard Palander somewhat resembled Strotz, MacAnulty, and Naines (1953) who first explicitly showed chaotic dynamics in an economics model, but did not understand mathematically what they had done other than that the dynamic pattern was "irregular" in nature. Although it is possible to observe complex dynamics in models with single equilibria, given sufficiently great nonlinearities induced by lag effects or learning mechanisms, most of the models of complex oligopoly dynamics involve multiple equilibria where such dynamics more readily arise.

2 Basic Chaotic Oligopoly Dynamics

It is appropriate that the first consciously generated model of chaotic dynamics in economics used the Cournot model, given his initiation of the use of calculus in economics. This effort was carried out by David Rand (1978) who assumed that firm reaction functions were nonlinearly non-monotonic with multiple equilibria arising, although May (1976) had previously suggested that the logistic map could generate chaotic dynamics in a variety of economic models without actually developing any of them.

In Rand's (1978) model, two agents A and B produce $x \in [0,1]$ and $y \in [0,1]$ respectively, which maximize the respective agent utility functions, $u_A(x,y)$ and $u_B(x,y)$. Given an initial point (x_0, y_0), A adjusts x to a local maximum x_1, assuming y_0 fixed, and B adjusts y to a local maximum y_1, assuming that x_0 is fixed. This defines a dynamic evolutionary process, $\phi(x,y)$, from

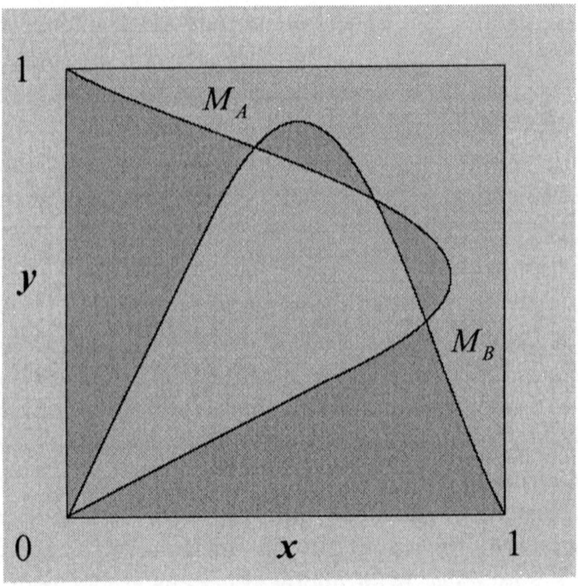

Figure 1. Cournot-Rand Chaotic Duopoly Reaction Functions

which reaction functions M_A and M_B can be derived. The reaction functions for one of Rand's smooth examples with all fixed points locally unstable are depicted in Figure 1.

Rand proceeded to prove that if C^r are rth order critical points, then there exists a non-empty open subset U of $C^r(I^2,\mathbf{R}) \times C^r(I^2,\mathbf{R})$ such that $(u_A, u_B) \in U$ implies that ϕ has periodic orbits of every period and the non-wandering set of ϕ contains a Cantor set. This Cantor set represents a strange attractor for an infinite set of periodicities of possible orbits that are chaotic dynamics. Rand recognized that a continuous model would generate simpler dynamics, but that an oligopoly of three producers (triopoly) would produce similar results to the duopoly case. The problem of continuous oligopoly dynamics was investigated for a more general case by Chiarella and Khomin (1996) who showed that even with nonlinear cost or revenue functions and various distributed lags with an unstable equilibrium, the most complex dynamics to emerge will be limit cycles.

Unsurprisingly Rand's results inspired many followups and modifications. One due to Shaffer (1984) was to argue that chaotic duopoly dynamics depend on "sophisticated" reaction functions in that non-monotonicities arise only if firms account for such phenomena as interfirm externalities. "Naiveté" without such awareness tends to eliminate such non-monotonicities and thus also chaotic dynamics, although such phenomena as unstable cobweb cycles can still result. Shaffer proposed that if firms adopted "consistent conjectures" in which firms react to each other's reaction functions in a "higher" form of sophistication, rapid convergence to Nash equilibrium will result without chaos or cobwebs.

Dana and Montrucchio (1986) generalized Rand's model to infinite horizon game models via Markov-Perfect-Equilibria (MPE). General conditions that allow for the existence of MPE in the conditions of the extended Rand model are hard to demonstrate, although for a sufficiently small discount rate any pair of smooth reaction functions could be the MPE of some game. If an MPE exists, then an orbit of agents playing alternately will resemble a Cournot tâtonnement. As the discount rate approaches zero the set of MPEs converges on one-shot game best-reply dynamics similar to that studied by Rand. Dana and Montrucchio argue that the inability of agents to distinguish exogenous randomness from endogenous chaos can lead them to such Cournot-Rand dynamics rather than convergence on long-run optimal behavior. Bischi, Mammana, and Gardini (2000) study models with MPEs that also exhibit complex dynamics beyond chaos.

A weakness of the Rand model, even as generalized by Dana and Montrucchio, is that it implies a monopolist will produce zero because a duopolist will produce zero when the other firm produces zero. Van Witteloostuijn and van Lier (1990) corrected this problem by showing that the general result holds even with the assumption of positive output. They rationalized non-monotonic firm reaction functions by noting how strategic behavior can lead to firms raising output as a new competitor appears in order to deter entry, drawing on arguments due to Bulow, Geanakoplis, and Klemperer (1985). The latter noted that if there is a constant elasticity of demand, then each firm's marginal revenue function is increasing in the other's output.

Puu (1991) followed this up by showing for a discrete model of Cournot duopoly dynamics, if there is an iso-elastic market demand curve with price simply the reciprocal of the sum of the two firms' outputs and the firms face constant marginal costs, periodic and chaotic dynamics can easily arise. This is so even though the reaction functions are unimodal, intersecting only once. This case again implies zero output for the monopoly case and an infinite price in such a case.

3 Other Forms of Complex Oligopoly Dynamics

Puu (1996, 1998) then extends his model to the case of three oligopolists and especially considers the problem of Stackelberg leadership in which one firm becomes the leader by taking the reaction functions of the other firms into account. Puu studies the famous case where there is not agreement regarding which firm is the leader, known to be unstable in the standard duopoly case. Introducing a stochastic element, Puu finds a variety of complex dynamics beyond chaotic dynamics. In Puu (1997, Chap. 5; 2000, Chap. 7) he follows up on his analysis of the monopoly case by carrying out a more detailed analysis of the duopoly, and triopoly cases. As these cases give rise to multidimensional systems, he uses the methods of critical curves (or manifolds) and global bifurcations to show many varieties of complex dynamics that will be discussed below, including multistability in the form of coexistent attractors and fractal boundaries of basins of attraction.

The first to apply these methods to oligopoly theory were Kopel (1996) and Bischi, Stefanini, and Gardini (1998). Such methods were originated by Mira and Gumowski (1980) and were closely studied for the coupled logistic case by Gardini, Abraham, Record, and Fournier-Prunaret (1994), the case studied by Kopel (1996). These methods and various extensions and implications were further developed in Mira, Gardini, Barugola, and Cathala (1996), which has served as a prime influence for many of these more recent developments in complex oligopoly dynamics.

The method of critical manifolds or critical sets (Gumowski and Mira, 1980) is the multidimensional extension of the concept of a critical point in unidimensional dynamical systems, with critical curves being the two dimensional version, frequently used in the duopoly models studied in this literature. They arise particularly in noninvertible maps (of non-monotonic functions) and are loci characterized by the coincidence of at least two preimages. Similar to critical points, Jacobian determinants of the dynamical systems vanish on such manifolds of rank zero. Critical manifolds can be of varying ranks and are boundary lines at which images are folded. They also generally bound the absorbing areas that contain the attractors while in turn they are contained within the basins of attraction.

The method of critical manifolds especially involves analysis of the characteristics of the preimages and the images of the critical manifold or set. Thus for such a critical manifold there will be at least two preimages that are coincident in the preceding period, that is preimages of rank -1. The set on which they are located is a set of merging preimages. This coincidence of preimages is a natural implication of the degeneracy of the Jacobian determinant on the critical manifold. Initial application of this method to complex oligopoly dynamics was due to Kopel (1996) and Bischi, Stefanini, and Gardini (1998).

In turn, following Mira, Gardini, Barugola, and Cathala (1996), absorbing areas or regions are observed by analyzing the images of the critical manifold to ever higher ranks until a closed region appears. Absorbing areas trap trajectories of dynamical systems after finite iterations and are bounded by portions of the critical manifold and its images of increasing rank. A given absorbing area may contain smaller ones, with a minimal invariant absorbing area generally existing. Generally if the attractor is chaotic it will fill the minimal invariant absorbing area.

Furthermore, the characteristics of the basins and the basin boundaries can be studied using the method of critical curves. With regard to the basins, their structure changes when their boundaries cross a critical curve as some parameter is varied. Such a contact or global bifurcation can cause a connected basin to become disconnected. In the model studied by Kopel (1996) an interfirm externality on the cost side leads to duopolists' Cournot reaction functions generating a coupled logistic system in which the two control parameters are one that determines the strength of the coupling between the firms' behavior and a speed of adjustment parameter. Contact or global bifurcations occur as these parameters pass through certain value combinations. Agiza (1999) shows how control of chaos can occur in the Kopel model.

When there are multiple coexisting basins of attraction, the condition of multistability, then the peculiarities of their boundaries and interpenetrations become of interest. The coexisting attractors may be of very different types, e.g. cyclic and chaotic. Analyzing the coupled logistic duopoly game of Kopel (1996), Bischi, Mammana, and Gardini (2000) show that multiple coexisting attractors of rectangular shapes can arise that exhibit a fractal pattern.

Interesting phenomena have been observed in studying cases involving synchronized behavior by duopolists. Bischi, Stefanini, and Gardini (1998) showed the possibility of intermittency arising from contact bifurcations. Synchronous behavior alternates with explosions of oscillatory behavior.

The case of synchronicity, or more precisely of near-synchronocity, was studied by Bischi, Gallegati, and Naimzada (1999). They showed that if firms are slightly off from being synchronized in their initial parameter values that bubbling bifurcations can occur that lead to riddled basins where trajectories that are locally repelled from unstable manifolds may belong to basins of other attractors. In the case of a blowout bifurcation the transversely unstable trajectories outweigh the stable ones and the attractor becomes a chaotic saddle. This analysis draws heavily on the expanded concept of an attractor due to Milnor (1985).

Finally, when a bounded attractor contacts the basin of infinity, a final bifurcation can occur that completely destroys the attractor, observed for the duopoly game by Bischi, Stefanini, and Gardini (1998). This phenomenon was first identified by Abraham (1972, 1985) and variously labeled the blue-sky catastrophe or blue-bagel chaostrophe.

4 Adjustment and Learning Complexities

More recently a greater emphasis has been placed upon studying models with greater degrees of adjustment. Models in which firms are assumed to use adaptive expectations have been studied. Among the first to do this was Puu (1998, 2000) who found a variety of complex dynamics arising in the Cournot duopoly case with adaptive expectations, including the appearance of fractal attractors. Others studying this model with more firms and other modifications include Ahmed and Agiza (1998), Agiza (1998), Agiza, Bischi, and Kopel (1999), Ahmed, Agiza, and Hassan (2000), Agliari, Gardini, and Puu (2000), and Huang (2001).

Such efforts have been extended by Bischi and Kopel (2001), drawing on learning models such as those of Cox and Walker (1998) and Rasssenti, Reynolds, Smith, and Szidarovsky (2000) that reliled on experimental results. They study the model of Kopel (1996) which is characterized by both adjustment and linking parameters. Assuming homogeneous expectations among the firms they derive zones for the parameters and initial values that produce a variety of outcomes from convergence to a Nash equilibrium to the coexisting attractors situation with rectangular basins of attraction. Furthermore, they determine conditions of global bifurcations where basins become disconnected using methods similar to those discussed above.

They then consider the case of heterogeneous expectations by the firms and find that this introduces a variety of asymmetries and further complexities. However, in considering the possibility of consistent expectations equilibria where simple learning autoregressive learning rules can mimic underlying chaotic dynamics (Grandmont, 1998; Hommes, 1998; Hommes and Sorger, 1998) they were unable to find such phenomena in their model.

Tuinstra (2001, Chap. 5) considers in more detail the problem of heterogeneous expectations in Cournot duopoly dynamics from the standpoint of evolutionary games with learning such as studied by Vega-Redondo (1997). They derive replicator dynamics (Binmore and Samuelson, 1997) based on a discrete choice model of Anderson, De Palma, and Thisse (1992). This then is shown to follow arguments developed in Brock and Hommes (1997) based on homoclinic bifurcation theory (Palis and Takens, 1993) that imply a variety of complex dynamics can arise.

What evolves are population shares following certain strategies that learn based on past performance. Willingness to switch strategies is a crucial control parameter in this setup. Although in Brock and Hommes (1997) it is posited that there are many agents, the population shares in Tuinstra (2001, Chap. 5) reflect probabilities that given strategies will be selected by two firms in a mixed-strategy game theoretic context. In particular the contrasting strategies are one that requires less information but is more destabilizing and one that requires more information but is stabilizing. He examines the specific cases of best-reply versus perfect foresight and also best-reply versus imitator dynamics.

In both of these cases, variations of the parameter controlling the willingness to switch strategies triggers homoclinic bifurcations wherein tangencies arise between stable and unstable manifolds. There such relatively unsurprising outcomes as wild oscillations of the stable and unstable manifolds, the existence of a Cantor set containing infinitely many periodic points and an uncountable set of aperiodic points, as well as sensitive dependence on initial conditions. In addition there are the less frequently observed cases of Hénon-like strange attractors for an open interval, the coexistence of many stable cycles, and cascades of infinitely many period doubling and period halving bifurcations.

We note that all these possible complex dynamics are occurring in a model with a unique Nash equilibrium. But this simply reinforces an observation of Bischi and Kopel (2001) that there are large zones of initial conditions and parameter values in evolutionary game dynamics models that do not converge to the unique equilibrium, with many of these non-convergent trajectories following seriously irregular paths.

5 Some Further Speculations and Possibilities

At this point we shall move beyond established results, including those presented in the later chapters of this book, and speculate somewhat more freely and loosely regarding some possible future directions that this vein of research might follow. In some cases these involve relatively minor variations on themes already seen here. In others the differences are noticeably more substantial.

One somewhat obvious line of pursuit would be to consider a wider range of models that might exhibit the phenomenon of consistent expectations equilibria in models of Cournot duopoly learning dynamics. As noted above, Bischi and Kopel (2001) were unable to detect such an outcome for the coupled logistic model that they studied. This might seem to be expected in that the original models studied by Grandmont (1998), Hommes (1998), and Hommes and Sorger (1998) all involved piecewise linear tent maps, and the coupled logistic system is not piecewise linear. But Hommes and Rosser (2001) have shown that such outcomes can occur with smooth non-invertible maps. Thus, there may be other reaction function formulations of learning in Cournot duopoly models that might produce this outcome. They might also find the related learning to believe in chaos (Sorger, 1998) wherein if firms do not initially use coefficients in their autoregressive rule of thumb, they can learn to adjust the coefficients so that they do accurately mimic the underlying chaotic dynamics.

Yet another exotic possibility might arise by combining the approach of Tuinstra (2001) with that of Bischi and Kopel (2001). Thus one might have a coupled logistic system operating within a framework of discrete learning with Brock and Hommes (1997) style mean-field dynamics of interacting agents, interpretable as probabilities of strategies. Shibata and Kaneko (1998) have studied such a system and discovered that in the windows of cyclic dynamics within the chaotic zones, tongues can occur in which there collective fluctuations can arise with substantially greater amplitudes. One can well imagine not unreasonable models of oligopoly dynamics in which this sort of thing might transpire.

Finally there is an issue that is not addressed in the essays cited in this chapter, but which easily could have been and arguably should be. This is the matter of transformations of industrial structure from competition to oligopoly or monopoly and back again. It was the father of conventional American industrial organization, Joe Bain (1951), who argued that in any industry there is a critical level of concentration at which a qualitative shift occurs in the nature of the industry between being competitive and oligopolistic with the latter's possibility of collusive conduct. From the perspective of firm managers that point may be where that go from knowing who their competitors are and are able to think about the behavior of each and when they can no longer do so. The impersonal invisible hand rules the market.

Clearly this is not a new idea. Rosser (2000, Chap. 3) discusses this issue in terms of the sudden increases and decreases of control over the world oil market by OPEC, citing a proposed catastrophe theory model due to Woodcock and Davis (1978, Chap. 7) to explain this. Bradburd and Over (1982) provided empirical support for Bain's original proposition more generally across a wide array of U.S. industries.

It would seem that many of the techniques and approaches used in some of the literature discussed above might be applicable to this question. It may well be that the models with larger numbers of heterogeneous agents who cross thresholds of cooperation or competitiveness are what may apply here, with the old prisoner's dilemma example from game theory serving as an obvious metaphor. The appropriate approach would clearly involve evolutionary approaches to game theory, with a full accounting for mathematical complexity. Some approaches to this can incorporate the mean-field approach of Brock and Hommes (1997), with a study by Lindgren (1997) being especially promising and provocative. Albin with Foley (1998) also provide a complexity context for the examination of repeated prisoner's dilemma games. And, whereas some of the older models assume a fixed number of firms, the work of Axtell (1999) examines the splitting and forming of firms within an extended dynamic game theoretic context with local increasing returns and local instabilities. This observer thinks that there is much more that can be done with such models that could illuminate the problem of the evolution of market structure itself.

6 Final Observations

We have seen that since the seminal contributions of David Rand in 1978, a substantial and growing literature has arisen that presents a variety of models of complex oligopoly dynamics of many sorts. Most of the array of the modern mathematics of nonlinear dynamics and complexity theory has been brought to bear on the problem, with many fruitful and interesting results demonstrated. Indeed, just as when Cournot introduced calculus to economics to study duopoly and Rand introduced chaos theory to extend his study, we see the problem of oligopoly dynamics leading to some of the major applications of multidimensional chaos theory and its relatives and offshoots in economics. Much has been done, but much more will be done in the future.

I would like to conclude by quoting an observation made by Tönu Puu (2000, pp. 236-237) regarding the complexities that can arise in the analysis of monopoly behavior.

"It may be hard for economists, nourished with textbook monopoly theory, to digest that the monopolist does not know all about the market and may behave in a way seemingly erratic...

Should the monopolist become suspicious about the fact that he is behaving like a random number generator, the blame could always be on exogenous shifts in the demand function, due to changing prices of close substitutes or to general business cycles. Not even a regularly periodic behaviour need make him suspicious, because the imagery of economic phenomena is crowded with all kinds of regular periodic cycles."

Needless to say, this observation applies with even greater force and relevance to the more complex and complicated cases involving oligopolies and their conduct.

References

Abraham, R.H., 1972, *Hamiltonian Catastrophes* (Université Claude-Bernard, Lyons).

Abraham, R.H., 1985, "Chaostrophes, Intermittency, and Noise", in Fischer, P. and Smith, W.R. (eds.), *Chaos, Fractals, and Dynamics* (Marcel Dekker, New York): 3-22.

Agiza, H.N., 1998, "Explicit Stability Zones for Cournot Games with 3 and 4 Competitors", *Chaos, Solitons & Fractals* 9: 1955-1966.

Agiza, H.N., 1999, "On the Stability, Bifurcation, Chaos and Chaos Control of Kopel Map", *Chaos, Solitions & Fractals* 11: 1909-1916.

Agiza, H.N., Bischi, G.-I., and Kopel, M., 1999, "Multistability in a Dynamic Cournot Game with Three Oligopolists", *Mathematics and Computers in Simulation* 51: 63-90.

Agliari, A., Gardini, L., and Puu, T., 2000, "The Dynamics of a Triopoly Cournot Game", *Chaos, Solitons & Fractals* 11: 2531-2560.

Ahmed, E.and Agiza, H.N., 1998, "Dynamics of a Cournot Game with n-Competitors", *Chaos, Solitons & Fractals* 9: 1513-1517.

Ahmed, E., Agiza, H.N., and Hassan, S.Z., 2000, "On Modifications of Puu's Dynamical Duopoly", *Chaos, Solitons & Fractals* 11: 1025-1028.

Albin, P.S., edited with an introduction by Foley, D.K., 1998, *Barriers and Bounds to Rationality: Essays on Economic Complexity and Dynamics in Interactive Systems* (Princeton University Press, Princeton).

Anderson, S.A., de Palma, A., and Thisse, J.-F., 1992, *Discrete Choice Theory of Product Differentiation* (MIT Press, Cambridge).

Axtell, R., 1999, "The Emergence of Firms in a Population of Agents: Local Increasing Returns, Unstable Nash Equilibria, and Power Law Distributions", Santa Fe Institute Working Paper 99-03-019E (*American Economic Review*, forthcoming).

Bain, J.S., 1951, "Relation of Profit Rate to Industry Concentration, American Manufacturing, 1936-40", *Quarterly Journal of Economics* 65: 293-324.

Binmore, K., and Samuelson, L., 1997, "Muddling Through: Noisy Equilibrium Selection", *Journal of Economic Theory* 74: 235-265.

Bischi, G.-I. , Gallegatti, M., and Naimzada, A., 1999, "Symmetry-Breaking Bifurcations and Representative Firm in Dynamic Duopoly Games", *Annals of Operations Research* 89: 253-272.

Bischi, G.-I., and Kopel, M., 2001, "Equilibrium Selection in an Nonlinear Duopoly Game with Adaptive Expectations", *Journal of Economic Behavior and Organization* 46: 73-100.

Bischi, G.-I., Mammana, C., and Gardini, L., 2000, "Multistability and Cyclic Attractors in Duopoly Games", *Chaos, Solitons & Fractals* 11: 543-564.

Bischi, G.-I., Stefanini, L., and Gardini, L., 1998, "Synchronization, Intermittency and Critical Curves in a Duopoly Game", *Mathematics and Computers in Simulation* 44: 559-585.

Bonanno, G., 1987, "Monopoly Equilibrium and Catastrophe Theory", *Australian Economic Papers* 26: 197-215.

Bradburd, R.M., and Over, M., 1982, "Organizational Costs, 'Sticky Equilibria,' and Critical Levels of Concentration", *Review of Economics and Statistics* 64: 50-58.

Brock, W.A., and Hommes, C.H., 1997, "A Rational Route to Randomness", *Econometrica* 65: 1059-1095.

Bulow, J.L., Geanakoplis, J.D., and Klemperer, P.D., 1985, "Holding Idle Capacity to Deter Entry", *Economic Journal* 95: 178-182.

Chiarella, C., and Khomin, A., 1996, "An Analysis of the Complex Dynamic Behaviour of Nonlinear Oligopoly Models with Time Delays", *Chaos, Solitons & Fractals* 7: 2049-2065.

Cournot, A.A., 1838, *Recherches sur les Principes Mathématiques de la Théorie des Richesses* (Hachette, Paris).

Cox, J.C., and Walker, M., 1998, "Learning to Play Cournot Duopoly Strategies", *Journal of Economic Behavior and Organization* 36: 141-161.

Dana, R.-A., and Montrucchio, L., 1986, "Dynamic Complexity in Duopoly Games", *Journal of Economic Theory* 40: 40-56.

Gardini, L., Abraham, R., Record, R.J., and Fournier-Prunaret, D., 1994, "A Double-Logistic Map", *International Journal of Bifurcations and Chaos* 4: 145-176.

Grandmont, J.-M., 1998, "Expectations Formation and Stability in Large Socio-Economic Systems", *Econometrica* 66: 741-781.

Gumowski, I., and Mira, C., 1980, *Dynamique Chaotique* (Cepadues Éditions, Toulouse).

Hommes, C.H., 1998, "On the Consistency of Backward-Looking Expectations: The Case of the Cobweb", *Journal of Economic Behavior and Organization* 33: 333-362.

Hommes, C.H., and Rosser, Jr., J.B., 2001, "Consistent Expectations and Complex Dynamics in Renewable Natural Resource Markets", *Macroeconomic Dynamics* 5: 180-203.

Hommes, C.H., and Sorger, G., 1998, "Consistent Expectations Equilibria", *Macroeconomic Dynamics* 2: 287-321.

Huang, W., 2001, "Theory of Adaptive Adjustment", *Discrete Dynamics in Nature and Society* 5: 247-263.

Kopel, M., 1996, "Simple and Complex Adjustment Dynamics in Cournot Duopoly Models", *Chaos, Solitons & Fractals* 7: 2031-2048.

Lindgren, K., 1997, "Evolutionary Dynamics in Game-Theoretic Models", in W.B. Arthur, S.N. Durlauf, and D. Lane, (eds.), *The Economy as an Evolving Complex System II* (Addison-Wesley, Reading): 337-367.

von Mangoldt, V.K.E., 1863, *Grundriss der Volkswirtschaftslehre* (Engelhorm, Stuttgart).

May, R.M., 1976, "Simple Mathematical Models with Very Complicated Dynamics", *Nature* 261: 459-467.

Milnor, J., 1985, "On the Concept of an Attractor", *Communications in Mathematical Physics* 102: 517-519.

Mira, C., Gardini, L., Barugola, A., and Cathala, J.C., 1996, *Chaotic Dynamics in Two-Dimensional Invertible Maps* (World Scientific, Singapore).

Nash, J., 1951, "Non-cooperative Games", *Annals of Mathematics* 54: 286-295.

Palander, T.F., 1939, "Konkurrens och Marknadsjämvikt vid Duopol och Oligopol", *Ekonomisk Tidskrift* 41: 124-145 and 222-250.

Palis, J., and Takens, F., 1993, *Hyperbolicity and Sensitive Chaotic Dynamics at Homoclinic Bifurcations: Fractal Dimensions and Infinitely Many Attractors* (Cambridge University Press, Cambridge).

Puu, T., 1991, "Chaos in Duopoly Pricing", *Chaos, Solitons & Fractals* 1: 573-581.

Puu, T., 1995, "The Chaotic Monopolist", *Chaos, Solitons & Fractals* 5: 35-44.

Puu, T., 1996, "Complex Dynamics with Three Oligopolists", *Chaos, Solitons & Fractals* 7: 2075-2081.

Puu, T., 1997, *Nonlinear Economic Dynamics*, 4th Edition (Springer-Verlag, Berlin).

Puu, T., 1998, "The Chaotic Duopolists Revisited", *Journal of Economic Behavior and Organization* 33: 385-394.

Puu, T., 2000, *Attractors, Bifurcations, & Chaos: Nonlinear Phenomena in Economics* (Springer-Verlag, Berlin).

Rand, D., 1978, "Exotic Phenomena in Games and Duopoly Models", *Journal of Mathematical Economics* 5: 173-184.

Rassenti, S., Reynolds, S.S., Smith, V.L., and Svidarovsky, F., 2000, "Adaptation and Convergence of Behavior in Repeated Experimental Cournot Games", *Journal of Economic Behavior and Organization* 41: 117-146.

Robinson, J., 1933, *The Economics of Imperfect Competition* (Macmillan, London).

Rosser, Jr., J.B., 2000, *From Catastrophe to Chaos: A General Theory of Economic Discontinuities, Vol. I: Mathematics, Microeconomics, Macroeconomics, and Finance*, 2nd Edition (Kluwer Academic Publishers, Boston).

Shaffer, S., 1984, "Chaos, Naiveté, and Consistent Conjectures", *Economics Letters* 14: 155-162.

Shibata, T. and Kaneko, K., 1998, "Tongue-Like Bifurcation Structures of the Mean-Field Dynamics in a Network of Chaotic Elements", *Physica D* 124: 177-200.

Sorger, G., 1998, "Imperfect Foresight and Chaos: An Example of a Self-Fulfilling Mistake", *Journal of Economic Behavior and Organization* 33: 363-383.

Strotz, R.H., McAnulty, J.C., and Naines, Jr., J.B., 1953, "Goodwin's Nonlinear Theory of the Business Cycle: An Electro-Analog Solution", *Econometrica* 21: 390-411.

Tuinstra, J., 2001, *Price Dynamics in Equilibrium Models: The Search for Equilibrium and the Emergence of Endogenous Fluctuations* (Kluwer Academic Publishers, Boston).

Vega-Redondo, F., 1997, "The Evolution of Walrasian Behavior", *Econometrica* 65: 375-384.

Wald, A., 1936, "Über einige Gleichungssysteme der mathematischen Ökonomie", *Zeitschrift für Nationalökonomie* 7: 637-670.

van Witteloostuijn, A., and van Lier, A., 1990, "Chaotic Patterns in Cournot Competition", *Metroeconomica* 41: 161-185.

Woodcock, A., and Davis, M., 1978, *Catastrophe Theory* (E.P. Dutton, New York).

3 Some Methods for the Global Analysis of Dynamic Games Represented by Iterated Noninvertible Maps

Anna Agliari[1], *Gian-Italo Bischi*[2] *and Laura Gardini*[2]

[1]Istituto di Econometria e Matematica AEFA, Università Cattolica di Milano, Italy
[2]Istituto di Scienze Economiche, Università di Urbino, Italy

1 Introduction

The time evolution of an oligopoly system is often described as an n-players game which is played repeatedly, in the sense that at each discrete time period $t = 0, 1, \ldots, n$ producers choose their actions, $x_1(t), \ldots, x_n(t)$, by solving an optimization problem based on the knowledge of the actions observed in the past. For example, the classical Cournot adjustment (see Cournot, 1838, Teocharis, 1960, and many others) is obtained by assuming that at each period any player chooses its own production strategy which is a best response to the choices of the competitors in the previous period.

This is often expressed in the form of a discrete dynamical system defined in a given strategy space $S \subseteq \mathbb{R}^n$. Given an initial condition $\mathbf{x}(0) \in S$, the sequence of actions $\mathbf{x}(t)$, $t \in \mathbb{N}$, is obtained *inductively* by the iteration of a map $T : S \to S$ defined by

$$\mathbf{x}' = T(\mathbf{x}) \tag{1}$$

where $'$ denotes the unit-time advancement operator, that is, if the right hand side variables represent the actions at time period t then the left hand side represents the set of actions at time $(t+1)$.

A dynamic process defined by the iterated map (1) may converge to a given steady state (or equilibrium) or to a more complex attractor. Indeed, as shown in a pioneering paper by Rand, 1978, quite complex dynamics, with periodic and chaotic trajectories, may characterize the long run behavior of duopoly games (see also Poston and Stewart, 1978). Examples of economically interesting discrete time dynamic oligopoly games, showing complex dynamics, have been given by Dana and Montrucchio, 1986, Puu, 1991,

1997, 2000, Kopel, 1996, Agiza et al., 1999, Bischi et al., 2000a, 2001a, Agliari et al., 2000b, just to quote a few. In these papers it is shown that dynamic Cournot oligopoly games may have time evolutions which never settle to a steady state, and in the long run they exhibit bounded dynamics which may be periodic, or quasi-periodic or chaotic. In such cases, a delimitation of a bounded region of the strategy space where the system dynamics are ultimately trapped, despite of the complexity of the long-run time patterns, may be a useful information for practical applications. Moreover, as some parameters are varied, global bifurcations may cause sudden qualitative changes in the properties of the attracting sets (see the contact bifurcations in Mira et al., 1996 and the so called *crises* in Grebogi et al., 1983).

Another problem which often arises in the study of nonlinear maps which describe dynamic oligopoly games concerns the existence of several attracting sets, each with its own basin of attraction. In this case, a problem of equilibrium selection arises (see Van Huyck et al., 1994, Bischi and Kopel, 2001) because the dynamic process becomes path-dependent, i.e. which kind of long run dynamics is chosen depends on the starting condition of the game. This opens the question of the delimitation of the basins of attraction and their changes as the parameters of the model vary.

These two problems lead to two different routes to complexity, one related to the complexity of the attracting sets which characterize the long run time evolution of the dynamic process, the other one related to the complexity of the boundaries which separate the basins when several coexisting attractors are present. These two different kinds of complexity are not related in general, in the sense that very complex attractors may have simple basin boundaries, whereas boundaries which separate the basins of simple attractors, such as coexisting stable equilibria, may have very complex structures.

Both the questions outlined above require an analysis of the global dynamical properties of the dynamical system, that is, an analysis which is not based on the linear approximation of the map (1). When the map T is noninvertible (i.e. "many-to-one") the global dynamical properties can be usefully characterized by the method of critical sets, a powerful tool introduced in the seventies (see Gumovsky and Mira, 1980a,b and references therein, Mira et al. 1996, Abraham et al. 1997) but only recently employed in the study of dynamic modelling of economic and financial systems (see e.g. Gardini, 1992, Bischi et al., 1999b,c, 2000b, Agliari et al, 2000a, Puu 2000, Bischi and Kopel 2001, Bischi et al., 2001b, Dieci et al., 2001, Chiarella et al 2001a,b). Indeed, several dynamic models of oligopoly games are represented by the iteration of a noninvertible map , i.e. a point transformation T which maps distinct points into the same point. Loosely speaking, this can be expressed by saying that the map "folds and pleats" the state space. As we shall describe in the following, the folding action associated with the application of a noninvertible map, as well as the "unfolding" associated with the geometric action of the inverses, can be described by using the formalism of critical sets. The repeated application of a noninvertible map repeatedly folds the state space along the critical sets and their images, and often this allows one to define a bounded region where asymptotic dynamics are trapped. Instead, the repeated application of the inverses "repeatedly unfolds" the state space, so that a neighborhood of an attractor may have preimages far from it. This may give rise to complicated topological structures of the basins, which may be formed by the union of non connected portions.

The paper is organized as follows. In section 2 we recall some definitions and properties of noninvertible maps. In section 3 we describe the construction of absorbing regions and in section 4 we recall the main contact bifurcations which lead to complex basins of attraction. In section 5 we describe the properties of a Cournot duopoly game and in section 6 we consider the special case of with symmetric duopoly games (i.e. duopoly games with identical players) leading to chaos synchronization problems and riddled basins.

2 Noninvertible maps: Basic definitions and properties

In this section we give some basic definitions and properties, and a minimal vocabulary, about the theory of noninvertible maps of the plane and the method of critical sets. A map $T : S \to S$, $S \subseteq \mathbb{R}^n$, like the one defined in (1), transforms a point $\mathbf{x} \in S$ into a unique point $\mathbf{x}' \in S$. The point $\mathbf{x}'$ is called the *rank-1 image* of $\mathbf{x}$, and a point $\mathbf{x}$ such that $T(\mathbf{x}) = \mathbf{x}'$ is a *rank-1 preimage* of $\mathbf{x}'$. Starting from an *initial condition* $\mathbf{x}_0 \in S$, the repeated application (*iteration*) of T uniquely defines a *trajectory*

$$\tau(\mathbf{x}_0) = \left\{\mathbf{x}(t) = T^t(\mathbf{x}_0), t = 0, 1, 2, ...\right\}, \tag{2}$$

where T^0 is the identity map and $T^t = T(T^{t-1})$.

A set $A \subset \mathbb{R}^n$ is *trapping* if it is mapped into itself, $T(A) \subseteq A$, i.e. if $x \in A$ then $T(x) \in A$. A trapping set is *invariant* if it is mapped onto itself: $T(A) = A$, i.e. all the points of A are images of points of A. A closed invariant set A is an attractor if it is *asymptotically stable,* i.e. if a neighborhood U of A exists such that $T(U) \subseteq U$ and $T^t(\mathbf{x}) \to A$ as $t \to +\infty$ for each $\mathbf{x} \in U$.

The *Basin* of an attractor A is the set of all points that generate trajectories converging to A

$$\mathcal{B}(A) = \{\mathbf{x} | T^t(\mathbf{x}) \to A \text{ as } t \to +\infty\} \tag{3}$$

Starting from the definition of stability, let $U(A)$ be a neighborhood of an attractor A whose points converge to A. Of course $U(A) \subseteq \mathcal{B}(A)$, but also the points of the phase space which are mapped inside U after a finite number of iterations belong to $\mathcal{B}(A)$. Hence, the *total basin of* A (or briefly the basin of A) is given by

$$\mathcal{B}(A) = \bigcup_{n=0}^{\infty} T^{-n}(U(A)) \tag{4}$$

where $T^{-1}(\mathbf{x})$ represents the set of the rank-1 preimages of $\mathbf{x}$ (i.e. the points mapped into x by T), and $T^{-n}(x)$ represents the set of the rank-n preimages of x (i.e. the points mapped into x after n applications of T).

If $\mathbf{x} \neq \mathbf{y}$ implies $T(\mathbf{x}) \neq T(\mathbf{y})$ then T is an *invertible map*, because the inverse mapping that gives $\mathbf{x} = T^{-1}(\mathbf{x}')$ is uniquely defined; otherwise T is a *noninvertible map*. So, noninvertible means "many-to-one", that is, distinct points $\mathbf{x} \neq \mathbf{y}$ may have the same image, $T(\mathbf{x}) = T(\mathbf{y}) = \mathbf{x}'$. Hence, several rank-1 preimages may exist and the inverse relation $\mathbf{x} = T^{-1}(\mathbf{x}')$ may be multivalued. Geometrically, the action of a noninvertible map T can be expressed by saying that it "folds and pleats" the plane, so that the two distinct points p_1 and p_2 are mapped into the same point p. This is equivalently stated by saying that several inverses are defined in p, and these inverses "unfold" the plane.

For a noninvertible map (1) $\mathbb{R}^n$ can be subdivided into regions Z_k, $k \geq 0$, whose points have k distinct rank-1 preimages. Generally, as the point $\mathbf{x}'$ varies in $\mathbb{R}^n$, pairs of preimages appear or disappear as this point crosses the boundaries separating different regions. Hence, such boundaries are characterized by the presence of at least two coincident (merging) preimages. This leads to the definition of the critical sets, one of the distinguishing features of noninvertible maps (Gumovski and Mira, 1980, Mira et al., 1996):

Definition. The *critical set* CS of a continuous map T is defined as the locus of points having at least two coincident $rank-1$ preimages, located on a set CS_{-1} called *set of merging preimages*.

The critical set CS is the n-dimensional generalization of the notion of critical value (when it is a local minimum or maximum value) of a one-dimensional map[1], and of the notion of *critical curve* LC (from the French "Ligne Critique", following Gumowski and Mira, 1980), of a noninvertible two-dimensional map. The set CS_{-1} is the generalization of the notion of critical point (when it is a local extremum point) of a one-dimensional map, and of the fold curve LC_{-1} of a two-dimensional noninvertible map. The critical set CS is generally formed by $(n-1)$-dimensional hypersurfaces of $\mathbb{R}^n$, and portions of CS separate regions Z_k of the phase space characterized by a different number of $rank-1$ preimages, for example Z_k and Z_{k+2} (this is the standard occurrence).

From the definition given above it is clear that the relation $CS = T(CS_{-1})$ holds, and the points of CS_{-1} in which the map is continuously differentiable are necessarily points where the Jacobian determinant vanishes, so that if T is smooth we have :

$$CS_{-1} \subseteq J_0 = \{p \in \mathbb{R}^n | \det DT(p) = 0\} \tag{5}$$

In fact, in any neighborhood of a point of CS_{-1} there are at least two distinct points which are mapped by T in the same point. Accordingly, the map is not locally invertible in points of CS_{-1}.

In order to explain the geometric meaning of the critical sets, let us consider a portion of CS, say $\widehat{CS}$, which separates two regions Z_k and Z_{k+2} of the phase space, and let $\widehat{CS}_{-1}$ be the corresponding locus of merging preimages, i.e. $\widehat{CS} = T\left(\widehat{CS}_{-1}\right)$. This means that two inverses of T exist, say T_1^{-1} and T_2^{-1}, which are defined in the region Z_{k+2} (and have respective ranges in the regions R_1 and R_2 separated by $\widehat{CS}_{-1}$). Both inverses merge on $\widehat{CS}_{-1}$ (i.e. they give merging preimages on $\widehat{CS}_{-1}$) and no longer exist in the region Z_k. Now, let $U \subset \mathbb{R}^n$ be a ball which intersects $\widehat{CS}_{-1}$ in $D = U \cap \widehat{CS}_{-1}$. Then $T(D) \subseteq \widehat{CS}$, and $T(U)$ is "folded" along $\widehat{CS}$ into the region Z_{k+2}. In fact, considering the two portions of U separated by $\widehat{CS}_{-1}$, say $U_1 \in R_1$ and $U_2 \in R_2$, we have that $T(U_1) \cap T(U_2)$ is a nonempty set included in the region Z_{k+2}, which is the region whose points p' have rank-1 preimages $p_1 = T_1^{-1}(p') \in U_1$ and $p_2 = T_2^{-1}(p') \in U_2$. This means that two points $p_1 \in U_1$ and $p_2 \in U_2$, located at opposite sides with respect to $\widehat{CS}_{-1}$, are mapped in the same side with respect to $\widehat{CS}$, in the region Z_{k+2}. This is also expressed by saying that the ball U is "folded" by T along CS on the side with more preimages (examples in two dimensions

[1] This terminology, and notation, originates from the notion of critical points as it is used in the classical works of Julia and Fatou.

are given below, see e.g. Fig. 7). The same concept can be equivalently expressed by stressing the "unfolding" action of T^{-1}, obtained by the application of the two distinct inverses in Z_{k+2} which merge along CS, because if we consider a ball $V \subset Z_{k+2}$, then the set of its $rank-1$ preimages $T_1^{-1}(V)$ and $T_2^{-1}(V)$ is made up of two balls $T_1^{-1}(V) \in R_1$ and $T_2^{-1}(V) \in R_2$, and these balls are disjoint if $V \cap \widehat{CS} = \emptyset$.

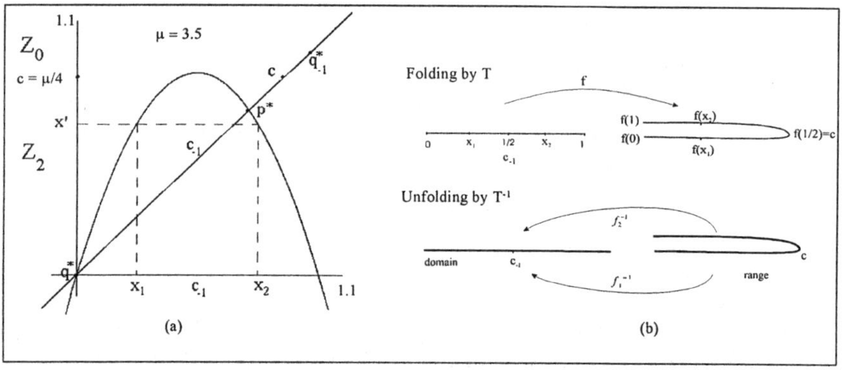

Figure 1: (a) *The logistic map.* (b) *Folding and unfolding actions.*

For one-dimensional continuous maps $x' = f(x)$, $x \in \mathbb{R}$, the critical set is formed by the local maximum or minimum values. For example, the well known logistic map (Fig.1a)

$$x' = f(x) = \mu x(1-x) \tag{6}$$

has a unique critical point $c = \mu/4$, which separates the real line into the two subsets: $Z_0 = (c, +\infty)$, where no inverses are defined, and $Z_2 = (-\infty, c)$, whose points have two rank-1 preimages, computed by the two inverses

$$x_1 = f_1^{-1}(x') = \frac{1}{2} - \frac{\sqrt{\mu(\mu - 4x')}}{2\mu}; \quad x_2 = f_2^{-1}(x') = \frac{1}{2} + \frac{\sqrt{\mu(\mu - 4x')}}{2\mu} \tag{7}$$

If $x' \in Z_2$, its two rank-1 preimages, computed according to (7), are located symmetrically with respect to the point $c_{-1} = 1/2 = f_1^{-1}(\mu/4) = f_2^{-1}(\mu/4)$, i.e. c_{-1} is the point where the two merging preimages of c are located. Of course, since the map (6) is differentiable, at c_{-1} the first derivative vanishes. However, we remark that in general the condition of vanishing derivative is not sufficient to define c_{-1}, because such condition may also be satisfied by points which are not local extrema (e.g. the inflection points with horizontal tangent). Moreover, a critical point may even be a point where the map is not differentiable, as it happens for continuous piecewise differentiable maps where critical points are located at the kinks where local maxima and minima are formed in the points at which two branches having slopes of opposite sign join, such as the well known tent map or other piecewise linear maps (see Fig.2a).

The importance of the critical points lies in the fact that they separate regions Z_k characterized by different number of preimages. We note however that this property is not a characteristic only of the critical points, because the boundary of a region Z_k may also be a particular set, called *prefocal set*, whose properties are associated with inverses not defined in the whole space, as shown in Bischi et al. 1999a.

In order to explain the action of a critical point, let us consider again the logistic map and let us notice that, as x moves from 0 to 1, the corresponding image $f(x)$ spans the interval $[0, c]$ twice, the critical point c being the turning point. In other words, if we consider how the segment $\gamma = [0, 1]$ is transformed by the map f we can say that it is *folded and pleated* to obtain the image $\gamma' = [0, c]$. This folding gives a geometric reason why two distinct points of γ, say x_1 and x_2, located symmetrically with respect to the point $c_{-1} = 1/2$, are mapped into the same point $x' \in \gamma'$ due to the folding action of f (Fig.1b). This is a geometric interpretation of the fact that (6) is a two-to-one map.

The same arguments can be explained by looking at the two inverse mappings f_1^{-1} and f_2^{-1} defined in $(-\infty, \mu/4]$ according to (7). We can consider the range of the map f formed by the superposition of two half-lines $(-\infty, \mu/4]$, joined at the critical point $c = \mu/4$ (Fig.1b), and on each of these half-lines a different inverse is defined. With other words, instead of saying that two distinct maps are defined on the same half-line we say that the range is formed by two distinct half lines on each of which a unique inverse map is defined. This point of view gives a geometric visualization of the definition of the critical point c of rank-1 as the point in which two distinct inverses merge. The action of the multivalued inverse, say $f^{-1} = f_1^{-1} \cup f_2^{-1}$, causes an unfolding of the range by mapping c into c_{-1} and by opening the two half-lines one on the right and one on the left of c_{-1}, so that the whole real line $\mathbb{R}$ is covered. Of course, a small segment γ' inside Z_2 and not including c, is splitted by the two inverses into two disjoint segments, located at opposite sides with respect to c_{-1}.

Up to now we have considered continuous maps, but the properties of critical points can easily be extended also to a discontinuous map T. In this case a point of discontinuity may be a critical point of T. This happens when the ranges of the map on the two sides of the discontinuity have an overlapping zone, so that at least one of the two limiting values of the function at the discontinuity separates regions having a different number of rank-1 preimages. The difference with respect to the continuous case is that now the number of distinct rank-1 preimages through a critical point differs generally by one (instead of two), that is, a critical value c (in general the critical set CS) separates regions Z_k and Z_{k+1}. An example is shown in Fig.2b, with a one-dimensional map. The discontinuity point is a critical point c_{-1}, and both the two limiting values of the function in c_{-1} are critical points, say c_a and c_b, associated with c_{-1}, as both c_a and c_b separate regions Z_1 and Z_2. Notice that now the critical points have no merging rank-1 preimages. Consider for example the critical point c_b in Fig.2b. The two distinct rank-1 preimages of c_b are the critical point c_{-1} and the so called "excess rank-1 preimage" $c^e_{-1,b}$. More on the properties and bifurcations of discontinuous maps of the plane can be found in Mira et al., 1996. In the following sections we shall only consider, for the sake of simplicity, continuous maps.

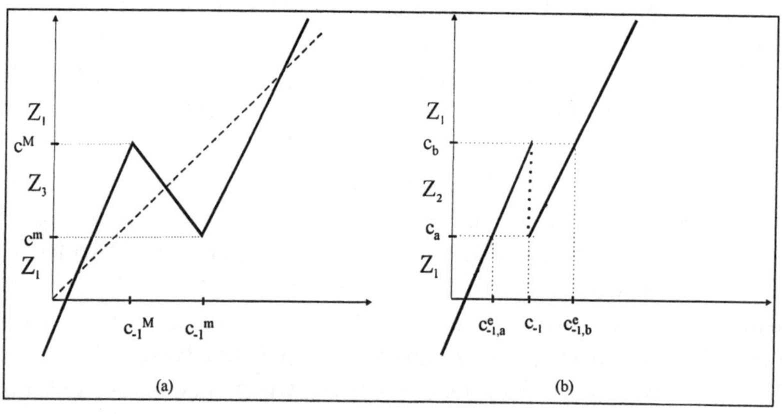

Figure 2: (a) *A piecewise linear noninvertible map.* (b) *A discontinuous noninvertible map.*

Another interpretation of the folding action of the unimodal map f is the following. Since $f(x)$ is increasing for $x \in [0, 1/2)$ and decreasing for $x \in (1/2, 1]$, then its application to a segment $\gamma_1 \subset [0, 1/2)$ is orientation preserving, whereas its application to a segment $\gamma_2 \subset (1/2, 1]$ is orientation reversing. This suggests that an application of f to a segment $\gamma_3 = [a, b]$ including the point $c_{-1} = 1/2$ preserves the orientation of the portion $[a, c_{-1}]$, i.e. $f([a, c_{-1}]) = [f(a), c]$, whereas it reverses the portion $[c_{-1}, b]$, i.e. $f([c_{-1}, b]) = [f(b), c]$, so that $\gamma_3' = f(\gamma_3)$ is folded, the folding point being the critical point of rank-1 c.

The extension of these concepts to the case of two-dimensional noninvertible maps leads us to the notion of critical curves. We present here some geometric characterizations of the action of the critical curves, because they present some new features with respect to the one dimensional case, and their properties can often be used as a useful visualization for the analogous properties of n-dimensional cases, with $n > 2$. Moreover, dynamic duopoly games, represented by a continuous map of the plane into itself $T : (x_1(t), x_2(t)) \to (x_1(t+1), x_2(t+1))$ defined as

$$T : \begin{cases} x_1' = T_1(x_1, x_2) \\ x_2' = T_2(x_1, x_2) \end{cases} , \tag{8}$$

constitute the simplest oligopolies, and are often studied in the literature.

If we solve the system of the two equations (8) with respect to the unknowns x_1 and x_2, then, for a given (x_1', x_2'), we may have several solutions, representing rank-1 preimages (or backward iterates) of (x_1', x_2'), say $(x_1, x_2) = T^{-1}(x_1', x_2')$, where T^{-1} is in general a multivalued relation. In this case we say that T is noninvertible, and the critical curves LC are the boundaries which separate regions of the plane characterized by a different number of rank-1 preimages. Along LC at least two inverses give merging preimages, located on the set denoted by LC_{-1}. For a continuous and (at least piecewise) differentiable noninvertible map of the plane, the study of the sign of the Jacobian determinant can help one to find the critical curves, because the set LC_{-1} is included in the set where $\det DT(x_1, x_2)$ changes sign, since T is locally an orientation preserving map near points (x_1, x_2) such that $\det DT(x_1, x_2) > 0$ and orientation reversing if $\det DT(x_1, x_2) < 0$, and $LC = T(LC_{-1})$.

In order to understand this point, let us recall that when a linear transformation $\mathbf{x}' = A\mathbf{x}$ of the plane onto itself

$$G : \begin{cases} x_1' = a_{11}x_1 + a_{12}x_2 \\ x_2' = a_{21}x_1 + a_{22}x_2 \end{cases} \tag{9}$$

is applied to a plane figure F, then the area of the transformed figure $F' = G(F)$ grows, or shrinks, by a factor $\rho = |\det A|$ with respect to the area of

F, being A the Jacobian matrix DG, and if $\det A > 0$ then the orientation of the figure on which (9) is applied is preserved, whereas if $\det A < 0$ then the orientation is reversed. This property of a linear two-dimensional map can be applied to the linear approximation of (8) in a neighborhood of a point $\mathbf{p} = (x_1, x_2)$, given by an affine map, the Jacobian matrix DT evaluated at the point $\mathbf{p}$:

$$\mathbf{DT}(\mathbf{p}) = \begin{bmatrix} \partial T_1/\partial x_1 & \partial T_1/\partial x_2 \\ \partial T_2/\partial x_1 & \partial T_2/\partial x_2 \end{bmatrix} \tag{10}$$

In fact, a small area around a point $\mathbf{p} = (x_1, x_2)$ is reduced (or expanded) by a factor $\rho(x_1, x_2) = |\det DT(x_1, x_2)|$, and if in a neighborhood of $\mathbf{p}$ the Jacobian determinant is positive (negative), then the map (8) is locally orientation preserving (reversing). This gives an intuitive visualization of the relation between the locus of points where the Jacobian determinant changes sign and the *folding properties* of continuous two-dimensional noninvertible maps (see Fig.3).

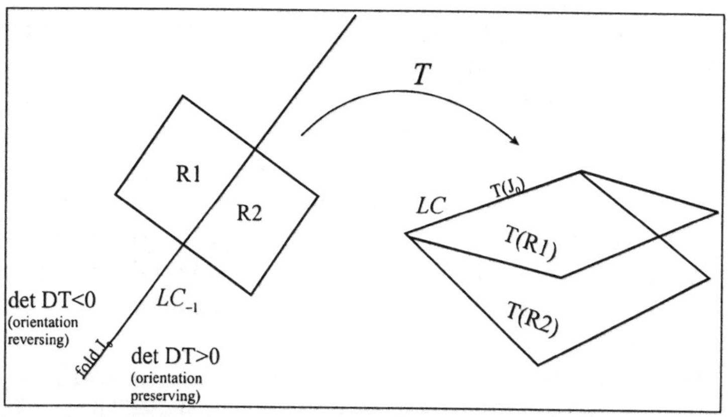

Figure 3: *Folding properties of a continuous two-dimensional map.*

This implies that if the map (8) is continuously differentiable in the whole plane, then

$$LC_{-1} \subseteq J_0 = \{(x_1, x_2) \in \mathbb{R}^2 | \det DT(x_1, x_2) = 0\} \tag{11}$$

Let us remark, however, that condition of vanishing Jacobian is necessary for differentiable maps, but not sufficient to detect a critical point of LC_{-1} as defined above (i.e. the inclusion in (11) may be strict).

In order to give a geometrical interpretation of the action of the multi-valued inverse relation T^{-1}, it is useful to consider a region Z_k as the superposition of k sheets, each one associated with a different inverse. Such a representation is known as *Riemann foliation* of the plane (see e.g. Mira et al., 1996). Different sheets are connected by folds joining two sheets, and the critical curve LC belongs to the projections of such folds on the phase plane (note however that the *vice-versa* is not necessarily true, due to the properties associated with *prefocal sets*, see Bischi et al., 1999a).

We illustrate these concepts by some examples. Let us consider the quadratic map (see Mira et al., 1996, Abraham et al., 1997) defined by

$$T : \begin{cases} x' = ax + y \\ y' = b + x^2 \end{cases} \tag{12}$$

This is a noninvertible map. In fact, given x' and y', if we try to solve the algebraic system with respect to the unknowns x and y we get two solutions, given by

$$T_1^{-1} : \begin{cases} x = -\sqrt{y' - b} \\ y = x' + a\sqrt{y' - b} \end{cases} \quad ; \qquad T_2^{-1} : \begin{cases} x = \sqrt{y' - b} \\ y = x' - a\sqrt{y' - b} \end{cases} \tag{13}$$

if $y' \geq b$, and no solutions if $y' < b$. In other words, (12) is a $Z_0 - Z_2$ noninvertible map, where Z_0 (region whose points have no preimages) is the half plane $Z_0 = \{(x, y) \,|y < b\}$ and Z_2 (region whose points have two distinct rank-1 preimages) is the half plane $Z_2 = \{(x, y) \,|y > b\}$. The line $y = b$, which separates these two regions, is the locus of points having two merging rank-1 preimages, located on the line $x = 0$. Hence the line $y = b$ is LC and the line $x = 0$ is LC_{-1} (see Fig.4.). Being (12) a continuously differentiable map, the points of LC_{-1} necessarily belong to the set of points at which the Jacobian determinant vanishes, i.e. $LC_{-1} \subseteq J_0$, where

$$J_0 = \{(x, y) \,|\, \det DT(x, y) = 0\}$$

and, since for the map (12) we have

$$\det\left(DT(x, y)\right) = \det\left(\begin{bmatrix} a & 1 \\ 2x & 0 \end{bmatrix}\right) = -2x$$

in this case LC_{-1} coincides with J_0 (the y axis $x = 0$). The critical curve LC is the image by T of LC_{-1}

$$LC = T(LC_{-1}) = T\left(\{x = 0\}\right) = \{(x, y) \,|y = b\}$$

We can consider the region Z_2 of the two-dimensional noninvertible map (12) as formed by the superposition of two sheets, and on each of these sheets a different inverse map is defined. The two sheets join along the critical line LC, where the two inverses merge (see Fig.4). This is the Riemann foliation for the $Z_0 - Z_2$ map (12). The action of the inverses, $T^{-1} = T_1^{-1} \cup T_2^{-1}$, causes an unfolding of the sheets by opening these sheets at opposite sides with respect to the line LC_{-1}. That is, given a point $(x', y') \in Z_2$ the preimages $T_1^{-1}(x', y')$ and $T_2^{-1}(x', y')$ are located on the right and on the left of LC_{-1} respectively.

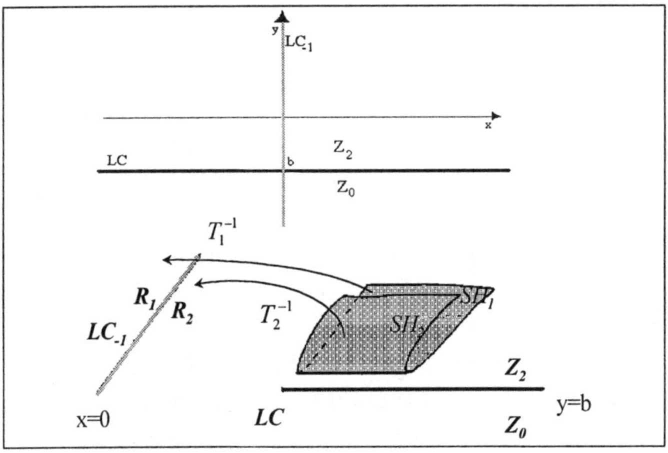

Figure 4: *Riemann foliation associated with the map T in (12).*

We propose another example, given by the map $T : (x, y) \to (x', y')$ defined by

$$T : \begin{cases} x' = \mu_1 y(1 - y) + \varepsilon_1 (y - x) \\ y' = \mu_2 x(1 - x) + \varepsilon_2 (x - y) \end{cases} \tag{14}$$

This is a $Z_0 - Z_2 - Z_4$ noninvertible map. In fact, given x' and y', if we solve the fourth degree algebraic system with respect to the unknowns x and y we may get four, two or no real solutions. The inverse maps are not easy to write by an elementary analytic form, since they are obtained by solving a fourth degree algebraic system. Nevertheless, it is easy to draw critical curves in order to obtain the boundaries separating the regions Z_k characterized by different numbers of inverses. In fact, from the Jacobian matrix

$$DT(x,y) = \begin{bmatrix} -\varepsilon_1 & \mu_1(1-2y)+\varepsilon_1 \\ \mu_2(1-2x)+\varepsilon_2 & -\varepsilon_2 \end{bmatrix}$$

the equation of LC_{-1}, defined by det $DT(x,y)=0$, is given by

$$4\mu_1\mu_2xy - 2\mu_2(\mu_1+\varepsilon_1)x - 2\mu_1(\mu_2+\varepsilon_2)y + \mu_1\mu_2 + \mu_1\varepsilon_2 + \mu_2\varepsilon_1 = 0.$$

Hence LC_{-1} is formed by the two branches, say $LC_{-1}^{(a)}$ and $LC_{-1}^{(b)}$, of an equilateral hyperbola (see Fig.5a). Also the critical set of rank-1 LC =

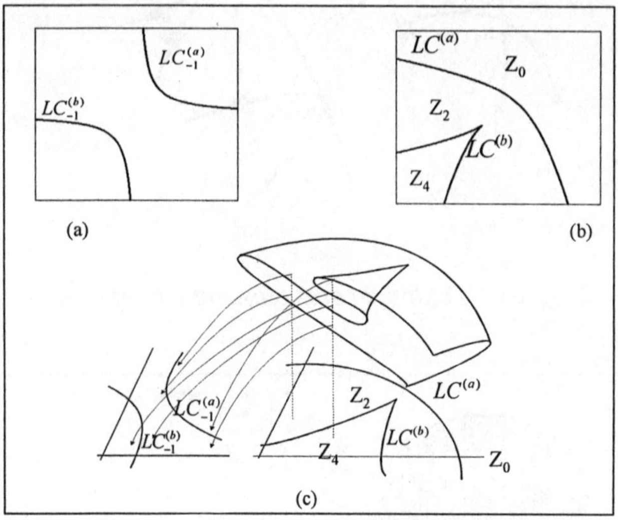

Figure 5: (a) LC_{-1}. (b) *LC and the Zones* $Z_2 - Z_4 - Z_0$. (c) *Riemann foliation for the map (14).*

$T(LC_{-1})$, obtained by taking the image by T of LC_{-1}, is formed by two disjoint branches: $LC = LC^{(a)} \cup LC^{(b)}$, where $LC^{(a)} = T(LC_{-1}^{(a)})$ and $LC^{(b)} = T(LC_{-1}^{(b)})$ (see Fig.5b). The branch $LC^{(a)}$ separates the regions Z_0 and Z_2, the branch $LC^{(b)}$ separates the regions Z_2 and Z_4. The Riemann foliation associated with the map (14) is qualitatively sketched in Fig.5c. Notice that the cusp point of $LC^{(b)}$ is characterized by three merging preimages at the junction of two folds.

An important property of the critical curves is that when an arc γ crosses LC_{-1} then its image $T(\gamma)$ is folded along LC, so that it is entirely included in the region characterized by a higher number of preimages. A qualitative picture is shown in Fig.6. The effect on a plane figure is shown in Fig.7, where the map (12) is applied to a circle crossing through LC_{-1}.

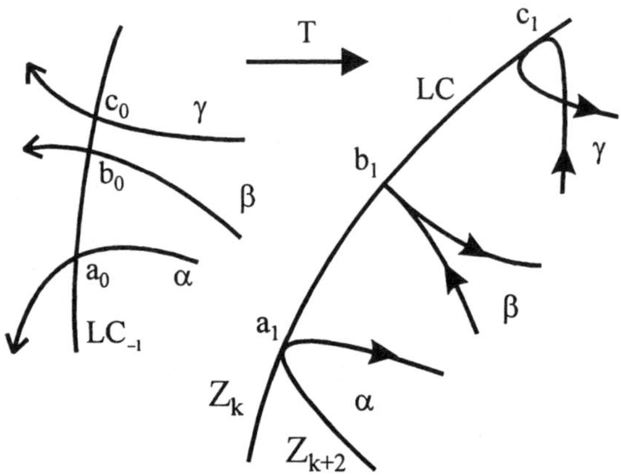

Figure 6: *Folding of arcs crossing through LC_{-1}.*

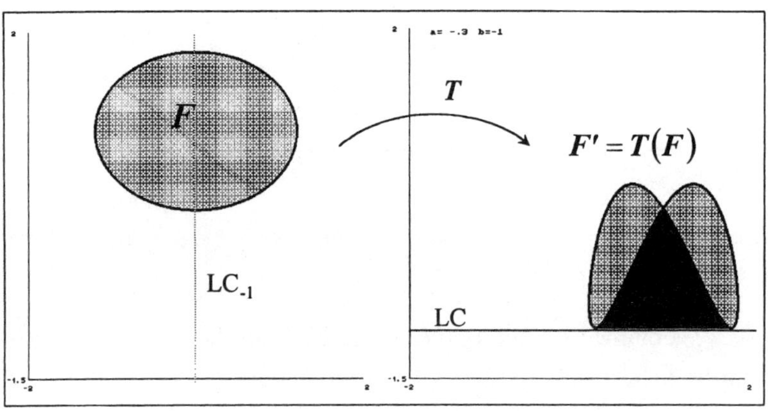

Figure 7: *Folding of a circle crossing through LC_{-1}.*

3 Absorbing regions, chaotic attractors and their delimitation

Portions of the critical set CS and its images $CS_k = T^k(CS)$ can be used to obtain the boundaries of trapping regions where the asymptotic dynamics of the iterated points of a noninvertible map are confined. This can be easily explained for a one-dimensional noninvertible map, for example the quadratic map (6). In fact, it is quite evident that if we iterate the logistic map for $2 < \mu < 4$ starting from an initial condition inside the interval $[c_1, c]$, with $c_1 = f(c)$, no images can be obtained out of this interval (see Fig.8), i.e. the interval formed by the critical point c and its rank-1 image c_1 is trapping. Moreover, any trajectory generated from an initial condition in

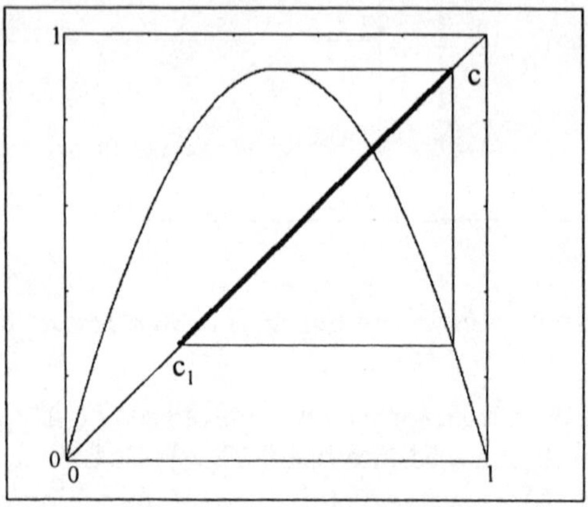

Figure 8: *Absorbing interval*

$(0, 1)$, enters $[c_1, c]$ after a finite number of iterations. Following the terminology introduced in Mira et al., 1996, the interval $[c_1, c]$ is called absorbing. In general, for an n-dimensional map, an *absorbing region* $\mathcal{A}$ (intervals in R, areas in R^2, volumes in R^3, ...) is defined as a bounded set whose boundary is given by portions of the critical set CS and its images of increasing order $CS_k = T^k(CS)$, such that a neighborhood $U \supset \mathcal{A}$ exists whose point enter $\mathcal{A}$ after a finite number of iterations and then never escape it, since $T(\mathcal{A}) \subseteq \mathcal{A}$, i.e. $\mathcal{A}$ is trapping (see e.g. Mira et al., 1996 for more details). Loosely speaking, we can say that the iterated application of a noninvertible map, folding and folding again the space, defines trapping regions bounded by critical sets of increasing order.

Sometimes, smaller absorbing regions are nested inside a bigger one. This can be illustrated, again, for the logistic map (6), as shown in Fig.9a, where inside the absorbing interval $[c_1, c]$ a trapping subset is obtained by higher rank images of the critical point, given by $\mathcal{A} = [c_1, c_3] \cup [c_2, c]$. In Fig.9b it is shown that, for the same parameter value $\mu = 3.61$ as in Fig.9a, the iteration of the logistic map gives points which never escape $\mathcal{A}$.

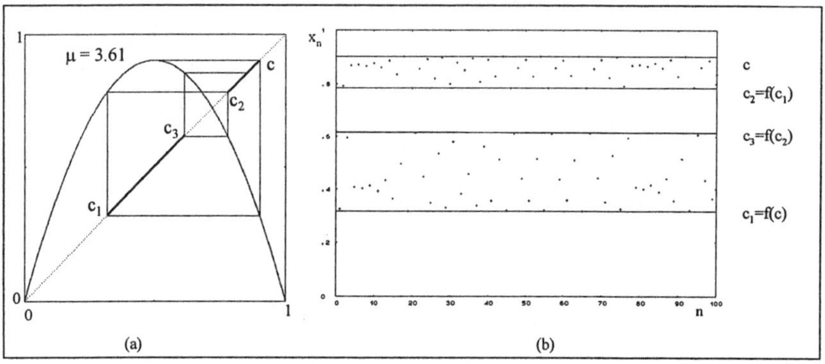

Figure 9: *Absorbing intervals nested inside* $[c_1, c]$.

An absorbing region $\mathcal{A}$, for which the property $T(\mathcal{A}) \subseteq \mathcal{A}$ holds, may be invariant, i.e. exactly mapped into itself, $T(\mathcal{A}) = \mathcal{A}$, or strictly mapped into itself, $T(\mathcal{A}) \subset \mathcal{A}$. Moreover, several invariant absorbing regions may exist, one embedded into the other. In these cases the concept of *minimal invariant absorbing region* is often useful, where *minimal* means the smallest one, in the sense that no other smaller absorbing regions are nested inside it.

Inside an absorbing region one or more attractors may exist. However, if a chaotic attractor exists which fills up the absorbing region then it is also called chaotic region, and the boundary of the chaotic attractor is formed by portions of critical sets. This is the situation shown in Fig.9, where the absorbing interval $\mathcal{A} = [c_1, c_3] \cup [c_2, c]$ is invariant and filled up by a chaotic trajectory, as shown in Fig.9b. To better illustrate this point, we also give a two-dimensional example, obtained by using the map (12). In Fig.10a a chaotic trajectory is shown, and in Fig.10b an absorbing area around it is obtained by the union of a segment of LC and three iterates $LC_i = T^i(LC)$, $i = 1, 2, 3$.

Indeed, following Mira et al., 1996, a practical procedure can be outlined in order to obtain the boundary of an absorbing area (although it is difficult to give a general method). Starting from a portion of LC_{-1}, approximately taken in the region occupied by the area of interest, its images by T of increasing rank are computed until a closed region is obtained. When such a region is mapped into itself, then it is an absorbing area $\mathcal{A}$. The length of the initial segment is to be taken, in general, by a trial and error method, although several suggestions are given in the books referenced above. Once an absorbing area $\mathcal{A}$ is found, in order to see if it is invariant or not the same procedure must be repeated by taking only the portion

$$\gamma = \mathcal{A} \cap LC_{-1} \tag{15}$$

as the starting segment. Then one of the following two cases occurs:

(case I) the union of m iterates of γ (for a suitable m) covers the whole boundary of $\mathcal{A}$; in which case $\mathcal{A}$ is an invariant absorbing area, and

$$\partial\mathcal{A} \subset \bigcup_{k=1}^{m} T^k(\gamma) \tag{16}$$

(case II) no natural m exists such that $\bigcup_{i=1}^{m} T^i(\gamma)$ covers the whole boundary of $\mathcal{A}$; in which case $\mathcal{A}$ is not invariant but strictly mapped into itself. An invariant absorbing area is obtained by $\cap_{n>0} T^n(\mathcal{A})$ (and may be obtained by a finite number of images of $\mathcal{A}$).

The application of this procedure to the problem of the delimitation of the chaotic area of Fig.10 by portions of critical curves suggests us, on the basis of Fig.10b, to take a smaller segment γ and to take an higher number of iterates in order to obtain also the inner boundary. The result is shown in Fig.11. By five iterates we get the outer boundary, shown in Fig.11a, by more iterates also the inner boundary of the chaotic area is get, as shown in Fig.11b. As it can be clearly seen, and as clearly expressed by the strict inclusion in (16), the union of the images also include several arcs internal to the invariant area $\mathcal{A}$. Indeed, the images of the critical arcs which are mapped inside the area play a particular role, because these curves represent the "foldings" of the plane under forward iterations of the map, and this is the reason why these inner curves often denote the portions of the region which are more frequently visited by a generic trajectory inside it (compare Fig.10a and Fig.11b); many examples are given in the literature on noninvertible maps, see e.g. Mira et al., 1996. The points close to a critical arc LC_i, $i \geq 0$, are more frequently visited because there are several distinct parts of the invariant area which are mapped in the same region (close to LC_i) in $i+1$ iterations.

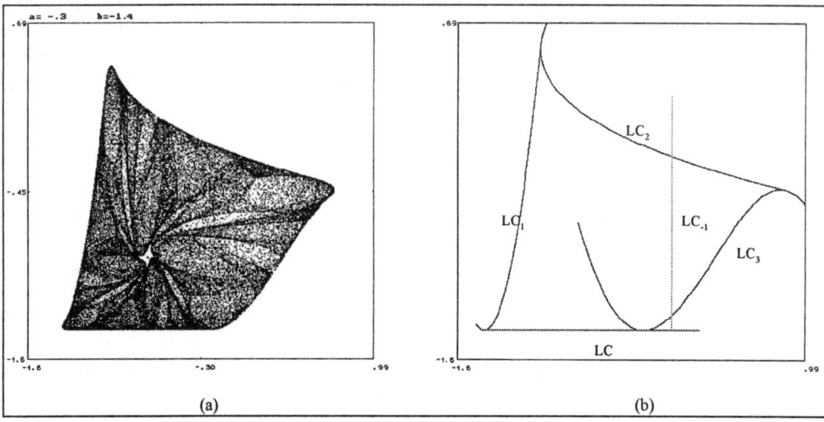

Figure 10: (a) *Chaotic trajectory for a two-dimensional map.* (b) *The outer boundary of the chaotic area obtained by the segments of* LC_k, $k = 1, 2, 3$.

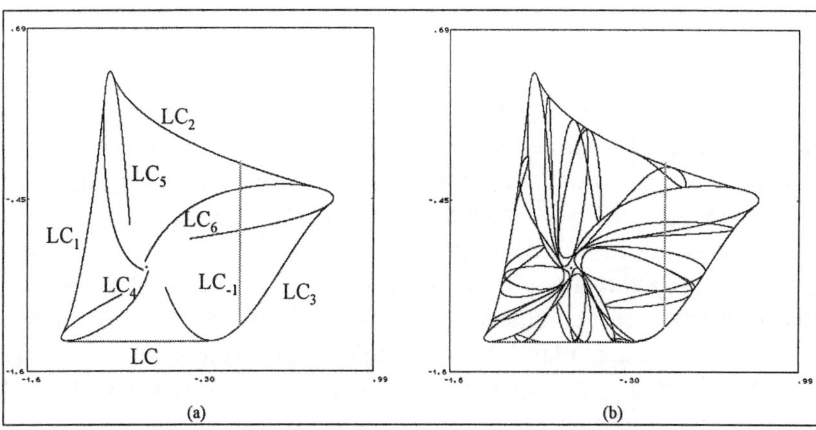

Figure 11: (a) *Five iterates of LC give the outer boundary of the chaotic area shown in of Fig.10.* (b) *More iterates also give the inner boundary.*

4 Contact bifurcations and the creation of complex basins

From (4) it is clear that the properties of the inverses are important in order to understand the structure of the basins and the main bifurcations which change their qualitative properties. In the case of noninvertible maps, the multiplicity of preimages may lead to basins with complex structures, such as multiply connected or non connected sets, sometimes formed by infinitely many non connected portions (see Mira et al., 1994, Mira and Rauzy, 1995, Mira et al., 1996, ch.5, Abraham et al., 1997,ch.5). In the context of noninvertible maps it is useful to define the *immediate basin* $\mathcal{B}_0(A)$, of an attracting set A, as the widest connected component of the basin which contains A. Then the total basin can be expressed as

$$\mathcal{B}(A) = \bigcup_{n=0}^{\infty} T^{-n}(\mathcal{B}_0(A))$$

where $T^{-n}(\mathbf{x})$ represents the set of all the rank-n preimages of $\mathbf{x}$, i.e. the set of points which are mapped in $\mathbf{x}$ after n iterations of the map T. The backward iteration of a noninvertible map *repeatedly unfolds* the phase space, and this implies that the basins may be non-connected, i.e. formed by several disjoint portions.

Also in this case, we first illustrate this property by using a one-dimensional map[2] In Fig.12 the graph of a $Z_1 - Z_3 - Z_1$ noninvertible map is shown, where two stable fixed points exist, denoted by O and p, whose basins $\mathcal{B}(O)$ and $\mathcal{B}(p)$ are represented along the diagonal by light and dark grey thick lines respectively. In Fig.12a the two basins are connected sets, bounded by the unstable fixed points q and r. As a parameter is varied, the critical point c (the local minimum), which separates the lower region Z_1 from the region Z_3, moves downwards, until it has a contact with the basin boundary q, and a crossing occurs if the parameter is further changed. Before the contact $c > q$ (see the enlargement in Fig.12b), whereas after the contact we have $c < q$. This implies that a portion of $\mathcal{B}(O)$ enters Z_3, i.e. the segment $[c, q) = H_0$, thus causing the appearance of non connected portions of $\mathcal{B}(O)$ nested inside $\mathcal{B}(p)$. Indeed, infinitely many non connected portions (or *holes*) are suddenly created at the contact bifurcation, given by the preimages of any rank of the portion H_0 of $\mathcal{B}(O)$ included into Z_3, see $H_{-i} = T^{-i}(H_0)$, $i = 1, 2, ...$, in Fig.12c. Before the bifurcation $\mathcal{B}(p) = (q, r)$ (Fig.12a), after the bifurcation the basin of p is given

[2]The example is taken from an evolutionary game proposed in Bischi et al., 2001b.

by the immediate basin $\mathcal{B}_0(p) = (q, q_{-1,1})$ (Fig.12c), and all its preimages, $\mathcal{B}(p) = \bigcup_{n=0}^{\infty} T^{-n}(\mathcal{B}_0)$, given by infinitely many segments that have the unstable fixed point r as limit point. So, the contact between the critical point c and the basin boundary q marks the transition from simple connected to non connected basins.

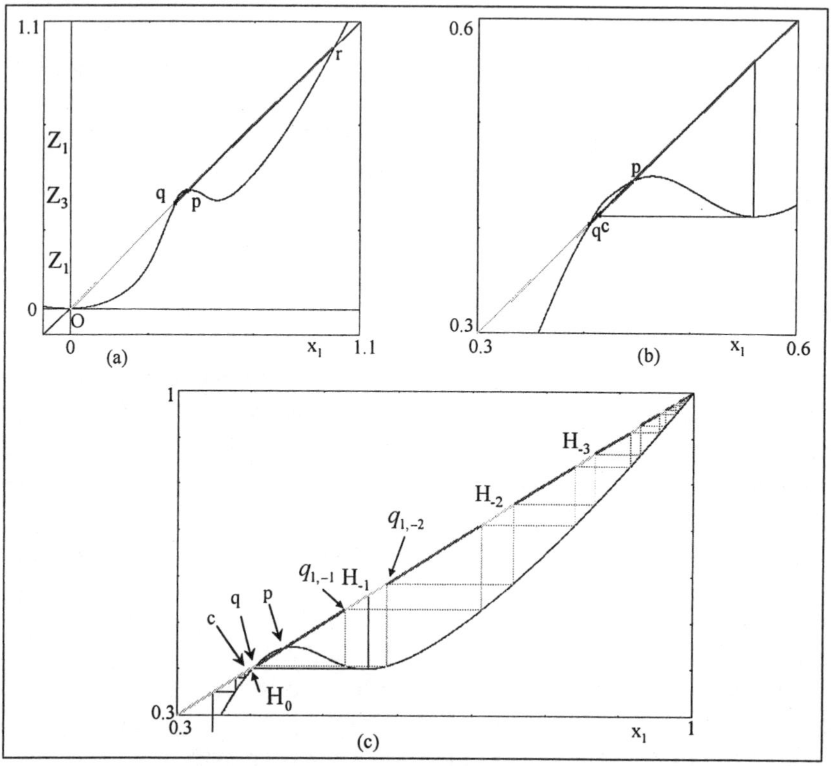

Figure 12: (a) *Graph of a* $Z_1 - Z_3 - Z_1$ *noninvertible map and basins of attraction of* p *and* O*, before the contact between the critical point* c *and the basin boundary* q. (b) *Enlargement.* (c) *The basins of* p *and* O *after the contact.*

Similar global bifurcations, due to contacts between critical sets and basin boundaries, also occur in higher dimensional maps. In general, the origin of complex topological structures of the basins, like those formed by non connected sets, can be easily explained on the basis of the geometrical properties of a noninvertible map. In fact, suppose that p is a fixed point of

T, i.e. $T(p) = p$. Hence one of the preimages of p is p itself, but if T^{-1} is multivalued in p, i.e. $p \in Z_k$, with $k \geq 2$, then other preimages of the fixed point p exist. If the fixed point is stable, and $\mathcal{B}_0(p)$ is its immediate basin, then the total basin of p must also include all the rank-1 preimages of the points of $\mathcal{B}_0(p)$, which may be also far (i.e. disjoint) from $\mathcal{B}_0(p)$. Whenever such disjoint preimages belong to regions where many inverses exist, higher rank preimages of $\mathcal{B}_0(p)$ must be included in the basin of p and so on[3]. Such behavior of the multivalued inverse of T may be better visualized by using the concept of Riemann foliation (see Fig.13).

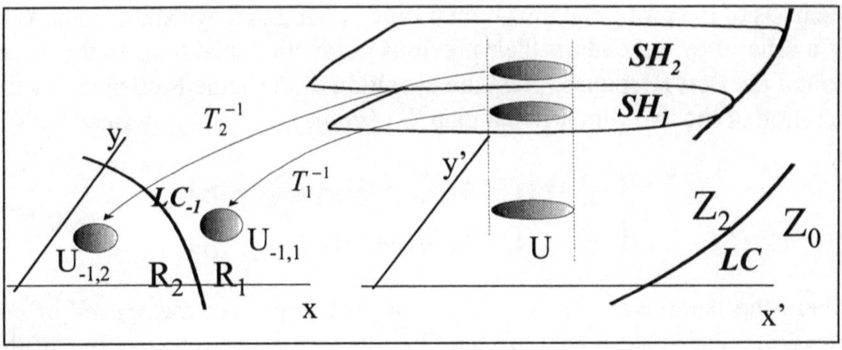

Figure 13: *Riemann foliation which visualizes the basic mechanism for the creation of disconnected basins*

Also in higher dimensional cases, the global bifurcations which give rise to complex topological structures of the basins, like those formed by non connected sets, can be explained in terms of contacts of basins boundaries and critical sets. In fact, if a parameter variation causes a crossing between a basin boundary and a critical set which separates different regions Z_k so that a portion of a basin enters a region where a higher number of inverses is defined, then new components of the basin may suddenly appear at the contact. However, for maps of dimension greater than 1, such kinds of bifurcations can be very rarely studied by analytical methods, since the analytical equations of such singularities are not known in general. Hence these studies are mainly performed by geometric and numerical methods.

[3] A similar reasoning applies to any kind of attractor, such as a periodic orbit (a cycle) or a chaotic set.

4.1 Basin bifurcations in a duopoly game with two stable Nash equilibria. We consider a Cournot duopoly game proposed in Kopel, 1996, for which the structure of the basins and their qualitative changes are analyzed in Bischi et al., 1999b. This game describes a market where, at each time period t, two firms decide their productions for the next period on the basis of best reply functions expressed as $q_i(t+1) = r_i(q_j(t))$, $i,j = 1,2$, $i \neq j$, where the reaction functions r_i assume the form of logistic maps $r_i(q_j) = \mu_i q_j (1 - q_j)$ (in Kopel, 1996, these functions are derived as Best Responses). The dynamic duopoly game is obtained by assuming that competitors do not immediately adjust to the optimal quantity they computed on the basis of the profit maximization problem, but that they exhibit some kind of inertia: they only adjust their previous production quantities in the direction of the Best Response. The time evolution of the game is obtained by the iteration of the two-dimensional map $T : (q_1, q_2) \to (q_1', q_2')$ defined by

$$T : \begin{cases} q_1' = (1 - \lambda_1) q_1 + \lambda_1 \mu_1 q_2 (1 - q_2) \\ q_2' = (1 - \lambda_2) q_2 + \lambda_2 \mu_2 q_1 (1 - q_1) \end{cases} . \qquad (17)$$

where the parameters $\lambda_i \in [0,1]$, $i = 1,2$, represent the *speeds of adjustment*. The fixed points of map (17), located at the intersections of the two reaction curves, coincide with the Nash equilibria of the duopoly game (see Kopel, 1996). As shown in Bischi et al., 1999b, under the assumption $\mu_1 = \mu_2 = \mu$, the fixed points of (17) can be expressed by simple analytical expressions, and a range of parameters μ, λ_1, λ_2 exist such that two of them are both stable. For example, for the set of parameters used in Fig.14a, given by $\mu = 3.5$, $\lambda_1 = 0.6$, $\lambda_2 = 0.8$, four equilibria exist, indicated by O, S, E_1 and E_2 in the figure: O and S are saddle points, whereas E_1 and E_2 are both stable, each with its own basin of attraction, say $\mathcal{B}(E_1)$ and $\mathcal{B}(E_2)$, represented by white and light grey respectively (the dark grey region represents the basin of infinity, i.e. the set of initial conditions which generate unbounded trajectories).

In the presence of multiple stable Nash equilibria the problem of equilibrium selection arises, and this naturally leads to the question of the delimitation of the basins of attraction. As argued in section 2, the properties of the inverses of the map become important in order to understand the structure of the basins and their qualitative changes. Indeed, the map (17) is a noninvertible map, as it can be easily deduced from the fact that given a point $\mathrm{q}' = (q_1', q_2') \in \mathbb{R}^2$, its rank-1 preimages $T^{-1}(\mathrm{q}')$, computed by solving the fourth degree algebraic system (17) with respect to the quantities q_1 and

q_2, may be up to four. It is easy to realize that the map (17) is a noninvertible map of $Z_0 - Z_2 - Z_4$ type, its *Riemann* foliation is similar to the one associated with the map (14) analyzed in section 2, see Fig. 5.

Being the map T continuously differentiable, LC_{-1} coincides with the set of points on which $\det DT = 0$, which gives

$$\left(q_1 - \frac{1}{2}\right)\left(q_2 - \frac{1}{2}\right) = \frac{(1-\lambda_1)(1-\lambda_2)}{4\lambda_1\lambda_2\mu_1\mu_2} \tag{18}$$

This equation represents an equilateral hyperbola, whose two branches are denoted by $LC_{-1}^{(a)}$ and $LC_{-1}^{(b)}$ in Fig.14. It follows that also $LC = T(LC_{-1})$ is the union of two branches, say $LC^{(a)} = T(LC_{-1}^{(a)})$ and $LC^{(b)} = T(LC_{-1}^{(b)})$. The branch $LC^{(b)}$ separates the region Z_0, whose points have no preimages, from the region Z_2, the other branch $LC^{(a)}$ separates the region Z_2 from Z_4, whose points have four distinct preimages (see Fig.14).

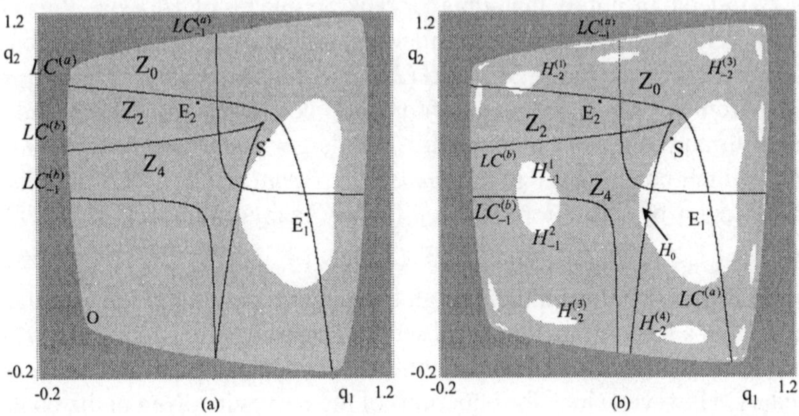

Figure 14: (a) *Simply connected basin of* E_1. (b) *Disconnected basin of* E_1 *due to a contact bifurcation.*

In order to understand how complex basin structures are obtained, we start from a situation in which the basins are connected sets, like the one shown in Fig 14a. In this situation, the smaller basin $\mathcal{B}(E_1)$ is a simply connected set, and the boundary which separates the basins $\mathcal{B}(E_1)$ and $\mathcal{B}(E_2)$ is given by the whole stable set $W^s(S)$ of the saddle S. In Fig.14a, $W^s(S)$ is entirely included inside the regions Z_2 and Z_0. However, the fact that a portion of $W^s(S)$ is close to $LC^{(b)}$ suggests the occurrence of a global bifurcation. If the parameters are changed, so that a contact between $W^s(S)$ and LC occurs, this contact will mark a bifurcation which causes qualitative changes in the structure of the basins. If a portion of $\mathcal{B}(E_1)$ enters Z_4 after a contact with $LC^{(b)}$, new rank-1 preimages of that portion will appear near $LC_{-1}^{(b)}$, and such preimages must belong to $\mathcal{B}(E_1)$. Indeed, this is the situation shown in Fig.14b, obtained after a small change of λ_1. The portion of $\mathcal{B}(E_1)$ inside Z_4 is denoted by H_0. It has two rank-1 preimages, denoted by $H_{-1}^{(1)}$ and $H_{-1}^{(2)}$, which are located at opposite sides with respect to $LC_{-1}^{(b)}$ and merge on it (in fact, by definition, the rank-1 preimages of the arc of $LC^{(b)}$ which bound H_0 must merge along $LC_{-1}^{(b)}$). The set $H_{-1} = H_{-1}^{(1)} \cup H_{-1}^{(2)}$ constitutes a non connected portion of $\mathcal{B}(E_1)$. Moreover, since H_{-1} belongs to the region Z_4, it has four rank-1 preimages, denoted by $H_{-2}^{(j)}$, $j = 1, ..., 4$ in Fig.14b, which constitute other four "islands" of $\mathcal{B}(E_1)$, or "holes" of $\mathcal{B}(E_2)$. Points of these "islands" are mapped into H_0 in two iterations of the map T. Indeed, infinitely many higher rank preimages of H_0 exist, thus giving infinitely many smaller and smaller disjoint "islands" of $\mathcal{B}(E_1)$. Hence, at the contact between $W^s(S) = \partial\mathcal{B}(E_1)$ and LC the basin $\mathcal{B}(E_1)$ is transformed from a simply connected into a non connected set, constituted by infinitely many disjoint components. The larger connected component of $\mathcal{B}(E_1)$ which contains E_1 is the *immediate basin* $\mathcal{B}_0(E_1)$, and the whole basin is given by the union of the infinitely many preimages of $\mathcal{B}_0(E_1)$: $\mathcal{B}(E_1) = \bigcup_{n\geq 0} T^{-n}(\mathcal{B}_0(E_1)) = \mathcal{B}_0(E_1) \cup \left[\bigcup_{n\geq 0} T^{-n}(H_0)\right]$. So, also in this two-dimensional example, the global bifurcation which causes a transformation of a basin from connected set into the union of infinitely many non connected portions, is caused by a contact between a critical set and a basin boundary. However, since the equations of the curves involved in the contact cannot be analytically expressed in terms of elementary functions, the occurrence of contact bifurcations can only be revealed numerically. This happens frequently in nonlinear dynamical systems of dimension greater than one, where the study of global bifurcations is generally obtained through an interplay between theoretical and numerical methods, and the occurrence of these bifurcations is shown by computer-assisted proofs, based on the knowledge of the properties of the critical curves and their graphical representation (see e.g. Mira et al., 1996, for many examples). This "modus operandi" is typical in the study of the global bifurcations of nonlinear two-dimensional maps.

This implies that an extension of such methods to the study of higher dimensional noninvertible maps is not easy in general. Indeed, some nontrivial practical problems arise, related to the obvious reason that the computer screen is two-dimensional, so the visualization of objects in a phase spaces of dimension greater than two, and the detection of contacts among these objects as their shapes change, may become a very difficult task. In other words, the extension to higher-dimensional systems of the results on contact bifurcations, which gave so many interesting and promising results in the study of two-dimensional noninvertible maps, may become a very hard and challenging task, due to the difficulties met in the computer-assisted graphical visualization. For example, in Agiza et al., 1999, a repeated Cournot game is considered, whose time evolution is obtained by the iteration of the three-dimensional map $T : (q_1, q_2, q_3) \to (q_1', q_2', q_3')$

$$T : \begin{cases} q_1' = (1 - \lambda_1)\, q_1 + \lambda_1 \mu_1 \left[q_2 \left(1 - q_2\right) + q_3 \left(1 - q_3\right)\right] \\ q_2' = (1 - \lambda_2)\, q_2 + \lambda_2 \mu_2 \left[q_3 \left(1 - q_3\right) + q_1 \left(1 - q_1\right)\right] \\ q_3' = (1 - \lambda_3)\, q_3 + \lambda_3 \mu_3 \left[q_1 \left(1 - q_1\right) + q_2 \left(1 - q_2\right)\right] \end{cases} \tag{19}$$

which can be seen as the extension of the game illustrated above to the case of three players. In Agiza et al., 1999, two-dimensional sections are employed in order to visualize the basins of coexisting attractors, but this method is not useful to detect the occurrence of qualitative changes in the structure of the basins and the contact bifurcations which cause such changes. The same game has been re-examined in Bischi et al., 2001c, where enhanced graphical methods have been used to modulate the opacity of the outer objects in order "to see through", in order to visualize objects which are nested inside other objects . Moreover, the critical sets, which are now two-dimensional surfaces embedded in a three-dimensional phase space, have been represented like semi-transparent veils, so that their contacts with portions of basin boundaries, also given by two-dimensional surfaces, can be detected. Some example of the kind of graphical results obtained are shown in Fig.15, which can be considered just as snapshots of animated sequences which allow interactive rotation of the three dimensional figures. In Fig.15 the basins of four coexisting stable Nash equilibria are represented by different colors, and the region outside is the outer boundary constitutes the basin of infinity. The sequence of figures shown in Fig.15 clearly show that a contact bifurcation occurs at which one of the basins is transformed from connected into a non connected set. The occurrence of this global bifurcations is caused by a contact between a basin boundary and a critical surface, after which a portion of the immediate basin enters a zone characterized by a higher number of preimages, as explained above. In the figures the critical surfaces are not shown, because a proper visualization requires the usage of several grades of colors. The animated sequences which emphasize the contacts can the seen in the web page associated with the paper by Bischi et al., 2001c[4].

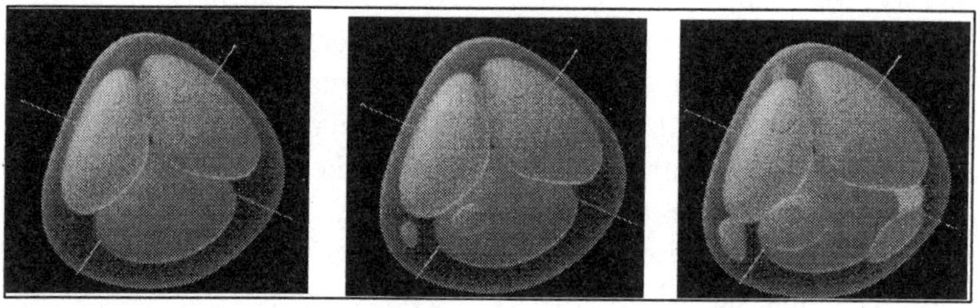

Figure 15: *Example of contact bifurcation of basins of attraction in a three-dimensional map.*

5 An important class of Cournot duopoly games

In this section we consider a repeated Cournot duopoly model with naive expectations, i.e. a classical discrete-time Cournot tâtonnement modeled by the iteration of the two dimensional map

$$\Phi : \begin{cases} x' = r_1(y) \\ y' = r_2(x) \end{cases} \tag{20}$$

where r_1 and r_2 are the reaction functions (or best reply functions) defined in X and Y respectively (so that Φ is defined in the rectangle $X \times Y$). In a famous paper by Rand, 1978, it is proved that quite complex dynamics, with periodic and chaotic trajectories, can emerge from the iteration of (20). Other peculiar properties of this kind of Cournot duopoly games are given in Dana and Montrucchio, 1986, where it is shown that the properties of the two-dimensional map (20) can be deduced from the properties of one-dimensional maps obtained by the composition of the reaction functions, while peculiar properties of the bifurcations associated with the two-dimensional maps (20) are given in Lupini et al., 1997. Starting from these papers, Bischi et al., 2000a, show that, in general, maps of the form (20) are characterized by *multistability*, i.e. coexistence of many distinct attractors,

[4]See the URL `http://bandviz.cg.tuwien.ac.at/basinviz/disjoint/`

that may be stable periodic cycles or cyclic chaotic attractors, and study the structure of their basins of attraction.

We now recall some general properties of the dynamic duopoly games (20), and we stress the peculiar structure of the critical sets and the basins.

5.1 General properties of maps Φ: $(x,y) \to (r_1(y), r_2(x))$. A trajectory of the map Φ represents the Cournot tâtonnement of a duopoly game in which the producers simultaneously update their productions at each discrete time period. Moreover, as already noticed in Dana and Montrucchio, 1986, among the possible sequences generated by the iteration of (20) there are also the so called *Markov-Perfect-Equilibria* (MPE henceforth) processes, where at each discrete time only one player moves, that is, the two players move alternatively, each choosing the best reply to the previous action of the other player. This occurs if the phase point (x_t, y_t) belongs alternatively to the graphs of the reaction curves $y = r_2(x)$ and $x = r_1(y)$. This condition is satisfied if the initial condition (i.c. henceforth) (x_0, y_0) belongs to a reaction curve, i.e. $y_0 = r_2(x_0)$ (player 1 moves first) or $x_0 = r_1(y_0)$ (player two moves first). This follows from the fact that the set

$$R_{12} = R_1 \cup R_2 \tag{21}$$

with $R_1 = \{(r_1(y), y) \,|y \in Y\}$ and $R_2 = \{(x, r_2(x)) \,|x \in X\}$, that represents the union of the graphs of the two reaction functions, is a trapping set for Φ, i.e. $\Phi(R_{12}) \subseteq R_{12}$. In fact, it is easy to realize that the image of a point belonging to a reaction curve belongs to the other reaction curve, so any i.c. $(x_0, y_0) \in R_{12}$ generates a trajectory entirely belonging to R_{12}, $\Phi^t(x_0, y_0) \in R_{12}\ \forall t \geq 0$. We shall call such a trajectory an *MPE trajectory*.

A particular MPE trajectory is a fixed point of the map Φ. In fact, (x^*, y^*) is a fixed point of Φ iff $x^* = r_1(y^*)$ and $y^* = r_2(x^*)$, i.e. a point of intersection of the graphs R_1 and R_2 of the two reaction functions,

$$(x^*, y^*) \in R_{12}\,. \tag{22}$$

While an i.c. $(x_0, y_0) \in R_{12}$ generates an MPE trajectory, a "generic" i.c. $(x_0, y_0) \notin R_{12}$ shall give rise to a Cournot tâtonnement, with (x_t, y_t) not belonging, in general, to R_{12}. Note, however, that a trajectory starting with an i.c. $(x_0, y_0) \notin R_{12}$ may enter the trapping set R_{12} after a finite number of steps, since a point of the set R_{12} can have preimages out of R_{12}.

Let us turn now to the generic dynamics. We first recall some properties of the map (20), that will be used in the following. Let

$$F(x) = r_1 \circ r_2(x)\ ,\ x \in X, \quad \text{and} \quad G(y) = r_2 \circ r_1(y)\ ,\ y \in Y \tag{23}$$

where we assume that the sets X and Y are such that the maps F and G are well defined. Then the following three properties hold (see Dana and Montrucchio, 1986)

Property 1. $\Phi^{2k}(x,y) = \left(F^k(x), G^k(y)\right)$ *for each integer* $k \geq 1$.

This property easily follows from the fact that the square map Φ^2 (the second iterate of Φ) is a decoupled map, since $\Phi^2(x,y) = \Phi(r_1(y), r_2(x)) = (r_1(r_2(x)), r_2(r_1(y))) = (F(x), G(y))$.

Property 2. *For each* $n \geq 1$ *the two one-dimensional maps* F *and* G *satisfy*:

$$r_1 \circ G^n(y) = r_1 \circ r_2 \circ r_1 \circ ... \circ r_2 \circ r_1(y) = F^n \circ r_1(y)$$
$$r_2 \circ F^n(x) = r_2 \circ r_1 \circ r_2 \circ ... \circ r_1 \circ r_2(x) = G^n \circ r_2(x)$$

From Property 2 we deduce that the cycles of the maps F and G (and their stability properties), are strictly related. In particular, a correspondence between the cycles of the two maps is defined by the following

Property 3. *If* $\{x_1, ..., x_n\}$ *is an n-cycle of* F *then* $\{y_1, ..., y_n\} = \{r_2(x_1), ..., r_2(x_n)\}$ *is an n-cycle of* G.
If $\{y_1, ..., y_n\}$ *is an n-cycle of* G *then* $\{x_1, ..., x_n\} = \{r_1(y_1), ..., r_1(y_n)\}$ *is an n-cycle of* F

Such kinds of cycles of F and G shall be called *conjugate*. That is, for each cycle of F (resp. G), a conjugate one of G (resp. F) exists, and the two conjugate cycles have the same stability property (both are stable or both are unstable). In fact, due to the chain-rule for the derivative of composite functions, the cycle $\{x_1, ..., x_n\}$ of F and the conjugate cycle $\{y_1 = r_2(x_1), ..., y_n = r_2(x_n)\}$ of G have the same eigenvalue $\lambda = \prod_{i=1}^n DF(x_i) = \prod_{i=1}^n DG(y_i) = \prod_{i=1}^n Dr_1(y_i) Dr_2(x_i)$.

These properties show that the cycles of the Cournot map Φ are related to the cycles of the one-dimensional maps F and G defined in (23). Now we answer to the following questions: if F and G have cycles $C_F^{(p)} = \{x_1^*, ..., x_p^*\}$ and $C_G^{(q)} = \{y_1^*, ..., y_q^*\}$, of period p and q respectively, with eigenvalues $\lambda_F = \prod_{i=1}^p DF(x_i^*)$ and $\lambda_G = \prod_{i=1}^q DG(y_i^*)$ how many cycles of Φ are generated by these cycles ? and how can the points of the cycles of Φ be obtained ? and what are their stability properties ?.

We first consider the case $p = q = 1$, that is, the case of fixed points of F and G. Due to the correspondence defined by property 3, F and G have the same number of fixed points, say m, and if $X^* = \{x_1^*, ..., x_m^*\}$ is the set of fixed points of F then $Y^* = \{y_1^*, ..., y_m^*\}$, with $y_i^* = r_2(x_i^*)$, is the set of fixed points of G. Let us consider the m^2 points of the phase space of Φ obtained by the Cartesian product $X^* \times Y^*$. Among these points there are exactly m fixed points of the Cournot map Φ, belonging to $R_1 \cap R_2$, given by

$$p_i^* = (x_i^*, y_i^*) = (x_i^*, r_2(x_i^*)) \quad i = 1, ..., m \tag{24}$$

whereas the remaining $m^2 - m$ points of the form $\left(x_i^*, y_j^*\right)$, $i \neq j$, belong to 2-cycles of the map Φ. In fact

$$\begin{aligned}\Phi\left(x_i^*, y_j^*\right) &= \Phi\left(x_i^*, r_2\left(x_j^*\right)\right) = \left(r_1 \circ r_2(x_j^*), r_2\left(x_i^*\right)\right) \\ &= \left(F(x_j^*), r_2\left(x_i^*\right)\right) = \left(x_j^*, y_i^*\right)\end{aligned}$$

and, analogously,

$$\Phi\left(x_j^*, y_i^*\right) = \Phi\left(x_j^*, r_2\left(x_i^*\right)\right) = \left(x_i^*, y_j^*\right) .$$

Since each 2-cycle is formed by two points of $X^* \times Y^*$ not belonging to the trapping set R_{12} we have $N_2 = \left(m^2 - m\right)/2$ cycles of Φ of period 2, given by

$$C_\Phi^{(2)} = \left\{\left(x_i^*, r_2(x_j^*)\right), \left(x_j^*, r_2(x_i^*)\right)\right\} \;\; i < j, \; i, j = 1...m . \tag{25}$$

Thus the existence of m distinct fixed points of F (and consequently of G) implies the existence, for the Cournot map Φ, of
(a) m fixed points given by (24);
(b) $m\,(m-1)/2$ cycles of period two given by (25).

The stability of the fixed points and of the 2-cycles of Φ can be easily obtained. For the fixed points (24) we have

$$D\Phi\left(x_i^*, r_2\left(x_i^*\right)\right) = \begin{bmatrix} 0 & Dr_1\left(r_2\left(x_i^*\right)\right) \\ Dr_2\left(x_i^*\right) & 0 \end{bmatrix}$$

so the eigenvalues are $\lambda_{1,2} = \pm\sqrt{DF(x_i^*)} = \pm\sqrt{DG(y_i^*)}$, i.e. the fixed point is a stable or unstable node (focus) if the multiplier of F at the fixed point x_i^* is positive (negative) with modulus less or greater than one respectively. Note in particular that a fixed point of Φ cannot be a saddle.

For a 2-cycle, which is given by (25), we have

$$D\Phi^2\left(x_i^*, r_2\left(x_j^*\right)\right) = \begin{bmatrix} DF(x_j^*) & 0 \\ 0 & DF\left(x_i^*\right) \end{bmatrix}$$

so the eigenvalues are $\lambda_1 = DF(x_j^*)$ and $\lambda_2 = DF(x_i^*)$. Then a 2-cycle is a stable node if both the fixed points of F and G, whose coordinates give the points of the cycle, are stable, whereas it is an unstable node (saddle) if both the fixed points are unstable (one stable and one unstable). A 2-cycle of Φ cannot be a focus cycle. Note also that a 2-cycle of Φ can never be a MPE cycle, since if the points of $X^* \times Y^*$ used to form the 2-cycle belong to the trapping set R_{12} then they necessarily are fixed points of Φ. We now generalize these results to the case in which F has a cycle of odd period.

Cycles of the map Φ due to cycles of F of odd period.

Let F have a cycle of odd period $n = 2k+1$, say $C_F^{(n)} = \{x_1^*, ..., x_n^*\}$, with eigenvalue $\lambda = \prod_{i=1}^{n} DF(x_i^*)$ (consequently G has the conjugate cycle $C_G^{(n)} = \{y_1^*, ..., y_n^*\}$ with $y_i^* = r_2(x_i^*)$, $i = 1, ..., n$, with the same eigenvalue λ). By combining the points of the Cartesian product $C_F^{(n)} \times C_G^{(n)}$ the following distinct coexisting cycles of the Cournot map Φ are obtained:

(a) one cycle of the same odd period $n = 2k+1$ given by

$$C_\Phi^{(n)} = \left\{ \Phi^t \left(x_1^*, r_2 \left(x_{k+1}^* \right) \right), t = 1, ..., n \right\} \tag{26}$$

with eigenvalues $\lambda_{1,2} = \pm\sqrt{\lambda}$. Hence $C_\Phi^{(n)}$ is a stable or unstable node (or focus) depending on the modulus (and the sign) of λ.

(b) $k = (n-1)/2$ cycles of even period $2n$ given by

$$\begin{aligned} {}^1C_\Phi^{(2n)} &= \left\{ \Phi^t \left(x_1^*, r_2 \left(x_1^* \right) \right), t = 1, ..., 2n \right\} \\ {}^2C_\Phi^{(2n)} &= \left\{ \Phi^t \left(x_1^*, r_2 \left(x_2^* \right) \right), t = 1, ..., 2n \right\} \\ &\vdots \\ {}^kC_\Phi^{(2n)} &= \left\{ \Phi^t \left(x_1^*, r_2 \left(x_k^* \right) \right), t = 1, ..., 2n \right\} . \end{aligned} \tag{27}$$

with eigenvalues $\lambda_1 = \lambda_2 = \lambda$ coincident with the common eigenvalue of the conjugated n-cycles of F and G, so that ${}^iC_\Phi^{(2n)}$, $i = 1, ..., k$ are stable (unstable) star nodes if the cycles $C_F^{(n)}$ and $C_G^{(n)}$ are stable (unstable)[5]. Of these $(k+1)$ cycles only ${}^1C_\Phi^{(2n)}$ is formed by points belonging to the trapping set R_{12}, i.e. only ${}^1C_\Phi^{(2n)}$ gives an MPE periodic trajectory (*MPE cycle*).

(c) Now let us suppose, as frequently happens, that F has more than one distinct cycles of period $n = 2k+1$, and hence the same holds for G. In

[5] A star-node is a node-cycle with real eigenvalues having equal moduli (see Mira, 1987, p.194).

this case, in addition to the points of the phase plane of Φ obtained by the Cartesian product of the pairs of conjugate cycles, there are also the points obtained by combining the points of the non-conjugate ones. For example, let F have two n-cycles $CX_F^{(n)} = \{x_1^*, ..., x_n^*\}$ and $CZ_F^{(n)} = \{z_1^*, ..., z_n^*\}$ with eigenvalues $\lambda_x = \prod_{i=1}^n DF(x_i^*)$ and $\lambda_z = \prod_{i=1}^n DF(z_i^*)$ respectively. This implies that G has the conjugate cycles $CX_G^{(n)} = \{r_2(x_1^*), ..., r_2(x_n^*)\}$ and $CZ_G^{(n)} = \{r_2(z_1^*), ..., r_2(z_n^*)\}$ with the same eigenvalues. In this case, by combining the $4n^2$ points of the Cartesian product $\left(CX_F^{(n)} \cup CZ_F^{(n)}\right) \times \left(CX_G^{(n)} \cup CZ_G^{(n)}\right)$ the $2(k+1)$ cycles listed in (26) and (27) can be obtained by using the $2n^2$ points of $\left(CX_F^{(n)} \times CZ_G^{(n)}\right) \cup \left(CZ_F^{(n)} \times CX_G^{(n)}\right)$, and by the remaining $2n^2$ points further n cycles of Φ of period $2n$ are obtained, given by

$$
\begin{aligned}
{}^1M_\Phi^{(2n)} &= \left\{\Phi^t\left(x_1^*, r_2\left(z_1^*\right)\right), t = 1, ..., 2n\right\} \\
{}^2M_\Phi^{(2n)} &= \left\{\Phi^t\left(x_1^*, r_2\left(z_2^*\right)\right), t = 1, ..., 2n\right\} \\
&\vdots \\
{}^nM_\Phi^{(2n)} &= \left\{\Phi^t\left(x_1^*, r_2\left(z_n^*\right)\right), t = 1, ..., 2n\right\}
\end{aligned}
\qquad (28)
$$

The eigenvalues of the cycles (28) can be easily computed, since $D\Phi^{2n}$ is diagonal, and are given, for each cycle, by $\lambda_1 = \lambda_x$ and $\lambda_2 = \lambda_z$. We call these cycles, formed by combining the points of pairs of non-conjugate cycles of F and G, *cycles of mixed type*, in order to distinguish them from the cycles described in (a) and (b) (formed by combining the points of conjugate cycles of F and G), which shall be denoted as *cycles of homogeneous type*.

These results can be easily generalized to the case in which F has m distinct coexisting cycles of period $n = 2k+1$, say $X_{1F}^{(n)} = \{x_{11}^*, ..., x_{1n}^*\}$,..., $X_{mF}^{(n)} = \{x_{m1}^*, ..., x_{mn}^*\}$, and consequently G has the conjugate cycles $X_{1G}^{(n)} = \{r_2(x_{11}^*), ..., r_2(x_{1n}^*)\}$,..., $X_{mG}^{(n)} = \{r_2(x_{m1}^*), ..., r_2(x_{mn}^*)\}$. In this case we have:

(a) m cycles of Φ of odd period n given by
$C_\Phi^{(n)} = \left\{\Phi^t\left(x_{i_1}^*, r_2\left(x_{i_{(k+1)}}^*\right)\right), t = 1, ..., n\right\}, i = 1, ..., m;$

(b) mk cycles of Φ of even period $2n$, m of which are MPE cycles, of homogeneous type, i.e. made up of points belonging to the Cartesian products between pairs of conjugate cycles, according to (27);

(c) $m\,(m-1)\,n/2$ cycles of Φ of even period $2n$, of mixed type, according to (28).

Of course, if $n=1$ (i.e. $k=0$) these results coincide with those, already shown above, of the case of m fixed points of F.

Cycles of the map Φ due to cycles of F of even period.

Let us now consider the case in which F has a cycle of even period $n = 2k$, say $C_F^{(n)} = \{x_1^*, ..., x_n^*\}$, with eigenvalue $\lambda = \prod_{i=1}^n DF(x_i^*)$, and let $C_G^{(n)} = \{y_1^*, ..., y_n^*\}$, with $y_i^* = r_2(x_i^*)$, be the conjugate cycle of G, that, as usual, has the same eigenvalue. By combining the n^2 points of the Cartesian product $C_F^{(n)} \times C_G^{(n)}$, $n^2/(2n) = n/2 = k$ cycles of Φ of period $2n$, of homogeneous type, are obtained, given by

$$\begin{aligned} {}^1C_\Phi^{(2n)} &= \left\{\Phi^t\left(x_1^*, r_2(x_1^*)\right), t = 1, ..., 2n\right\} \\ {}^2C_\Phi^{(2n)} &= \left\{\Phi^t\left(x_1^*, r_2(x_2^*)\right), t = 1, ..., 2n\right\} \\ &\vdots \\ {}^kC_\Phi^{(2n)} &= \left\{\Phi^t\left(x_1^*, r_2(x_k^*)\right), t = 1, ..., 2n\right\} \end{aligned} \tag{29}$$

Of these $n/2$ cycles only ${}^1C_\Phi^{(2n)}$ is an *MPE cycle*. For each cycle ${}^iC_\Phi^{(2n)}$, $i = 1, ..., k$, the eigenvalues are $\lambda_1 = \lambda_2 = \lambda$. Hence all the coexisting $k = n/2$ cycles are star nodes with the same stability property as the conjugate cycles of F and G that generate them.

Also in this case cycles of mixed type can be obtained if F (and consequently G) has $m > 1$ coexisting cycles of period $n = 2k$. In fact, in addition to the m pairs of conjugated cycles, each generating k cycles of Φ of period $2n$ according to (29), the remaining $m^2 - m$ pairs of non-conjugate cycles generate $\left(m^2 - m\right) n/2$ further cycles of period $2n$ of mixed type, with eigenvalues $\lambda_1 = \prod_{i=1}^n F(x_i^*)$ and $\lambda_2 = \prod_{i=1}^n F(z_i^*)$, where x_i^* and z_i^* are points of distinct cycles of F. The periodic points of these cycles of Φ of mixed type are given by

$$\begin{aligned} {}^1M_\Phi^{(2n)} &= \left\{\Phi^t\left(x_1^*, r_2\left(z_1^*\right)\right), t = 1, ..., 2n\right\} \\ {}^2M_\Phi^{(2n)} &= \left\{\Phi^t\left(x_1^*, r_2\left(z_2^*\right)\right), t = 1, ..., 2n\right\} \\ &\vdots \\ {}^nM_\Phi^{(2n)} &= \left\{\Phi^t\left(x_1^*, r_2\left(z_n^*\right)\right), t = 1, ..., 2n\right\} \end{aligned} \tag{30}$$

and none of these is an *MPE cycle*. We notice that the only substantial difference between the case of cycles of F of odd period $n = 2k+1$ and that of cycles of even period $n = 2k$ is given by the presence, in the case of odd n, of the cycle of the same odd period of Φ generated by pairs of conjugated cycles by taking the "central point" of the cycles, according to (26).

Cycles of the map Φ due to coexisting cycles of F of different periods.

Let us now consider the cycles of the Cournot map Φ obtained from the combination of points of cycles of F and G having different periods. Of course such cycles are not conjugate and will give rise to cycles of Φ of mixed type. Let $C_F^{(p)} = \{x_1^*, ..., x_p^*\}$ and $C_F^{(q)} = \{z_1^*, ..., z_q^*\}$ be a p-cycle and a q-cycle of F, with eigenvalues $\lambda_p = \prod_{i=1}^p DF(x_i^*)$ and $\lambda_q = \prod_{i=1}^q DF(z_i^*)$ respectively. The map G has the conjugate cycles $C_G^{(p)} = \{r_2(x_1^*), ..., r_2(x_p^*)\}$ and $C_G^{(q)} = \{r_2(z_1^*), ..., r_2^*(z_q^*)\}$ with the same eigenvalues. Let s be the least common multiple between p and q and n_1, n_2 the two natural numbers such that $s = n_1 p = n_2 q$. Then $N = pq/s$ distinct cycles of Φ of period $2s$ are obtained by combining the pq points of $C_F^{(p)} \times C_G^{(q)} \cup C_F^{(q)} \times C_G^{(p)}$. These cycles are given by

$$\begin{aligned} {}^1M_\Phi^{(2s)} &= \{\Phi^t(x_1^*, r_2(z_1^*)), t = 1, ..., 2s\} \\ {}^2M_\Phi^{(2s)} &= \{\Phi^t(x_1^*, r_2(z_2^*)), t = 1, ..., 2s\} \\ &\vdots \\ {}^N M_\Phi^{(2s)} &= \{\Phi^t(x_1^*, r_2(z_N^*)), t = 1, ..., 2s\} \end{aligned} \tag{31}$$

All these cycles have the same eigenvalues, given by $\lambda_1 = \lambda_p^{n_1}$ and $\lambda_2 = \lambda_q^{n_2}$. Hence the cycles (31) are stable if and only if both the cycles $C_F^{(p)}$ and $C_F^{(q)}$ are stable. We also note that none of the cycles (31) is a MPE cycle.

From the arguments given above, the following propositions follow:

Proposition 1. *Cycles of even period $2n$ of the map Φ are obtained from:*

- *pairs of conjugate cycles of period n, odd or even, of F and G according to (27) or (29) respectively;*
- *pairs of non-conjugate cycles of period n of F and G, according to (28) or (30);*
- *pairs of cycles of F and G of different periods p and q, such that the least common multiple between p and q is n, according to (31).*

Proposition 2. *Cycles of odd period $2n+1$ of the map Φ are obtained from pairs of conjugated cycles of odd period $2n+1$ of F and G according to (26).*

We recall that MPE cycles can only be obtained in case (i.1) with $n > 1$ (as previously stated, MPE cycles of period 2 cannot exist). The stability properties of the cycles of the map (20) can be deduced from the stability properties of the cycles of F and G that generate them. This important feature can be expressed by the following general statement:

Proposition 3. *A cycle C of the map Φ is stable if and only if the two cycles of F and G, from which the periodic points of C are obtained, are both stable.*

As an example, let us consider reaction functions expressed by $r_1(y) = \mu_1 y(1-y)$ and $r_2(x) = \mu_1 x(1-x)$. In this case the functions F and G are given by the fourth degree functions

$$F(x) = r_1 \circ r_2(x) = \mu_1\mu_2 x(1-x)\left(1+\mu_2 x^2 - \mu_2 x\right) \qquad (32)$$

and

$$G(y) = r_2 \circ r_1(y) = \mu_1\mu_2 y(1-y)\left(1+\mu_1 y^2 - \mu_1 y\right) \qquad (33)$$

Let us consider the parameters $\mu_1 = 3.4$ and $\mu_2 = 3.6$, so that the function F has four fixed points: $x_0^* = 0$, $x_1^* = q^*$, $x_2^* = p^*$, $x_3^* = u^*$, all unstable. The same holds for G, whose fixed points have coordinates $y_i^* = r_2(x_i^*)$, $i = 0, ..., 3$. The Cartesian product $\{x_i^*\} \times \{y_i^*\}$ is formed by 16 points of $\mathbb{R}^2$ and includes the four fixed points of the map Φ, located at the four intersections between the reaction curves (these are homogenous 1-cycles belonging to R_{12}) and six 2-cycles with periodic points out of R_{12} (i.e. not MPE cycles) three of which are on the coordinate axes.

From the properties stated above, it follows that cycles of odd period of Φ can only come from cycles of the same odd period of F and G, while cycles of even period of Φ can come from several kinds of cycles of F and G, of even or odd period, with equal or different periods, conjugate or not. From Proposition 1 we can also deduce another property, which is characteristic of the Cournot maps (20):

Proposition 4. *If F has a stable cycle of period $n > 2$ then the two-dimensional map Φ is characterized by multistability, i.e. it has several distinct coexisting attracting sets.*

As an example let us consider again two logistic reaction functions, as given in (32) and (33). For $\mu_1 = 3.83$ and $\mu_2 = 3.84$ the map F has only one attracting cycle of period 3, say $\{x_1, x_2, x_3\}$, with eigenvalue $\lambda = 0.39$. Then from (26) and (27) we get the following coexisting attracting cycles of the map (20):

- a stable 3-cycle $\{\Phi^i(x_1, r_2(x_2)), i = 1, 2, 3\} \notin R_{12}$ (i.e. not MPE)
- a stable 6-cycle $\{\Phi^i(x_1, r_2(x_1)), i = 1, ..., 6\} \in R_{12}$ (MPE cycle)

all of homogeneous type.

The peculiar property of multistability of a Cournot map is even more evident when F has coexisting attracting cycles, because in this case also all the cycles of mixed type in (31) are attracting for Φ.

For example, consider $\mu_1 = 3.53$ and $\mu_2 = 3.55$. In this case F has a stable 2-cycle $\{x_1, x_2\}$ coexisting with a stable 4-cycle $\{z_1, ..., z_4\}$, and the same occurs for the conjugate cycles of G . Then the map Φ has five coexisting attracting cycles:

- one homogeneous cycle of period 4, with periodic points $\{\Phi^i(x_1, r_2(x_1)), i = 1, ..., 4\} \in R_{12}$, an MPE cycle (according to (29) for the 2-cycle);
- two homogeneous cycles of period 8, given by $C^1 = \{\Phi^i(z_1, r_2(z_1)), i = 1, ..., 8\} \in R_{12}$, another MPE cycle, and $C^2 = \{\Phi^i(z_1, r_2(z_2)), i = 1, ..., 8\}$ which is not MPE (according to (29) for the 4-cycle);
- two cycles of mixed type of period 8, given by $M^1 = \{\Phi^i(x_1, r_2(z_1)), i = 1, ..., 8\}$ and $M^2 = \{\Phi^i(x_1, r_2(z_2)), i = 1, ..., 8\}$, (according to (30)).

The periodic points of these cycles are represented in Fig. 16, where also their basins of attraction are shown, represented by different colors.

We close this section by noting that, due to Property 2, a cycle of G is always the image by r_2 of a cycle of F, thus we can state that the study of only one of the maps defined in (23) is sufficient to give a complete understanding of all the cycles of the two-dimensional map Φ and their stability properties.

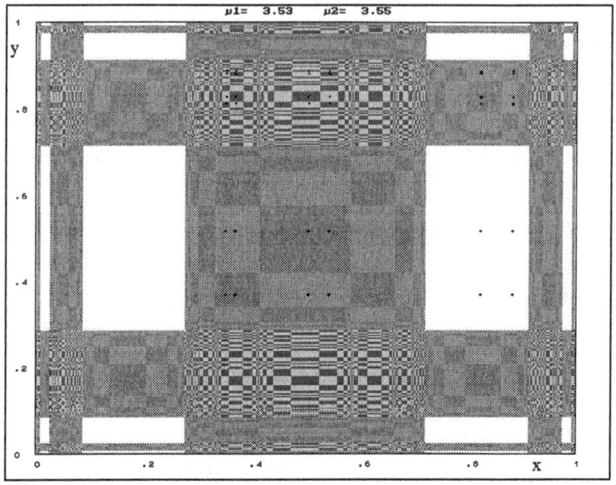

Figure 16: *Basins of different coexisting attracting cycles.*

5.2 Basins of attraction. As we have seen, the coexistence of attracting sets is a characteristic property of the class of maps (20), thus the structure of their basins of attraction becomes of particular interest in order to predict the asymptotic behavior of the games starting from a given i.c. (x_0, y_0). The peculiar structure of the basins of attraction, clearly visible in Fig.16, is another characteristic property of the class of maps (20), and we recall in this section the results presented in Bischi et al. 2000a.

From Propositions 1 and 2 of the previous section we know that any n-cycle $\mathcal{C}$ of Φ, of odd or even period n, is necessarily associated with a cycle of F and one of G, say $\mathcal{C}_x$ and $\mathcal{C}_y$ respectively. Let us denote by $\mathcal{B}(\mathcal{C})$ the total basin of an attractor $\mathcal{C}$. It is given by $\mathcal{B}(\mathcal{C}) = \bigcup_{n=0}^{\infty} \Phi^{-n}(\mathcal{B}_{im}(\mathcal{C}))$, where $\mathcal{B}_{im}(\mathcal{C})$ is the immediate basin of $\mathcal{C}$, made up of the connected components of the basin containing $\mathcal{C}$. Analogously, for the one-dimensional map F we have $\mathcal{B}(\mathcal{C}_x) = \bigcup_{n=0}^{\infty} F^{-n}(\mathcal{B}_{im}(\mathcal{C}_x))$, where $\mathcal{B}_{im}(\mathcal{C}_x)$ is the immediate 1-dimensional basin of $\mathcal{C}_x$ along the x axis. The following propositions hold (for a proof see Bischi et al. 2000):

Proposition 5. *Let* C *be an attracting cycle of* Φ *associated with the cycles* $\mathcal{C}_x$ and $\mathcal{C}_z$ of F, *then*

(i) $\mathcal{B}(\mathcal{C}) \subseteq [\mathcal{B}(\mathcal{C}_x) \cup \mathcal{B}(\mathcal{C}_z)] \times r_2([\mathcal{B}(\mathcal{C}_x) \cup \mathcal{B}(\mathcal{C}_z)])$

(ii) $\mathcal{B}_{im}(C) \subseteq [\mathcal{B}_{im}(\mathcal{C}_x) \cup \mathcal{B}_{im}(\mathcal{C}_z)] \times r_2([\mathcal{B}_{im}(\mathcal{C}_x) \cup \mathcal{B}_{im}(\mathcal{C}_z)])$ *and* $\mathcal{B}_{im}(C)$ *is made up of rectangles which include the points of C.*

Proposition 6. *Let* Φ *be a map of the form (20). Then:*
(i) *The image of a horizontal segment is a vertical segment and vice-versa.*
(ii) *The preimages of a horizontal segment, if any, are vertical segments and vice-versa.*

Proposition 7. *For any periodic point* $P=(x_1, y_1)$ *of the map T of period* $n \geq 1$, *the horizontal and vertical lines* $y = y_1$ *and* $x = x_1$, *issuing from P, are trapping sets for the map* T^n.

From the propositions stated above it follows that *any saddle cycle of T has stable and unstable sets formed by the union of segments which are parallel to the coordinate axes.*

5.3 Chaotic attractors and their basins. As stressed in the seminal paper of Rand, attractors which are more complex than periodic cycles can be easily observed in the long-run dynamics of Cournot maps (20). The transition from regular (i.e. periodic) to chaotic (i.e. aperiodic, or erratic) behavior of the Cournot tâtonnement, modeled by map (20), is marked by sequences of local bifurcations that create an increasing number of cycles, both stable and unstable. As it is well known, a chaotic regime is characterized by the presence of infinite unstable periodic points and, as remarked in Rand (1978), it is interesting to know what kind of bifurcations cause the creation of such cycles as some parameter is changed.

Since the dynamical properties of the Cournot map Φ are strictly linked to those of the one-dimensional maps F and G, these local and global bifurcations are expected to extend also to the map Φ. In fact, whenever a bifurcation occurs that creates (eliminates) cycles of the map F, and thus also of G, many cycles of the Cournot map are simultaneously created (eliminated) at the same parameter's value. Such bifurcations of the map Φ are often of particular type, due to the presence of two eigenvalues that simultaneously cross the unit circle, in which case we say that the cycles of Φ undergo a *degenerate bifurcation*, whose effects are generally different from those of a generic local bifurcation. In particular, a standard fold or flip bifurcation for the one-dimensional map F (or equivalently for G), is always associated with a degenerate bifurcation of Φ of *fold-type* (the eigenvalues cross the unit circle with $\lambda_1 = \lambda_2 = 1$), or of *flip-type* ($\lambda_1 = \lambda_2 = -1$), or of *saddle-type* ($\lambda_1 = -1$ and $\lambda_2 = 1$).

We can extend our propositions to attractors which are more complex than point cycles. The existence of such attractors occurs when the reaction functions are non monotonic functions. In fact, if the reaction functions $r_1(y)$ and $r_2(x)$ are invertible, i.e. increasing or decreasing functions, then the dynamical behavior of the Cournot map Φ is very simple, because in this case also the maps F and G are monotone. Instead, when r_1 and/or r_2 are noninvertible maps also the functions F and G are noninvertible maps. The attractors of F can be, besides k-cycles, also k-cyclic chaotic intervals or Cantor sets (a Cantor set is an attractor in Milnor's sense (Milnor, 1985), that can occur at particular bifurcation values, as the Feigenbaum points). Let us call by $\mathcal{A}_x$ (resp. $\mathcal{A}_y$) any one of the possible attractors of F (resp. G). Then results similar to those given in Propositions 1 and 5 still hold:

Proposition 8. *Let A be an attracting set of Φ. Then attracting sets of F exist, say A_x and A_z ($A_z = A_x$ or $A_z \neq A_x$), such that*

(i) $A \subseteq (A_x \cup A_z) \times (r_2(A_x \cup A_z))$

(ii) $\mathcal{B}(A) \subseteq [\mathcal{B}(A_x) \cup \mathcal{B}(A_z)] \times r_2([\mathcal{B}(A_x) \cup \mathcal{B}(A_z)])$

(iii) $\mathcal{B}_{im}(A) \subseteq [\mathcal{B}_{im}(A_x) \cup \mathcal{B}_{im}(A_z)] \times r_2([\mathcal{B}_{im}(A_x) \cup \mathcal{B}_{im}(A_z)])$ *and* $\mathcal{B}_{im}(A)$ *is made up of the rectangles which include the elements of* A.

More generally, the structure related to the Cartesian products for the attracting sets of Φ holds for any invariant set of Φ, also repelling, that is:

Proposition 9. *Let S be any invariant set of Φ (i.e. $\Phi(S)=S$), then there exist invariant sets of F, say S_x and S_z ($S_z = S_x$ or $S_z \neq S_x$), such that*
$S \subseteq (S_x \cup S_z) \times (r_2(S_x \cup S_z))$.

As we have seen, when the attracting sets of F include something more complex then a cycle, for example a chaotic set made up of k-cyclic chaotic intervals, then also the attracting sets of Φ are more complex. However, also in this case the asymptotic sets of Φ must belong to Cartesian products of attracting sets of F and G, and such two-dimensional sets may include *segments* and *rectangles*. For example, if the map F has cyclic chaotic intervals then, following the procedure indicated in Proposition 8, cyclic chaotic attractors of the Cournot map Φ can be obtained by the Cartesian product of the cyclic chaotic intervals of F and the conjugate ones of G, so that chaotic rectangles are obtained in the phase plane of Φ.

We again consider the particular Cournot game (20) with logistic reaction functions. Consider $\mu_1 = 2.8131$ and $\mu_2 = 3.85$. The map F has 3-cyclic chaotic intervals $\{I_1, I_2, I_3\}$ inside which the generic dynamics are aperiodic. Then $\{J_1, J_2, J_3\} = \{r_2(I_1), r_2(I_2), r_2(I_3)\}$ are the conjugate chaotic intervals of G. In this case the nine rectangles of the Cartesian product $\{I_1, I_2, I_3\} \times \{J_1, J_2, J_3\} = \{I_i \times J_j,\ i,j = 1,2,3\}$ include an attracting set of Φ made up of 3-cyclic rectangles (see Fig.17a) coexisting with an attracting set made up of 6-cyclic rectangles (see Fig.17b), inside which the dynamics are chaotic. The two distinct basins of attraction are shown in Fig.17c.

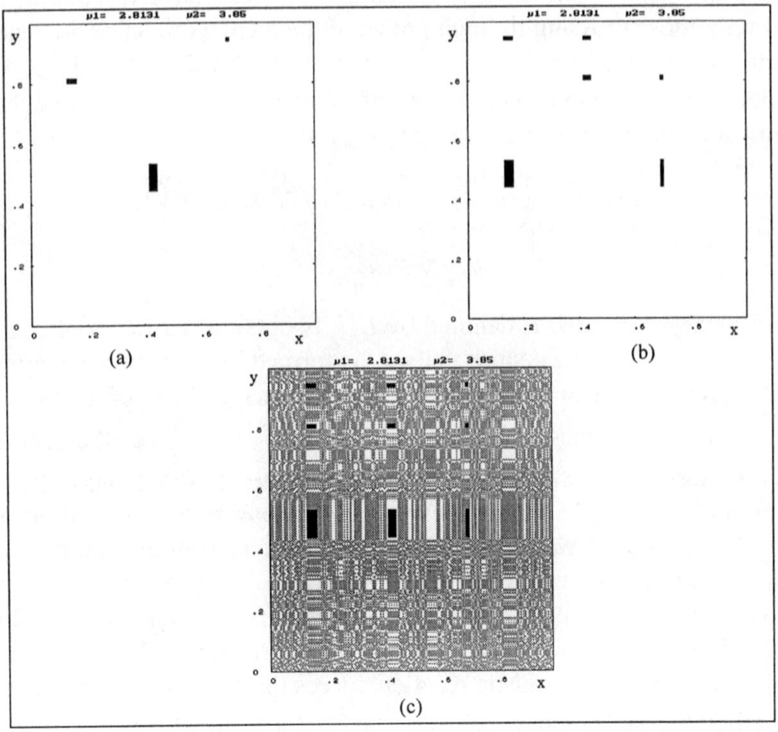

Figure 17: (a) *3-cycling chaotic rectangles.* (b) *6-cycling chaotic rectangles.* (c) *Their basins of attraction.*

As the cyclic chaotic intervals of the map F are bounded by the critical points of F, say $c^{(F)} = F\left(c_{-1}^{(F)}\right)$, and their images $c_k^{(F)} = F^k(c^{(F)})$, the sides of the chaotic rectangles of the two-dimensional map Φ are formed by segments of lines through these critical points and parallel to the coordinate axes. Hence the boundaries of these cyclic chaotic rectangles are completely known on the basis of the knowledge of the critical points $c_k^{(F)}$ of the one dimensional map F (and those of G, given by $c_{k+1}^{(G)} = r_2\left(c_k^{(F)}\right)$).

This fact can be also seen from the more general point of view of the delimitation of the absorbing and the chaotic areas of the noninvertible maps of the plane. The critical manifold of rank-0, denoted by LC_{-1}, belongs to the locus of points at which the Jacobian determinant $|D\Phi| = Dr_2(x) \cdot Dr_1(y)$ vanishes. Hence LC_{-1} is given by the union of vertical lines and horizontal lines related to $Dr_2(x)$ and $Dr_1(y)$ respectively. Among these lines only those crossing through points of the local extrema of the reaction functions are branches of LC_{-1}. In other words, let $\overline{x}^j_{-1}$, $j = 1, ..., N$, be the points of local maxima or minima of $r_2(x)$ and $\overline{y}^k_{-1}$, $k = 1, ..., M$, be the points of local maxima or minima of $r_1(y)$, then

$$LC_{-1} = \left\{(x,y) : x = \overline{x}^j_{-1}, j = 1, ..., N\right\} \cup \\ \cup \left\{(x,y) : y = \overline{y}^k_{-1}, k = 1, ..., M\right\}. \tag{34}$$

The critical set of rank-1, denoted by LC, is obtained as $LC = \Phi(LC_{-1})$. From Proposition 6 we deduce that LC is formed by segments belonging to horizontal and vertical lines. In particular, the images by Φ of the lines $x = \overline{x}^j_{-1}$ belong to the lines of equation $y = \overline{y}^j = r_2\left(\overline{x}^j_{-1}\right)$ i.e. the horizontal lines through the maximum and minimum values of the reaction function $r_2(x)$, and the images of the lines $y = \overline{y}^k$ belong to the lines of equation $x = \overline{x}^k = r_1\left(\overline{y}^k_{-1}\right)$ which are the vertical lines through the maximum and minimum values of the reaction function $r_1(y)$.

The branches of the critical set LC separate the phase plane into regions whose points have different numbers of preimages. For example, for the Cournot map with logistic reaction curves the critical curve of rank-0 is formed by the two branches $LC_{-1} = LC^{(a)}_{-1} \cup LC^{(b)}_{-1}$, where

$$LC^{(a)}_{-1} = \left\{(x,y) \mid y = \frac{1}{2}\right\} \quad and \quad LC^{(b)}_{-1} = \left\{(x,y) \mid x = \frac{1}{2}\right\}. \tag{35}$$

Also LC is formed by two branches, $LC = LC^{(a)} \cup LC^{(b)}$, where $LC^{(a)} = \Phi\left(x, \frac{1}{2}\right)$ is the half-line defined by

$$x = \frac{1}{4}\mu_1, \text{ with } y \leq \frac{1}{4}\mu_2 \tag{36}$$

and $LC^{(b)} = \Phi\left(\frac{1}{2}, y\right)$ is the half-line defined by

$$y = \frac{1}{4}\mu_2 \text{ with } x \leq \frac{1}{4}\mu_1 \tag{37}$$

In this case the two branches of LC separate a region, denoted by Z_0 in Fig.18, whose points have no real preimages, from a region, denoted by Z_4, whose points (x, y) have four real preimages, given by $x_{-1}^{(1,2)} \times y_{-1}^{(1,2)}$, where

$$x_{-1}^{(1,2)} = \frac{1}{2\mu_2}\left(\mu_2 \pm \sqrt{\mu_2^2 - 4\mu_2 y}\right) \; ; \; y_{-1}^{(1,2)} = \frac{1}{2\mu_1}\left(\mu_1 \pm \sqrt{\mu_1^2 - 4\mu_1 x}\right).$$

Critical sets of higher rank i, $i \geq 1$, defined as $LC_i = \Phi^{i+1}(LC_{-1})$ are important because generally the absorbing areas and the chaotic areas of a noninvertible map are bounded by segments of critical curves. This is true also for the absorbing and chaotic rectangles and segments of the map (20). For example in the situation shown in Fig.18 we have a chaotic attractor whose boundary is given by segments of LC, LC_1, and LC_2. As usual, segments of critical curves of higher rank bound zones inside the chaotic area where the points are more dense, i.e. are more frequently visited by the phase point of a generic trajectory.

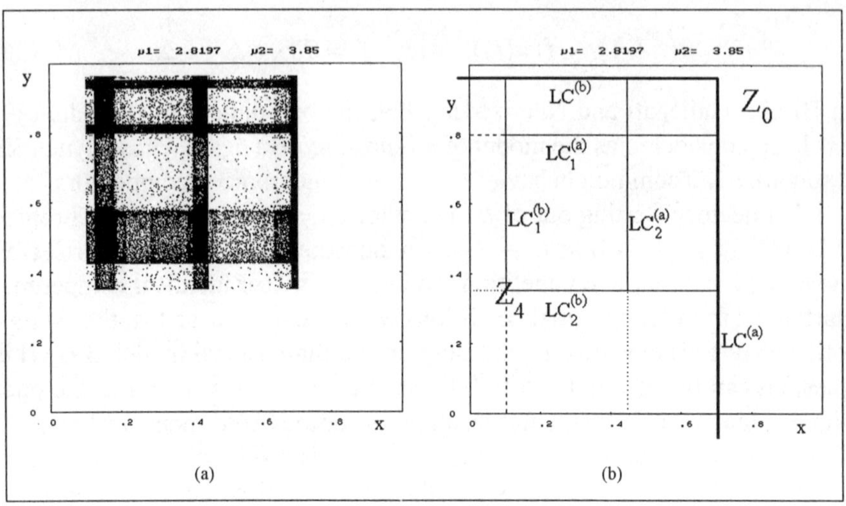

Figure 18: (a) *Chaotic attractor of the Cournot map with logistic reaction curves.* (b) *Boundary of the chaotic attractor obtained by segments of* LC_k, $k = 1, 2$.

6 Duopoly games with identical players: synchronization, riddling and intermittency phenomena

In this section we consider a dynamic duopoly game, whose time evolution is represented by the iteration of a two-dimensional map (8), in the case of identical players. This means that the dynamical system must remain the same if the variables x_1 and x_2 are interchanged, i.e. $T \circ P = P \circ T$, where $P : (x_1, x_2) \to (x_2, x_1)$ is the reflection through the diagonal

$$\Delta = \left\{(x_1, x_2) \in \mathbb{R}^2 | x_1 = x_2\right\} . \tag{38}$$

This symmetry property implies that the diagonal is mapped into itself, i.e., $T(\Delta) \subseteq \Delta$, which corresponds with the obvious statement that, in a deterministic framework, identical competitors, starting from identical initial conditions, behave identically for each time. The trajectories embedded into Δ, i.e. characterized by $x_1(t) = x_2(t)$ for every t, are called *synchronized trajectories*, and they are governed by the one-dimensional map given by the restriction of T to the invariant submanifold Δ

$$\mathbf{x}(t+1) = f(\mathbf{x}(t)) \quad \text{with} \quad f = T|_\Delta : \Delta \to \Delta. \tag{39}$$

In Bischi, Gallegati and Naimzada, 1999, the one-dimensional model (39) has been considered as the model of a *representative agent* whose dynamics summarize the common behavior of the two synchronized competitors.

A trajectory starting out of Δ, i.e. with $x_0 \neq y_0$, is said to synchronize if $|x_1(t) - x_2(t)| \to 0$ as $t \to +\infty$. A question which naturally arises, in the case of symmetric competition models, is whether identical competitors starting from different initial conditions will synchronize, so that the asymptotic behavior is governed by the simpler one-dimensional model (39). This question can be reformulated as follows. Let A_s be an attractor of the one-dimensional map (39). Is it also an attractor for the two-dimensional map T? Of course, an attractor A_s of the restriction f is stable with respect to perturbations along Δ, so an answer to the question raised above can be given through a study of the stability of A_s with respect to perturbations transverse to Δ (*transverse stability*). If A_s is a cycle, then the study of the transverse stability is the usual one, based on the modulus of the eigenvalues of the cycle in the direction transverse to Δ, whereas the problem becomes more interesting when the dynamics restricted to the invariant submanifold are chaotic. Indeed, dynamical systems with chaotic trajectories embedded into an invariant submanifold of lower dimensionality than the total phase space have raised an increasing interest in the scientific community (see e.g. Ashwin et al., 1996, Buescu, 1997, because the phenomenon of *chaos synchronization* may occur (see e.g. Fujisaka and Yamada, 1983, Pecora and Carrol, 1990, Hasler and Maistrenko, 1997), i.e., the time evolution of the two competitors synchronize in the long run even if each of them behaves chaotically. Moreover, in this case, *Milnor attractors* (see Milnor, 1985) which are not stable in Lyapunov sense appear quite naturally in this context. To better understand the meaning of this point, we recall some definitions.

Definition. *A is an asymptotically stable attractor (or topological attractor) if it is Lyapunov stable, i.e. for every neighborhood U of A there exists a neighborhood V of A such that* $T^t(V) \subset U \; \forall t \geq 0$, *and* $B(\mathcal{A})$ *contains a neighborhood of A.*

In other words, If $\mathcal{A}$ is a topological attractor then a neighborhood $W \supset \mathcal{A}$ exists such that $T^t(\mathbf{x}) \to \mathcal{A}$ as $t \to +\infty$ for any $\mathbf{x} \in W$. In this case the stable set $\mathcal{B}(\mathcal{A})$, also called basin of attraction, is an open set given by $\mathcal{B}(\mathcal{A}) = \bigcup_{t \geq 0} T^{-t}(W)$.

Definition. *A closed invariant set A is said to be a weak attractor in Milnor sense (or simply Milnor attractor) if its stable set* $B(\mathcal{A})$ *has positive Lebesgue measure.*

Note that a topological attractor is also a Milnor attractor, whereas the converse is not true. Really the more general notion of Milnor attractor has been introduced to evidence the existence of invariant sets which "attract" many points even if they are not attractors in the usual topological sense.

We now recall some definitions and results related to the problem of chaos synchronization, see Buescu, 1997, for a more complete treatment. Let T be a map of the plane, Δ a one-dimensional trapping subspace and A_s a chaotic attractor (with absolutely continuous invariant measure on it) of the restriction (39) of T to Δ. The key property for the study of the transverse stability of A_s is that it includes infinitely many periodic orbits which are unstable in the direction along Δ. For any of these cycles it is easy to compute the associated eigenvalues. In fact, due to the symmetry of the map, the Jacobian matrix of T computed at any point of Δ, say $DT(x,x) = \{T_{ij}(x)\}$, is such that $T_{11} = T_{22}$ and $T_{12} = T_{21}$. The two orthogonal eigenvectors of such a symmetric matrix are one parallel to Δ, say $\mathbf{v}_{\parallel} = (1,1)$, and one perpendicular to it, say $\mathbf{v}_{\perp} = (1,-1)$, with related eigenvalues given by

$$\lambda_{\parallel}(x) = T_{11}(x) + T_{12}(x) \quad \text{and} \quad \lambda_{\perp}(x) = T_{11}(x) - T_{12}(x)$$

respectively. Of course, $\lambda_{\|}(x) = f'(x)$. Since the product of matrices with the structure of $DT(x,x)$ has the same structure as well, a k-cycle $\{s_1, ..., s_k\}$ embedded into Δ has eigenvalues $\lambda_{\|}^k = \prod_{i=1}^k \lambda_{\|}(s_i)$ and $\lambda_{\perp}^k = \prod_{i=1}^k \lambda_{\perp}(s_i)$, with eigenvectors $\mathbf{v}_{\|}$ and $\mathbf{v}_{\perp}$ respectively.

In the recent literature on chaos synchronization, stability statements are given in terms of the *transverse Lyapunov exponents*, by which the "average" local behavior of the trajectories in a neighborhood of the invariant set A_s can be understood, and new kinds of bifurcations can be detected , such as the *riddling bifurcation* or the *blowout bifurcation*. For a chaotic set $A_s \subset \Delta$, infinitely many transverse Lyapunov exponents can be defined as

$$\Lambda_{\perp} = \lim_{N\to\infty} \frac{1}{N} \sum_{i=0}^{N} \ln |\lambda_{\perp}(s_i)| \tag{40}$$

where $\{s_i = f^i(s_0), i \geq 0\}$ is a trajectory embedded in A_s.

If x_0 belongs to a k-cycle then $\Lambda_{\perp} = \ln |\lambda_{\perp}^k|$, so that the cycle is transversely stable if $\Lambda_{\perp} < 0$, whereas if x_0 belongs to a generic aperiodic trajectory embedded inside the chaotic set A_s then $\Lambda_{\perp}$ is the *natural transverse Lyapunov exponent* $\Lambda_{\perp}^{nat}$, where by the term "natural" we mean the Lyapunov exponent associated to the natural, or SBR (Sinai-Bowen-Ruelle), measure, i.e., computed for a typical trajectory taken in the chaotic attractor A_s. Since infinitely many cycles, all unstable along Δ, are embedded inside a chaotic attractor A_s, a spectrum of transverse Lyapunov exponents can be defined, see Buescu, 1997,

$$\Lambda_{\perp}^{\min} \leq ... \leq \Lambda_{\perp}^{nat} \leq ... \leq \Lambda_{\perp}^{\max} \tag{41}$$

The meaning of the inequalities in (41) can be intuitively understood on the basis of the property that $\Lambda_{\perp}^{nat}$ expresses a sort of "weighted balance" between the transversely repelling and transversely attracting cycles (see Nagai and Lai, 1997). If $\Lambda_{\perp}^{\max} < 0$, i.e. all the cycles embedded in A_s are transversely stable, then A_s is asymptotically stable, in the usual Lyapunov sense, for the two-dimensional map T. However, it may occur that some cycles embedded in the chaotic set A_s become transversely unstable, i.e. $\Lambda_{\perp}^{\max} > 0$, while $\Lambda_{\perp}^{nat} < 0$. In this case, A_s is no longer Lyapunov stable, but it continues to be a *Milnor attractor,* i.e. it attracts a positive (Lebesgue) measure set of points of the two-dimensional phase space. So, if $\mathcal{A} \subset \Delta$ is a chaotic attractor of $T|_{\Delta}$ with absolutely continuous invariant measure, then a sufficient condition for a $\mathcal{A}$ be a Milnor, but not topological, attractor for the two-dimensional map T, is that

(a) at least one k-cycle embedded in $\mathcal{A}$ is transversely repelling, i.e. $\left|\lambda_{\perp}^{(k)}\right| > 1$, and

(b) the Lyapunov exponent $\Lambda_{\perp}^{nat}$ is negative.

This means that the majority of the trajectories on $\mathcal{A}$ are transversely attracting, but some (even infinitely many) trajectories inside $\mathcal{A}$ can exist whose transverse Lyapunov exponent is positive. In other words, transversely repelling trajectories can be embedded into a chaotic set which is attracting only "on average". In this case we have *weak stability* or *stability in Milnor sense*, but not asymptotic stability.

The transition from asymptotic stability to attractivity only in Milnor sense, marked by a change of sign of $\Lambda_{\perp}^{\max}$ from negative to positive, is denoted as the *riddling bifurcation* in Lai and Grebogi, 1996, (or *bubbling bifurcation* in Venkataramani et al., 1996). Even if the occurrence of such bifurcations is detected through the study of the transverse Lyapunov exponents, their effects depend on the action of the non linearities far from Δ, that is, on the global properties of the dynamical system. In fact, after the riddling bifurcation two possible scenarios can be observed according to the fate of the trajectories that are locally repelled along (or near) the local unstable manifolds of the transversely repelling cycles:

(L) they can be reinjected towards Δ, so that the dynamics of such trajectories are characterized by some bursts far from Δ before synchronizing on it (a very long sequence of such bursts, which can be observed when $\Lambda_{\perp}$ is close to zero, has been called *on-off intermittency* in Ott and Sommerer, 1994);

(G) they may belong to the basin of another attractor, in which case the phenomenon of *riddled basins* (Alexander et al., 1992) is obtained.

Some authors call *local riddling* the situation (L) and, by contrast, *global riddling* the situation (G) (see Ashwin et al. 1996, Maistrenko et al., 1997, 1998a). When also $\Lambda_{\perp}^{nat}$ becomes positive, due to the fact that the transversely unstable periodic orbits embedded into A_s have a greater weight as compared with the stable ones, a *blowout bifurcation* occurs, after which A_s is no longer a Milnor attractor, because it attracts a set of points of zero measure, and becomes a *chaotic saddle*, see Buescu, 1997. In particular, for $\lambda_{\perp}^{\min} > 0$ all the cycles embedded into Δ are transversely repelling, and A_s is called *normally repelling chaotic saddle*. Also the macroscopic effect of a blowout bifurcation is strongly influenced by the behavior of the dynamical system far from the invariant submanifold Δ: The trajectories starting close to the chaotic saddle may be attracted by some attracting set far from Δ

or remain inside a two-dimensional compact set located around the chaotic saddle A_s, thus giving on-off intermittency.

As noticed by many authors, (see e.g. Ashwin et al., 1996, Buescu, 1997, Hasler and Maistrenko, 1997, Maistrenko et al., 1998a,b), even if the occurrence of riddling and blowout bifurcations is detected through the transverse Lyapunov exponents, i.e. from a local analysis of the linear approximation of the map near Δ, their effects are determined by the global properties of the map. In fact, the effect of these bifurcations is related to the fate of the trajectories which are locally repelled away from a neighborhood of the Milnor attractor A_s, since they may reach another attractor or they may be folded back toward A_s by the action of the non linearities acting far from Δ. When T is a noninvertible map, as generally occurs in problems of chaos synchronization[6], the global dynamical properties can be usefully described by the method of *critical curves* and the reinjection of the locally repelled trajectories can be described in terms of their folding action.

This idea has been recently proposed in Bischi et al., 1998, for the study of symmetric maps arising in game theory, and in Bischi et al., 1999c, for the study of the effects of small asymmetries due to parameters mismatches. In these two papers the critical curves have been used to obtain the boundary of a compact *absorbing area* inside which intermittency and blowout phenomena are confined. In other words, the critical curves are used to bound a compact region of the phase plane that acts as a trapping bounded vessel inside which the trajectories starting near S are confined. In particular, in Bischi and Gardini, 1998, the concept of *minimal invariant absorbing area* is used in order to give a global characterization of the different dynamical scenarios related to riddling and blowout bifurcations. In order to give an example, let us consider the map (14) in the symmetric case

$$\varepsilon_1 = \varepsilon_2 = \varepsilon \quad , \qquad \mu_1 = \mu_2 = \mu \tag{42}$$

so that the map (14) becomes

$$T_s : \begin{cases} x' = \mu y(1-y) + \varepsilon(y-x) \\ y' = \mu x(1-x) + \varepsilon(x-y) \end{cases} \tag{43}$$

The restriction $T_s|_\Delta$ to the invariant diagonal Δ can be identified with the one-dimensional logistic map

$$x' = f_\mu(x) = \mu x(1-x). \tag{44}$$

[6] In fact the one-dimensional restriction f must be a noninvertible map in order to have chaotic motion along the invariant subspace Δ.

The eigenvalues of the symmetric Jacobian matrix $DT(x,x)$ are

$$\rho_{\|}(x) = \mu - 2\mu x \quad , \quad \rho_{\perp}(x) = 2\mu x - \mu - 2\varepsilon.$$

with eigenvectors which are parallel to Δ ($\mathbf{v}_{\|} = (1,1)$) and orthogonal to Δ ($\mathbf{v}_{\perp} = (1,-1)$) respectively. It is important to note that the coupling parameter ε only appears in the transverse eigenvalue $\lambda_{\perp}$, i.e. ε is a *normal parameter*: it has no influence on the dynamics along the invariant submanifold Δ, and only influences the transverse stability. This allows us to consider fixed values of the parameter μ, such that a chaotic attractor $A_s \subset \Delta$ of the map (44) exists, with an absolutely continuous invariant measure on it. So, we can study the transverse stability of A_s as the coupling between the two components, measured by the parameter ε, varies. Suitable values of the parameter μ, at which chaotic intervals for the restriction (44) exist, are obtained from the well known properties of the logistic map (see e.g. Collet and Eckmann, 1980, Mira, 1987). For example, at the parameter value $\overline{\mu}_2 = 3.5748049387592...$ the period-4 cycle of the logistic map undergoes the homoclinic bifurcation, at which four cyclic chaotic intervals are obtained by the merging of 8 cyclic chaotic intervals. By using $\overline{\mu}_2$ we get a four-band chaotic set A_s along the diagonal Δ, as shown in Fig.19a. In this case, for $\varepsilon = 0.24$ we have $\Lambda_{\perp}^{\max} > 0$ and $\Lambda_{\perp}^{nat} = -4.7 \times 10^{-3} < 0$. Hence, A_s is a Milnor attractor and local riddling occurs. The generic trajectory starting from initial conditions taken in the white region of Fig.19a leads to asymptotic synchronization. In Fig.19a the asymptotic part of a trajectory is shown, after a transient of $15,000$ iterations has been discarded. Indeed, if also the transient is represented, Fig.19b is obtained. During the transient, the time evolution of the system is characterized by several bursts away from Δ before synchronization occurs, as shown in Fig.20, where the difference $x_t - y_t$, computed along the trajectory of Fig.19, is represented versus time. It it worth to note the intermittent behavior of the trajectory: sometimes it seems to synchronize for a quite long number of iterations, then a sudden burst occurs. This phenomenon is also called *on-off intermittency.*

The Milnor attractor A_s is included inside a minimal invariant absorbing area whose boundary can be easily obtained by five iterations of an arc of LC_{-1}, as shown in Fig.21a. This absorbing area, obtained by the procedure outlined in section 3, constitutes a trapping region inside which the bursts observed during the transient are contained. This means that, even if it is difficult to predict the sequence of times at which asynchronous bursts occur, an estimate of their maximum amplitude can be obtained by the construction of the minimal invariant absorbing area which includes the Milnor attractor

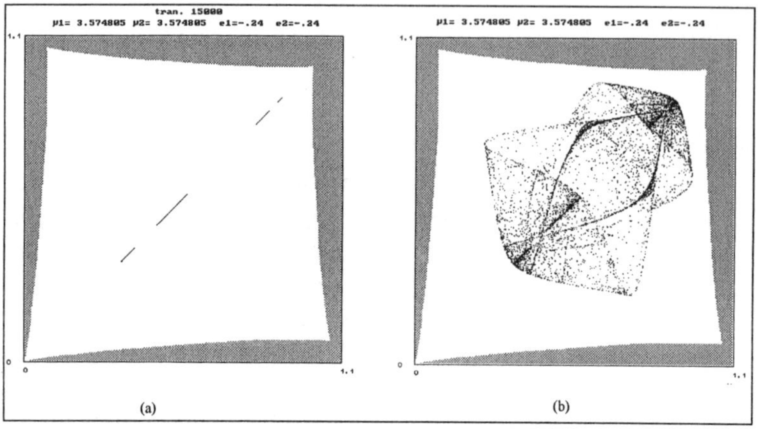

Figure 19: (a) *The four-band chaotic attracting set along the diagonal.* (b) *The transient part of a trajectory converging to the four band chaotic set*

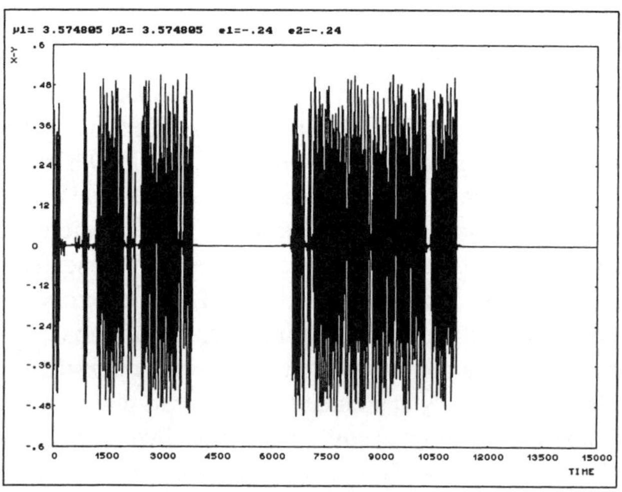

Figure 20: *Bursts away from the diagonal before synchronization occurs.*

on which synchronized dynamics take place. In such a situation, a method to obtain a trajectories which never synchronize, so that the bursts never stop and the iterated points fill up the whole minimal absorbing area, consists in the introduction of a small parameters' mismatch (see e.g. Bischi and Gardini, 1998), such as ε_1 slightly different from ε_2 or μ_1 slightly different from μ_2, so that the symmetry is broken. This implies that the invariance of Δ is lost, and consequently the one-dimensional Milnor attractor embedded in no longer exists. The study of the effects of small parameters' mismatches may be important in economic dynamic modelling, as stressed in Bischi et al., 1999c and Kopel et al., 2000.

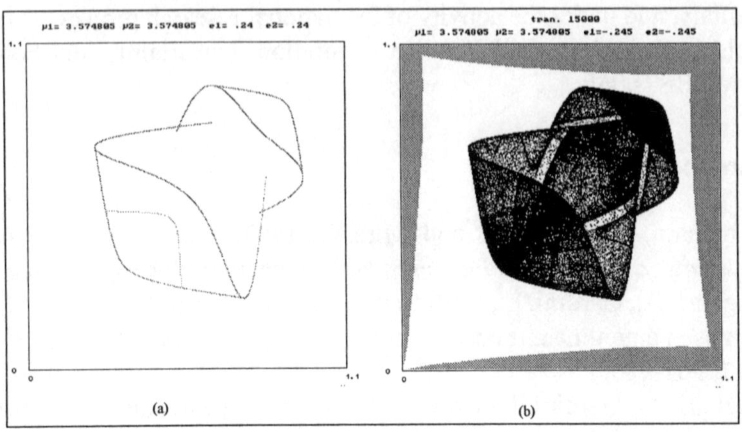

Figure 21: (a) *Minimal invariant absorbing area, obtained by iteration of LC, including the Milnor attractor.* (b) *A trajectory filling up the absorbing area, after the introduction of a small parameters' mismatch.*

A similar effect is obtained even in the symmetric case, if the value of the coupling parameter ε is increased so that $\Lambda_{\perp}^{nat}$ increases until it becomes positive, i.e. a blowout bifurcation occurs. After this bifurcation the bursts which characterize the first part of the trajectory of Fig.s 19 and 20, never stop, i.e. the firms never synchronize. A_s is now a chaotic saddle, and on-off intermittency is observed. This is what happens in the situation shown in Fig.21b, obtained for $\varepsilon = 0.245$, at which $\Lambda_{\perp}^{nat} = 2.2 \times 10^{-2} > 0$. Now the point of a generic trajectory starting from the white region fill the whole absorbing area, still bounded by segments of critical arcs.

We end this section noticing that in the case

$$\varepsilon_1 = \varepsilon_2 = 0 \tag{45}$$

we obtain the Cournot map with logistic reaction curves

$$T_{\varepsilon=0} : \begin{cases} x' = \mu_1 y(1-y) \\ y' = \mu_2 x(1-x) \end{cases} \tag{46}$$

whose properties have been analyzed in the previous section.

Acknowledgments. This work has been performed under the auspices of CNR, Italy, and under the activity of the national research project "Dynamic Models in Economics and Finance: Evolution, Uncertainty and Forecasting", MURST, Italy.

References

Abraham, R., Gardini, L. and Mira, C., 1997, *Chaos in discrete dynamical systems (a visual introduction in two dimension)*, Springer-Verlag.

Agliari, A., Gardini, L., Delli Gatti, D. and Gallegati, M., 2000a, "Global dynamics in a nonlinear model for the equity ratio", *Chaos, Solitons & Fractals*, **11**, 961-985.

Agliari, A., Gardini, L. and Puu, T., 2000b, "The dynamics of a triopoly game", *Chaos, Solitons & Fractals,* **11**, 2531-2560.

Agiza, H.N., Bischi, G.I. and Kopel, M., 1999, "Multistability in a Dynamic Cournot Game with Three Oligopolists", *Mathematics and Computers in Simulation*, **51**, 63-90.

Alexander, J.C., Yorke, J.A., You, Z. and Kan, I., 1992, "Riddled basins", *Int. Jou. of Bif. & Chaos*, **2**, 795-813

Ashwin, P., Buescu, J. and Stewart, I., 1996, "From attractor to chaotic saddle: a tale of transverse instability", *Nonlinearity*, **9**, 703-737.

Bischi, G.I. and Gardini, L., 1998, "Role of invariant and minimal absorbing areas in chaos synchronization", *Physical Review E*, **58**, 5710-5719.

Bischi, G.I., Stefanini, L. and Gardini, L., 1998 "Synchronization, intermittency and critical curves in duopoly games", *Mathematics and Computers in Simulations,* **44**, 559-585.

Bischi, G.I., Gardini, L. and Mira, C., 1999a, "Plane maps with denominator. Part I: some generic properties", *International Journal of Bifurcation and Chaos,* **9(1)**, 119-153.

Bischi, G.I., Gardini, L. and Kopel, M., 1999b, "Noninvertible maps and complex basin boundaries in dynamic economic models with coexisting attractors", Proceedings SICC99, Verbania, Italy, 1999.

Bischi, G.I., Gallegati, M. and Naimzada, A., 1999c, "Symmetry-Breaking bifurcations and representative firm in dynamic duopoly games", *Annals of Operations Research*, **89**, 253-272.

Bischi, G.I., Mammana, C. and Gardini, L., 2000a, "Multistability and cyclic attractors in duopoly games", *Chaos, Solitons & Fractals*, **11**, pp. 543-564

Bischi, G.I., Gardini, L. and Kopel, M., 2000b, "Analysis of Global Bifurcations in a Market Share Attraction Model", *Journal of Economic Dynamics and Control*, **24**, 855-879.

Bischi, G.I. and Kopel, M., 2001, "Equilibrium Selection in a Nonlinear Duopoly Game with Adaptive Expectations", *Journal of Economic Behavior and Organization*, **46(1)**, 73-100.

Bischi, G.I., Kopel, M. and Naimzada, A., 2001a, "On a rent-seeking game described by a non-invertible iterated map with denominator", *Nonlinear Analysis, Theory, Methods & Applications*, **47(8)**, 5309-5324.

Bischi, G.I., Dawid, H. and Kopel, M., 2001b, "Spillover Effects and the Evolution of Firm Clusters", *Journal of Economic Behavior and Organization* (to appear).

Bischi, G.I., Mroz, L. and Hauser, H., 2001c, "Studying basin bifurcations in nonlinear triopoly games by using 3D visualization" *Nonlinear Analysis, Theory, Methods & Applications,* **47(8)**, 5325-5341.

Buescu, J., 1997, *Exotic Attractors*, Birkhäuser, Boston.

Chiarella, C., R. Dieci and Gardini, L., 2001a, "Asset price dynamics in a financial market with fundamentalists and chartists " *Discrete Dynamics in Nature and Society*, **6**, 69-99.

Chiarella, C., Dieci, R. and Gardini, L., 2001b, "Speculative Behaviour and Complex Asset Price Dynamics: A Global Analysis", *Journal of Economic Behavior and Organization* (to appear).

Collet, P. and Eckmann, J.P., 1980, *Iterated maps on the interval as dynamical systems*, Birkhäuser, Boston.

Cournot, A., 1938, *Recherches sur les principes matematiques de la theorie de la richesse*, Hachette, Paris.

Dana, R.A. and Montrucchio, L., 1986, "Dynamic Complexity in duopoly games", *Journal of Economic Theory*, **40**, 40-56.

Dieci, R., Bischi, G.I. and Gardini, L., 2001, "From bi-stability to chaotic oscillations in a macroeconomic model", *Chaos, Solitons & Fractals* , **5(12)**, 805-822.

Fujisaka, H. and Yamada, T., 1983, "Stability theory of synchronized motion in coupled-oscillator systems", *Progress of Theoretical Physics*, **69 (1)**, 32-47.

Gardini, L., 1992, "Some global bifurcations of two-dimensional endomorphisms by use of critical lines", *Nonlinear Analysis, Theory, Methods & Applications*, **18**, 361-399.

Grebogi, C., Ott, E. and Yorke, J.A., 1983, "Crises, sudden changes in chaotic attractors and transient chaos", *Physica* 7D, 81-200.

Gumowski, I. and Mira, C., 1980a, *Dynamique Chaotique,* Cepadues Editions, Toulose.

Gumowski, I. and Mira, C., 1980b, *Recurrences and Discrete Dynamical Systems*, Springer Verlag, Berlin.

Hasler, M. and Maistrenko, Yu., 1997, "An introduction to the synchronization of chaotic systems: coupled skew tent maps", *IEEE Trans. Circuits Syst.*, **44 (10),** 856-866.

Kopel M., 1996, "Simple and Complex Adjustment Dynamics in Cournot Duopoly Models", *Chaos, Solitons & Fractals*,.**7(12)**, 2031-2048.

Kopel, M., Bischi, G.I. and Gardini, L., 2000, "On new phenomena in dynamic promotional competition models with homogeneous and quasi-homo-geneous firms" in *Interaction and Market Structure. Essays on Heterogeneity in Economics* D. Delli Gatti, M. Gallegati and A.P. Kirman (Eds.), Springer-Verlag, pp. 57-87.

Lai, Y.C.and Grebogi, C., 1996, "Noise-induced riddling in chaotic systems", *Physical Review Letters* **77**, 5047-5050.

Lupini R., Lenci, S. and Gardini, L., 1997, "Bifurcations and multistability in a class of two dimensional endomorphisms", *Nonlinear Analysis T. M. & A.* **28(1)** 61-85.

Maistrenko, Yu., Kapitaniak, T. and Szuminski, P., 1997, "Locally and globally riddled basins in two coupled piecewise-linear maps", *Physical Review E*, **57 (3)**, 6393-6399.

Maistrenko, Yu., Maistrenko, V., Popovich, A. and Mosekilde, E., 1998a, "Role of the Absorbing Area in Chaotic Synchronization", *Physical Review Letters*, **80 (8)**, 1638-1641.

Maistrenko, Yu., Maistrenko, V., Popovich, A. and Mosekilde, E., 1998b, "Transverse instability and riddled basins in a system of two coupled logistic maps", *Physical Review E*, **57 (3)**, 2713-2724.

Milnor, J., 1985, “On the concept of attractor”, *Commun. Math Phys*, **99**, 177-195.

Mira, C., 1987, *Chaotic Dynamics*, Word Scientific, Singapore.

Mira, C., Fournier-Prunaret, D., Gardini, L., Kawakami, H. and Cathala, J.C., 1994, “Basin bifurcations of two-dimensional noninvertible maps: fractalization of basins”, *International Journal of Bifurcation and Chaos*, **4**, 343-381.

Mira, C. and Rauzy, C., 1995, “Fractal aggregation of basin islands in two-dimensional quadratic noninvertible maps”, *International Journal of Bifurcations and Chaos*, **5(4)**, 991-1019.

Mira, C., Gardini, L., Barugola, A. and Cathala, J.C., 1996, *Chaotic Dynamics in Two-Dimensional Noninvertible Maps*, World Scientific, Singapore.

Nagai, Y. and Lai, Y.-C., 1997, “Periodic-orbit theory of the blowout bifurcation”, *Physical Review E*, **56 (4)**, 4031-4041.

Ott, E. and Sommerer, J.C., 1994, “Blowout bifurcations: the occurrence of riddled basins and on-off intermittency”, *Phys. Lett. A*, **188**, 39-47.

Pecora, L.M. and Carrol, T.L., 1990, “Synchronization in chaotic systems”, *Physical Review Letters*, **64 (8)** pp. 821-824.

Poston T. and Stewart, I.,1978, *Catastrophe Theory and its Applications*, Pitman.

Puu, T., 1991, “Chaos in Duopoly Pricing”, *Chaos, Solitons & Fractals*, **1(6)**, 573-581.

Puu, T., 1997, *Nonlinear Economic Dynamics*, Springer Verlag, Berlin.

Puu, T., 2000, *Attractors, Bifurcations and Chaos*, Springer Verlag, Berlin.

Rand, D., 1978, “Exotic phenomena in games and duopoly models”, *J. Math. Econ.*, **5**, 173-184.

Teocharis, R.D., 1960 “On the stability of the Cournot solution on the oligopoly problem” *Rev. Econ. Studies*, **27**, 133-134.

Van Huyck, J.B., Cook, J.P. and Battalio, R.C., 1994, “Selection dynamics, asymptotic stability, and adaptive behavior”, *Journal of Political Economy*, **102**, 975-1005.

Venkataramani, S.C., Hunt, B.R. and Ott, E., 1996, “Bubbling transition”, *Physical Review E*, **54**, 1346-1360.

4 The Role of Competition, Expectations and Harvesting Costs in Commercial Fishing

Gian-Italo Bischi[1] *and Michael Kopel*[2]

[1]Istituto di Scienze Economiche, Università di Urbino, Italy
[2]Department of Managerial Economics and Industrial Organization, University of Technology, Vienna, Austria

1 Introduction

In this chapter oligopoly theory and population dynamics are combined to describe the exploitation of a living resource under imperfect competition. This requires an interdisciplinary approach, because profit maximization arguments must be combined with the biological laws which regulate the natural growth of living resources in order to determine the long-run behavior of the natural system. The results obtained are often characterized by complex behavior and bifurcation phenomena, because the interaction of human economic decision making with ecological dynamics are highly non linear (see e.g. Rosser Jr., 2001). Indeed, resource economics is a very important field of application of dynamic analysis. In the literature on the economics of renewable resources, a plethora of questions has been studied using the tools from optimal control theory, dynamic programming and the theory of nonlinear dynamical systems. Renewable resources, e.g. grass, trees or fish, have the capacity for reproduction and growth over time and their stock is diminished by the harvesting activities of a sole owner or several individuals. Among the problems which have been investigated extensively are: How do (optimal) harvesting paths look like? Under which conditions is it more likely to observe conservation or extinction of the resource? What is the influence of the market structure? In the last 50 years considerable progress has been made, in particular, in fishery economics (for an overview, see Conrad 1995). To appreciate the breadth of topics which are the focus in the economics of fisheries and for a presentation of the main economic insights, the interested reader is advised to consult the book by Clarke (1990). Furthermore, the texts of Conrad and Clarke (1987) and Conrad (1999) can be recommended. In this chapter, we will introduce a bioeconomic model of commercial fishing to study the evolution of the stock of a fish population which is subject to harvesting over time. We will focus on three points: (i) the influence of the market structure, in particular, the role of strategic effects; (ii) the problem of extinction or conservation of the resource and the role of harvesting costs; (iii) the influence of errors in predicting the fish stock when determining future harvesting activities.

In order to address the first problem, we will analyze the model dynamics under imperfect competition in a duopoly framework and compare it with the dynamics under the assumption that the rights to harvest the resource are held by a sole owner. Equivalently, we can interpret this as a situation where the competitors form a cooperative venture. In many papers on the dynamics of renewable resources it has been assumed that the sea is open access, i.e. the fish stock is harvested by a large number of unregulated, competitive fishermen with no barrier to entry or exit. Due to perfectly competitive markets for harvested fish, the price for fish has been taken to be constant. Here, however, we assume that due to some form of regulation, e.g. limited entry, access to the fishery is restricted. Furthermore, we assume that the resource is offered on two distinct markets with downward sloping demand. The issue of international trade has been introduced recently into commercial fishing models by Okuguchi (1998) and Szidarovszki and Okuguchi (1998). Following their terminology, we will refer to the two markets in our model as the home market and the foreign market.

With respect to problem (ii), we will try to provide some insights under which circumstances conservation of the resource is more likely to be observed and how this depends on the harvesting costs. The costs of harvesting are crucial since they have a direct impact on the (profit-maximizing) behavior of the fishermen. For example, a regulator can influence the costs of harvesting by such methods as restricting the length of the fishing season, setting total catch limitations, and regulating the type of fishing gear used (see Clark 1990). Furthermore, costs can be reduced by giving R&D subsidies (see Okuguchi 1998). This in turn determines the level of the (optimal) harvesting activities of the fishermen. Again, the duopoly case will be compared to the cooperative venture case to see the effects of harvesting costs and competition on conservation of the resource.

In dealing with (iii), we are studying the problem that fishermen are only boundedly rational. When they determine their optimal level of future harvesting activities, they do not know the future fish stock. Instead, all they have is an imperfect prediction, which is revised as new information becomes available. To be a little bit more precise, let $X(t)$ denote the biomass or number of individuals in a fish stock. Then the actual evolution of the fish stock over time in the absence of fishing is determined by a so-called growth function $G(X)$, which is often expressed as $G(X) = XR(X)$, where R is the specific (or unitary) growth rate. A widely used form for R is the logistic growth

$$R(X) = (\alpha - \beta X)\,. \tag{1}$$

The parameter α is referred to as the intrinsic growth rate and $K = \alpha/\beta$ is called the carrying capacity. If an extra mortality term due to harvesting $h(t)$ is included, the dynamics of the fish stock is governed by $X(t+1) - X(t) = X(\alpha - \beta X) - h(t)$. In resource economics it is usually assumed that fishermen, when they determine the level of harvesting $h(t)$, have precise knowledge of this relation. However, in the real world this will hardly be the case. More realistically, economic agents have access to a collection of past data about the harvested amounts of fish and some other indicators of the size of the fish population, from which they then try to derive an estimate of the future fish stock, $X^e(t)$. With every new piece of information, the estimate $X^e(t)$ will be updated and be used to determine the future harvesting activities. Obviously, there are many ways to model how agents derive an estimate from past data. We will assume a simple learning rule called adaptive expectations, which states that the new estimate is a weighted average of the previous estimate and current data about the actual fish stock, where the weight on the past estimate (or belief) is a measure of the inertia of the agents. This leads to a two-dimensional dynamical system, where the dynamic variables are the actual value and the expected value of the fish stock. The study of the global properties of this two-dimensional system gives us some interesting insights on the combinations of true and expected values of the fish stock which lead to survival or extinction in the long run. As we shall see, this information is obtained through the study of the basins of attraction, which reveals another source of complexity, related to the complex structure of the basins' boundaries. This can be studied by the method of critical curves (see Mira et al., 1996, or the chapter 3 of this book). In fact, the two dimensional map, whose iteration gives the time evolution of the duopoly model with adaptive expectations, is noninvertible. This feature causes the creation of complex topological structures of the basins, such as non-connected or multiply connected sets, even if the attracting sets are very simple, e.g. stable fixed points (on this topic see also Bischi and Kopel, 2001).

This chapter is organized as follows. First, we make some remarks on the logistic growth model for different values of the intrinsic growth rate, since this has been an important topic in the literature. We then investigate the dynamics of the fish stock under imperfect competition. We are considering the effects of harvesting costs and adaptive expectations on the possibility of conservation and compare the results with the case of a sole owner. We end the chapter with some concluding remarks.

2 The logistic growth model

To begin with, we summarize some results on the dynamics of the unharvested fish population. As mentioned above, in the absence of any harvesting, the stock of the fish population in period t is determined by the discrete time logistic equation

$$X(t+1) = X\left(1+\alpha-\beta X\right). \tag{2}$$

The dynamical behavior of this equation has been studied extensively by May (1976, 1987) and May and Oster (1976). See also Conrad and Clark (1987).

The map in (2) is conjugate to the standard logistic map $z' = \mu z(1-z)$ with parameter $\mu = 1+\alpha$ through the linear transformation $X = (1+\alpha)z/\beta$. For any $\alpha > 0$ there are two fixed points

$$X_0^* = 0 \quad \text{and} \quad X_1^* = \frac{\alpha}{\beta}. \tag{3}$$

The first represents a particular biological equilibrium, known as extinction of the species, the second is called "carrying capacity" of the species when no harvesting occurs. The equilibrium point $X_0^* = 0$ is unstable for each $\alpha > 0$, and the positive equilibrium X_1^* is stable for $0 < \alpha < 2$. For $2 < \alpha < 3$, even if X_1^* is unstable, a bounded positive attractor exists around it, characterized by oscillatory dynamics (periodic or chaotic) and trapped inside the absorbing interval $I = \left[(1+\alpha)^2\left(3+2\alpha-\alpha^2\right)/16\beta, (1+\alpha)^2/4\beta\right]$. For each $0 < \alpha < 3$, the basin of attraction of the positive attractor is the interval

$$\mathcal{B} = \left(0, \frac{1+\alpha}{\beta}\right).$$

Note that any initial condition out of this interval would generate a trajectory with negative values, i.e. it leads to extinction of the fish population in finite time (see e.g. Clarke, 1990, p.13).

3 A Duopoly Model

In order to study the three questions stated in the introduction, namely the interplay between harvesting costs and extinction, the impact of competitive forces and the influence of adaptive expectations, we consider the following model of international commercial fishing. Two countries (the duopolists or players) harvest fish and sell it in their home market and in the foreign market. The inverse demand functions for the markets $i = 1, 2$ are given by $p_i = a_i - b_i(x_{1i} + x_{2i})$, where $x_{ki}(t)$ denotes the amount of fish harvested by player $k = 1, 2$ and sold in market i at time period t. Each player's harvesting costs depend on the harvest rate and, additionally, on the total fish stock. This latter assumption captures the fact that it is easier and less expensive to catch fish, if the fish population is large. Let $X(t)$ be the total fish biomass at time t in the common sea and $h_k(t) = x_{k1}(t) + x_{k2}(t)$ the amount of fish *harvested* by player k at time t. Then the cost function of player k is given by $C_k = c_k + \gamma_k h_k^2/X$, which satisfies the common assumptions that costs are convex in the fish stock and concave in harvest (see Clark 1990). Note that players might be heterogeneous with respect to their costs. In such a case the effect of cost leadership of one player on the resulting equilibrium can be investigated. Let $s_i(t) = x_{1i}(t) + x_{2i}(t)$ be the amount of fish *supplied (and sold)* in country i at time period t. We assume that the total fish harvested by the two competitors equals the total fish supplied in the two markets, i.e. $H(t) = h_1(t) + h_2(t) = s_1(t) + s_2(t)$.

3.1 Reaction functions and Nash equilibrium. Let $X^{e_j}(t)$ denote player j's expectation at time $t-1$ of the fish stock prevailing in the sea at time t. Furthermore, let $x_{ki}^{e_j}(t), j \neq k$ be player j's expectation at time $t-1$ of the amount of fish offered for sale by rival k in market i at time t. Then, the expected profits are

$$\pi_1^e(t) = [a_1 - b_1(x_{11} + x_{21}^{e_1})]\, x_{11} + [a_2 - b_2(x_{12} + x_{22}^{e_1})]\, x_{12} - c_1 - \gamma_1 h_1^2(t)/X^{e_1}(t)$$
$$\pi_2^e(t) = [a_1 - b_1(x_{11}^{e_2} + x_{21})]\, x_{21} + [a_2 - b_2(x_{12}^{e_2} + x_{22})]\, x_{22} - c_2 - \gamma_2 h_2^2(t)/X^{e_2}(t)$$

We assume that players are only boundedly rational. They try to determine their harvesting activities such that their current expected profit is maximized. The first order conditions for firm 1 are:

$$\frac{\partial \pi_1^e}{\partial x_{11}} = a_1 - 2b_1 x_{11} - b_1 x_{21}^{e_1} - 2\gamma_1 \frac{h_1(t)}{X^{e_1}(t)} = 0$$
$$\frac{\partial \pi_1^e}{\partial x_{12}} = a_2 - 2b_2 x_{12} - b_2 x_{22}^{e_1} - 2\gamma_1 \frac{h_1(t)}{X^{e_1}(t)} = 0$$

from which

$$x_{11} = \frac{a_1}{b_1} - (x_{11} + x_{21}^{e_1}) - 2\frac{\gamma_1}{b_1}\frac{h_1}{X^{e_1}}$$
$$x_{12} = \frac{a_2}{b_2} - (x_{12} + x_{22}^{e_1}) - 2\frac{\gamma_1}{b_2}\frac{h_1}{X^{e_1}}$$

follows. Accordingly, the optimal harvesting quantities for player 1, x_{1i}, for the markets $i = 1, 2$ depend on the predictions player 1 makes of the quantities offered by its rival and, additionally, on the expected fish stock. Equivalently, from the first order conditions for player 2 we get that the optimal quantities of player 2 depend on player 2's predictions of the quantities offered by its rival, $x_{11}^{e_2}$ and $x_{12}^{e_2}$, and the expected fish stock X^{e_2}.

For simplicity, we assume that the duopolists have homogeneous expectations and that the players are aware of this, i.e. $X^{e_1}(t) = X^{e_2}(t) = X^e(t)$. Furthermore, since we are interested in the equilibrium harvesting quantities, we assume that $x_{ki}^{e_j} = x_{ki}, i, j, k = 1, 2; j \neq k$. Given these assumptions, solving the system above would yield the optimal quantities of fish, $x_{ki}^*, i, k = 1, 2$, harvested by player k and sold in country i as a function of the (current) expectation X^e. Here instead, we focus on the total amount of harvest by player k. Adding the equations above yields

$$h_1 = 2A - (h_1 + h_2) - 2\frac{B\gamma_1}{X^e}h_1$$

where $A = 0.5\,(a_1/b_1 + a_2/b_2)$ and $B = (1/b_1 + 1/b_2)$. The parameter A is the average of the market volumes of the two markets. The parameter B is a measure of the (price-)sensitivity of demand. If prices in both markets drop by 1 unit, then total demand for fish increases by B units. From this equation, we can get the reaction function for player 1,

$$h_1 = \frac{A}{1 + \frac{B\gamma_1}{X^e}} - \frac{1}{2(1 + \frac{B\gamma_1}{X^e})}h_2.$$

Analogously, we get the reaction function for player 2,

$$h_2 = \frac{A}{1 + \frac{B\gamma_2}{X^e}} - \frac{1}{2(1 + \frac{B\gamma_2}{X^e})}h_1.$$

These two functions, represented by straight lines, are depicted in fig. 1. Obviously, in the duopoly case the harvesting quantities are strategic substitutes, i.e. if one player increases its quantity offered on the market, the other player's optimal reaction is to reduce its quantity.

The intersection point of the two lines defines the Nash equilibrium. From the graphical representation several properties of the Nash equilibrium quantities can be easily derived. For example, if the market volume A increases, then both reaction curves shift to the right. Accordingly, both Nash equilibrium harvesting quantities increase. Furthermore, strategic effects can also be studied. If the cost parameter γ_2 increases, then the intersection point of the reaction curve of player 2 with the vertical axis moves downwards. As a consequence, the harvesting quantity of player 2 in equilibrium (optimally) decreases. This is due to the direct cost effect. At the same time, player 1 op-

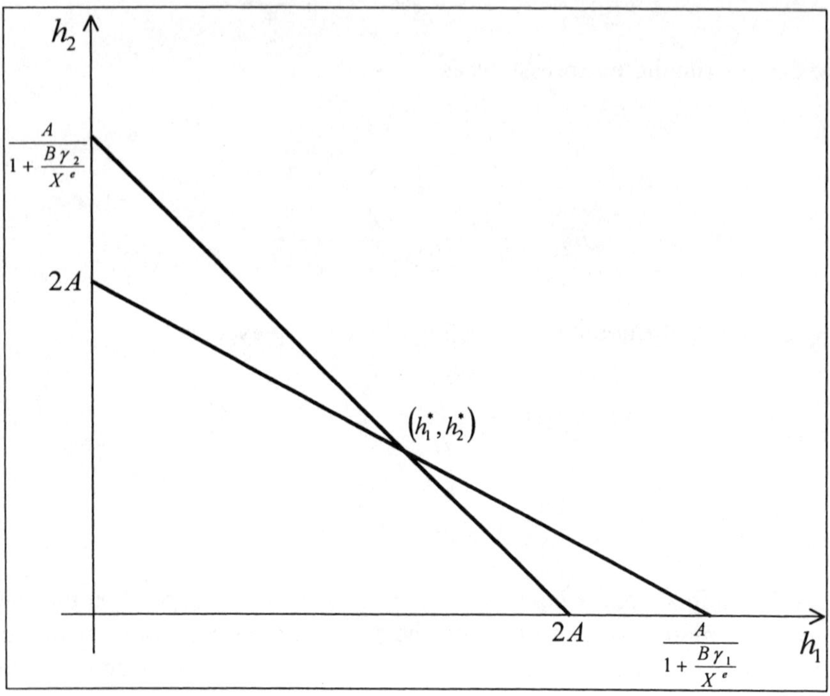

Figure 1: *Reaction curves*

timally increases the harvesting quantity, which is a consequence of a strategic effect. The harvesting quantities of the players in the Nash equilibrium can be calculated as

$$h_1^* = 2A\frac{1+\frac{2B\gamma_2}{X^e}}{4(1+\frac{B\gamma_1}{X^e})(1+\frac{B\gamma_2}{X^e})-1}$$
$$h_2^* = 2A\frac{1+\frac{2B\gamma_1}{X^e}}{4(1+\frac{B\gamma_1}{X^e})(1+\frac{B\gamma_2}{X^e})-1}.$$

By defining

$$f(x) = \frac{1}{1+\frac{2B\gamma_1}{x}} + \frac{1}{1+\frac{2B\gamma_2}{x}}.$$

we can rewrite these expressions as

$$h_1^* = \frac{2A}{(1+\frac{2B\gamma_1}{X^e})(1+f(X^e))} \tag{4}$$
$$h_2^* = \frac{2A}{(1+\frac{2B\gamma_2}{X^e})(1+f(X^e))}.$$

Hence, the total amount of harvesting of the two players is

$$H^*(t) = h_1(t) + h_2(t) = 2A\frac{f(X^e(t))}{1+f(X^e(t))} = \tag{5}$$
$$= 4A\frac{(1+\frac{B(\gamma_1+\gamma_2)}{X^e})}{4(1+\frac{B\gamma_1}{X^e})(1+\frac{B\gamma_2}{X^e})-1}.$$

The optimal harvesting quantities of player j in (4) depend on the average of the market volumes (A), on the sensitivity measure of demand (B), and on player j's cost parameter γ_j. Note that the equilibrium activities of the two players also depend on the player's expectation of the fish stock (X^e). This effect is usually absent in quantity-setting duopoly models and, hence, worth to be investigated in more detail. An increase in the (common) expectation of the fish stock affects both players. One might be tempted to argue that they both extend their harvesting activities. However, in a situation where the total amount of harvesting offered in the market is already very high, doing so would decrease the prices on the markets and would lower profits. Accordingly, it seems reasonable that the relation between (marginal) harvesting costs play a crucial role in determining the optimal reaction of the players. Calculating the derivative $\partial h_1^*/\partial X^e$, one notices that the sign depends on the sign of the expression

$$(2\gamma_1 - \gamma_2)X^2 + 4B\gamma_1\gamma_2 X + 4B^2\gamma_1\gamma_2^2.$$

Hence, for $\gamma_2 < 2\gamma_1$ we can be sure that the derivative is positive. On the other hand, for $\gamma_2 > 2\gamma_1$ it turns out that the derivative is positive only if the expected fish stock is sufficiently small (otherwise negative). An analysis of the derivative $\partial h_2^*/\partial X^e$ yields the same qualitative results with the indices of firm 1 and 2 exchanged, i.e. for $\gamma_2 > 0.5\gamma_1$ the derivative is positive; for $\gamma_2 < 0.5\gamma_1$ it is positive only if the expected stock is sufficiently small. Summing up, our analysis reveals that if the cost parameters of the duopolists do not differ too much, i.e. $0.5\gamma_1 < \gamma_2 < 2\gamma_1$, both players increase their harvesting activities if they predict a higher fish stock. Surprisingly, however, if the cost advantage of one firm is considerable ($\gamma_2 < 0.5\gamma_1$ or $\gamma_1 < 0.5\gamma_2$), then the cost leader increases its quantity only for sufficiently small values of the expected fish stock. If the expected fish stock is sufficiently high, then the cost leader reduces its harvesting effort! The reason is that the cost leader's harvesting is already high due to the high expected fish stock and due to its low marginal cost. In such a situation it is optimal, even if a higher fish stock is predicted, to reduce harvesting.

3.2 Perfect foresight for both countries. Taking into account (5), the evolution of the fish stock subject to harvesting by the duopolists is governed by

$$X(t+1) = X(t)\,(1 + \alpha - \beta X(t)) - \frac{2Af(X^e(t))}{1 + f(X^e(t))} \tag{6}$$

where the parameters γ_i, α, β, A and B are positive. By specifying how the players form their common expectation X^e, we obtain a dynamic model. To have a benchmark available, we start from the case where both players have perfect foresight, i.e.

$$X^e(t) = X(t) \text{ for each } t.$$

This means that the players are able to accurately predict the fish biomass which will prevail in the sea in the next period or, in other words, that the players know the dynamic equation which governs the evolution of the fish population. The model (6) becomes

$$X(t+1) = F(X(t))$$

where

$$F(x) = x(1+\alpha-\beta x) - H(x) = x(1+\alpha-\beta x) - 2A\frac{f(x)}{1+f(x)}$$

with

$$f(x) = 2x\frac{x+B(\gamma_1+\gamma_2)}{(x+2B\gamma_1)(x+2B\gamma_2)}.$$

The fixed point $X_0^* = 0$ always exists, and positive fixed points, if any, are obtained as solutions of the equation

$$\alpha - \beta x = \frac{H(x)}{x}. \tag{7}$$

With respect to the one-dimensional map of the logistic growth function, which describes the unharvested population, the presence of harvesting $H(x)$ makes the unimodal curve lower and an inflection point may exist. This implies that the map may have three fixed points, say $X_0^* = 0 < X_1^* < X_2^*$, with X_0^* and X_2^* stable and X_1^* unstable.

If the players have equal cost parameters, $\gamma_1 = \gamma_2 = \gamma_D$, then total harvesting becomes $H(x) = 4Ax/(3x+2B\gamma_D)$ and the equation to find the positive fixed points becomes

$$3\beta x^2 + (2\beta\gamma_D B - 3\alpha)x + 2(2A - B\alpha\gamma_D) = 0$$

Hence, if the common cost parameter is sufficiently large, namely

$$\gamma_D > \frac{2A}{B\alpha}, \tag{8}$$

the unique positive fixed point

$$X_{2,D}^* = \frac{3\alpha - 2\beta\gamma_D B + \sqrt{(3\alpha+2\beta\gamma_D B)^2 - 48A\beta}}{6\beta} \tag{9}$$

exists. Two positive solutions of the quadratic equation exist if $\gamma_D < 2A/B\alpha$ and $48A\beta < (3\alpha + 2\beta\gamma_D B)^2$. No positive fixed points exist if $48A\beta > (3\alpha + 2\beta\gamma_D B)^2$.

Given the results on the local stability of the steady states in the case of homogeneous players, we can derive some preliminary conclusions concerning the conservation of the renewable resource. Obviously, conservation or extinction of the resource depends on the relative magnitude of the intrinsic growth rate α of the renewable resource in comparison with the market volume, the price sensitivity of the consumers, and the cost parameter γ_D. From a regulator's point of view, α and β are biologically given, and A and B are the given market conditions (of course, they could be influenced by the firms' marketing activities). Accordingly, only the parameter γ_D can be influenced by a regulator in order to achieve conservation of the fish population. If the value of the cost parameter is sufficiently high ($\gamma_D > 2A/B\alpha$), extinction of the resource can be prevented (see fig. 2a). Note that the lower the intrinsic growth rate or the larger the market volume, the higher the cost parameter has to be in order to achieve conservation. If $\gamma_D < 2A/B\alpha$, extinction can occur. However, the situation depends on the interplay between the biological and the market parameters. If two fixed points X_1^* and X_2^* exist (see figs. 2b and 2d), then conservation results only if the stock of the renewable resource is in an intermediate range (larger than X_1^*, but smaller than $X_{1(-1)}^*$). For smaller or larger sizes of the population, extinction occurs either asymptotically (see fig. 2b) or in finite time (see fig. 2d). Observe that if the fish stock is larger than $X_{1(-1)}^*$, then natural mortality due to overpopulation together with the harvesting activities reduce the fish stock below the level X_1^*. finally, if the parameter of the harvesting costs is sufficiently small, then no fixed point exists, and the resource is driven extinct independent of the fish stock (see fig. 2c).

Several remarks should be made concerning our results. first, it makes sense that extinction does not occur if costs are sufficiently high. This effect has been observed before in bioeconomic models of the fishery (see Neher 1990, Pearce and Turner 1990, Clark 1990). It might be surprising, however, that for $\gamma_D < 2A/B\alpha$ extinction occurs despite the fact that harvesting costs increase for a decreasing fish stock. The reason is simply that in this situation for small sizes of the fish stock the optimal harvest (which is about $2A/B\gamma_D$) is greater than the internal growth rate α of the resource. finally, observe that a fish population can be viewed as a capital asset (see e.g. Dawid and Kopel, 1997, or Neher, 1990). The owners of the resource expect the asset to earn dividends at the normal rate of return; otherwise they would be tempted to dispose the asset. With respect to the fish stock, the owners compare the intrinsic growth rate of the resource with the rate of return of harvesting the resource and investing the proceeds. Extinction occurs, when the intrinsic growth rate is not sufficiently high. Looking at the local stability results, apparently the owners in our model behave according to this view (see figs. 2a-c). However, note that viewing the fish stock as a capital asset is an inherently intertemporal issue, i.e. the decision is made by considering the trade-off between harvesting now (and investing the returns) or let the resource grow (at the internal growth rate) and harvest in the future. The owners in our model are, in contrast to this, not forward-looking. They determine the harvest such that the expected profit of the current period is maximized. In doing this, they do not take into account the effects of their decision today on the fish stock and their opportunities tomorrow.

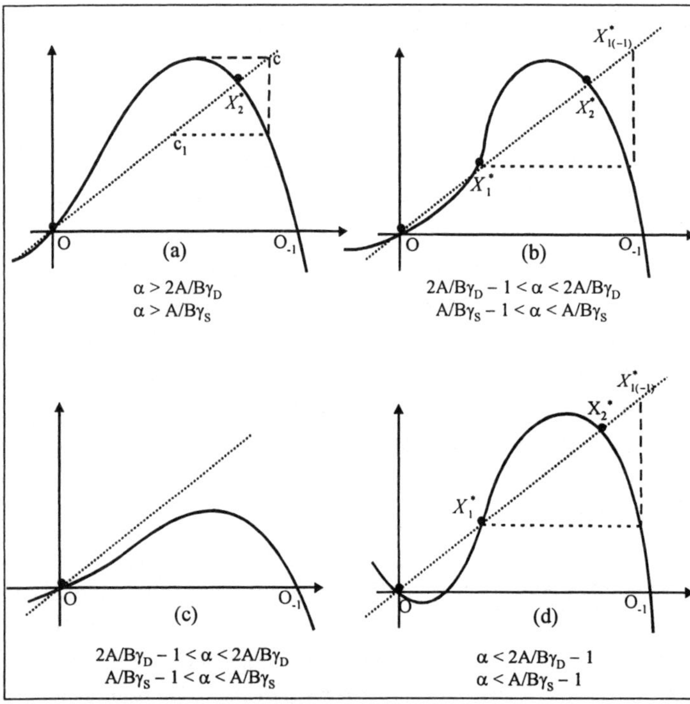

Figure 2: *One dimensional map F(x) whose iteration gives the dynamics with perfect foresight.*

If the players are heterogeneous with respect to their cost parameters, conditions for the existence of the positive fixed points are harder to come by. However, conditions for the stability of the origin can be easily found and a qualitative description of the resulting dynamics can be given. For $\alpha = A(\gamma_1 + \gamma_2)/B\gamma_1\gamma_2$, we have $F'(0) = 1$, and a transcritical bifurcation occurs. For

$$\alpha > A\frac{\gamma_1 + \gamma_2}{B\gamma_1\gamma_2},$$

we have $F'(0) > 1$. The fixed point X_0^* is unstable and only a positive fixed point X_2^* exists, where $X_2^* < \alpha/\beta$. The positive equilibrium may be stable or unstable and surrounded by a bounded attractor, as in the case of the unharvested resource. For

$$A\frac{\gamma_1 + \gamma_2}{B\gamma_1\gamma_2} - 1 < \alpha < A\frac{\gamma_1 + \gamma_2}{B\gamma_1\gamma_2}$$

we have $0 < F'(0) < 1$. The fixed point X_0^* is stable and two situations are possible. first, a pair of positive fixed points exists, $X_1^* < X_2^*$, unstable and stable, respectively. X_1^* is the boundary which separates the basin of X_0^* and that of X_2^*. Second, no positive fixed points exists and all the bounded trajectories converge to X_0^* (in finite or infinite time). For values of the intrinsic growth rate slightly smaller than $A(\gamma_1 + \gamma_2)/B\gamma_1\gamma_2$, we expect that the situation with three fixed points occurs. For $A(\gamma_1 + \gamma_2)/B\gamma_1\gamma_2 - 1 > \alpha$, there exists a neighborhood around X_0^* such that $F(X_0^*)$ is negative. This can be interpreted in saying that extinction is reached in finite time, as X_0^* is not an equilibrium in the biological sense, because the species goes extinct before reaching it (see also Clark 1990, p. 13). Observe that the set $\mathcal{B}$ of initial conditions which generate positive and bounded trajectories is given by $\mathcal{B} = (0, O_{-1})$,where $O_{-1} < (1+\alpha)/\beta$ is obtained by solving the equation

$$1 + \alpha - \beta x = \frac{H(x)}{x}.$$

4 Comparison with the cooperative venture case

For the sake of comparison, let us assume that the two players agree to form a cooperative venture. Equivalently, we might think of this as a situation where the property rights for the resource are in the hands of a sole owner, which may be imagined either as a private firm or a government agency that owns complete rights to the exploitation of the fish population. The sole owner harvests fish and sells it not only in its home market, but also in a foreign market. The inverse demand functions for the markets $i = 1, 2$ are given by

$p_i = a_i - b_i x_i$, where $x_i(t)$ denotes the amount of fish harvested and sold in market i at time period t. Let $X(t)$ be the total fish biomass at time t in the common sea and $h(t) = x_1(t) + x_2(t)$ the amount of fish *harvested* (and sold) at time t. Then the cost function is given by $C = c + \gamma_S h^2/X$. Furthermore, let $X^e(t)$ denote the owners prediction at time $t-1$ of the fish stock prevailing in the sea at time t. Then, the expected profit in period t reads

$$\pi^e(t) = [a_1 - b_1 x_1]\, x_1 + [a_2 - b_2 x_2]\, x_2 - c - \gamma_S \frac{h^2(t)}{X^e(t)}.$$

If we assume that the sole owner is only boundedly rational in the sense that he determines the harvesting activities such that the expected profit of the subsequent period is maximized, it is easy to see that the optimal harvesting quantity is

$$h^*(t) = \frac{A}{1 + \frac{B\gamma_S}{X^e(t)}} \tag{10}$$

where A and B are defined as before. The optimal harvesting quantity of a sole owner can be compared with the total harvesting quantity of the duopolists. If we assume that the cost parameters of both players are equal, i.e. $\gamma_1 = \gamma_2 = \gamma_D$, then the total amount of harvesting in equilibrium in the duopoly case is $H^*(X^e) = 4A/(3 + 2B\gamma_D/X^e)$, according to (4). It is easy to see that the following holds: If $\gamma_S \geq 0.5\gamma_D$, then for each level of X^e, $H^*(X^e) > h^*(X^e)$. In other words, competition has the effect that the resource is exploited more heavily for each level of expected fish stock. This has been noticed before in the literature, for example, in connection with open-access exploitation (see Clark 1990), and also in the context of a differential game model of fishing (see Levhari and Mirman 1982, and also Dutta and Sundaram 1993). Note, however, that this result does not give any information on the relation between the harvesting activities in the long run, i.e. in the long run steady state.

If we put the logistic growth function and the expression (10) for optimal harvesting together, then the dynamics of the fish stock subject to harvesting by a sole owner is determined by

$$X(t+1) = X(t)\,(1 + \alpha - \beta X(t)) - \frac{A}{1 + \frac{B\gamma_S}{X^e(t)}}. \tag{11}$$

If we assume that the owner can accurately predict the level of the fish stock in period t, i.e. $X^e(t) = X(t)\ \ \forall t$, we obtain a one-dimensional dynamical system. The time evolution of the fish population is then described by the iteration of the map

$$X' = F(X) = X(1+\alpha-\beta X) - \frac{AX}{X + B\gamma_S}. \tag{12}$$

The steady states of this dynamical equation are the non negative fixed points of the map (12). They are given by $X_0^* = 0$ and the positive solutions of the equation

$$\beta X^2 - (\alpha - \beta\gamma_S B)\,X + A - \alpha\gamma_S B = 0.$$

If

$$\alpha > \frac{A}{\gamma_S B}, \tag{13}$$

holds, then the slope at X_0^* is $DF(0) = 1+\alpha - A/\gamma_S B > 1$. Hence, X_0^* is unstable and there is a unique positive equilibrium given by

$$X_{2,S}^* = \frac{\alpha - \beta\gamma_S B + \sqrt{(\alpha+\beta\gamma_S B)^2 - 4A\beta}}{2\beta} \tag{14}$$

which may be stable or unstable (with an attractor around it). The equilibrium lies inside the absorbing interval $[c_1, c]$ (like in the situation shown in fig. 2a) with basin of attraction $(O, O_{(-1)})$. If

$$\alpha < \frac{A}{\gamma_S B} < \alpha + 1$$

then $0 < DF(0) < 1$, and no positive equilibria exist if $A > (\alpha+\beta\gamma_S B)^2/4\beta$. On the other hand, two positive equilibria $0 < X_1^* < X_2^*$ exist, where X_2^* is given by (14), if $A < (\alpha+\beta\gamma_S B)^2/4\beta$ and $\beta\gamma_S B < \alpha$. In this case, X_1^* is unstable. The other equilibrium, X_2^*, is stable with basin given by $\mathcal{B} = \left(X_1^*, X_{1(-1)}^*\right)$, where $X_{1(-1)}^*$ is the rank-1 preimage of X_1^* (compare fig. 2b). When X_2^* is unstable, then $\mathcal{B}$ is the basin of the absorbing interval $[c_1, c]$ around X_2^*, provided that $c < X_{1(-1)}^*$. finally, if

$$\alpha < \frac{A}{\gamma_S B} - 1, \tag{15}$$

then we have a right neighborhood of $X_0^* = 0$ where $F(X)$ is negative.

We can now make a comparison between the stability conditions for the duopoly case and the case of a sole owner in order to see the effects of harvesting costs and competition on the conservation of the fish population. first, note that the threshold (8) for the duopolists's common cost parameter γ_D resulting in conservation of the resource independent of the fish stock is twice as large as for the case of a sole owner, see (13). If we consider, for example, the existence and stability of the fixed points for $A = 5.5, B = 2, \beta = 1, \alpha = 3$, then conservation of the resource independent of the fish stock is achieved if $\gamma_1 = \gamma_2 = \gamma_D > 1.833$, whereas in the case of a sole owner, the value of the cost parameter only needs to be half as large. If we consider, for example, $\gamma_S = 1$, a sole owner with perfect foresight would achieve conservation of the resource, whereas competition between only two players leads to extinction of the resource. Interesting insights can be gained if we look at the long run evolution of the fish population in the two cases. Let us assume that $\gamma_S = 1$ and $\gamma_D = 2$. Recall, that in such a situation, for each level of X^e, $H^*(X^e) > h^*(X^e)$. In other words, a sole owner would harvest a smaller quantity than the duopolists together. However, for the particular set of parameters we are considering, in the steady states $X_{2,S}^*$ and $X_{2,D}^*$ the relation between the harvesting quantities is reversed: The sole owner achieves a higher permanent catch, $H^*(X_{2,D}^*) = 2 < h^*(X_{2,S}^*) = 2.24$, where $X_{2,S}^* = 1.37 > X_{2,D}^* = 1$. So, competition between the duopolists may lead to overconsumption, with less left for future periods.

With respect to conservation, in the case of homogeneous costs, it is easy to see that a more general statement can be made. If n firms harvest the resource, then their common cost parameter must be n times as large in order to achieve conservation of the resource, i.e. $\gamma_O > nA/B\alpha$ (the subscript 'O' stands for oligopoly). Hence, this is another indication that competition between players might have a detrimental effect on the resource, unless the costs of the rivals are considerably higher than the corresponding costs of a sole owner. This is of particular interest if we discuss the effect of market entry (see also Szidarovszki and Okuguchi 1998). Imagine a situation in which a sole owner harvests fish and offers it on the two markets. Due to economies of scale and learning effects, the sole owner's value of the cost parameter is quite low, e.g. $\gamma_S = 1$. Despite such low costs, no matter what the actual fish stock, conservation of the resource results. In this situation a potential entrant is considering market entry. If entry is accommodated by the incumbent (the sole owner), then we are dealing with a duopoly situation just analyzed. The remarkable effect is that even if the new firm has considerably higher costs, e.g. $\gamma_2 = 5$, extinction will result.

In the heterogeneous case, the cost parameters of *both* players play a crucial role, as the following simple example demonstrates. Let us assume that $A = 1, B = 1, \beta = 1, \alpha = 2$. Now consider the situation of a sole owner and assume that the value of the cost parameter is $\gamma_S = 1$. Since $\gamma_S > A/B\alpha = 0.5$, it follows that the fish population survives independent of the initial fish stock. Consider, on the other hand, a duopoly situation where the players are heterogeneous with respect to costs. Suppose player 1 has a value of the cost parameter of $\gamma_1 = 1$. Then $A(\gamma_1 + \gamma_2)/B\gamma_1\gamma_2 = 1 + 1/\gamma_2$, and conservation of the resource depends on the value of the cost parameter of player 2. If player 1 is the cost leader, i.e. $\gamma_2 > \gamma_1 = 1$, then $1 + 1/\gamma_2 < \alpha = 2$, and the resource is preserved no matter what the initial fish stock is. However, if player 2 is the cost leader, then conservation of the fish stock depends on the initial stock of fish. If player 2's cost advantage is considerable (γ_2 less than about 0.23), then extinction occurs for all initial stocks of the resource. The higher optimal harvesting activity by player results, in this case, in extinction of the resource no matter what the initial stock of fish is!

5 Duopoly and adaptive expectations

We now drop the assumption that the players can precisely predict the evolution of the fish stock. Instead, we assume that the duopolists form their common beliefs by using the adaptive expectations scheme. The model describing the evolution of the actual and the expected fish stock becomes two-dimensional:

$$T : \begin{cases} X(t+1) = X(t)\,(1+\alpha-\beta X(t)) - 2A\frac{f(X^e(t))}{1+f(X^e(t))} \\ X^e(t+1) = (1-\lambda)\,X^e(t) + \lambda X(t) \end{cases} \tag{16}$$

where

$$f(x) = 2x\frac{x + B\,(\gamma_1+\gamma_2)}{(x+2B\gamma_1)\,(x+2B\gamma_2)}.$$

Hence, for each t the duopolists have homogeneous expectations, where $\lambda \in (0,1]$ denotes the inertia in revising expectations. A higher value of λ coincides with a higher willingness to take new information into account. Note that the limiting case $\lambda = 1$ means that the competitors have naive expectations. The time evolution of realized and expected values is obtained by the iteration of the two-dimensional map $T : (X(t), X^e(t)) \rightarrow (X(t+1), X^e(t+1))$. On the horizontal axis of the state space the actual (or realized) values of the fish stock are measured, whereas along the vertical axis the expected values are measured at each time period. Hence, points in a neighborhood of the diagonal $X = X^e$ represent good estimates of the fish stock, whereas the points in the region above (below) the diagonal represent situations where the fish stock is over- (under-) estimated. From the second component of (16) it follows that the fixed points must be located along the diagonal $X^e = X$. This implies that in equilibrium the expectations coincide with the actual size of the fish stock. In other words, in equilibrium expectations are fulfilled and are, hence, rational. Accordingly, we have "Rational Expectations Equilibria" (REE), whose coordinates, obtained from the first equation in (16) with $X^e = X$, are the same as for the model with perfect foresight, namely the fixed point $O = (0,0)$ and those obtained by solving equation (7).

For the local stability of the REE, we now have to consider the two-dimensional Jacobian matrix

$$DT(X, X^e) = \begin{bmatrix} 1+\alpha-2\beta X & -2A\frac{f'(X^e)}{[1+f'(X^e)]^2} \\ \lambda & 1-\lambda \end{bmatrix}$$

computed at the fixed points. For example, for the equilibrium $O = (0,0)$, the range of stability is

$$\alpha < A\frac{\gamma_1+\gamma_2}{B\gamma_1\gamma_2} < 1-\alpha\frac{1-\lambda}{\lambda}$$

and, hence, smaller with respect to the case of perfect foresight ($\alpha < A(\gamma_1 + \gamma_2)/B\gamma_1\gamma_2$). This set is non-empty only if $\alpha < 1 - \alpha(1-\lambda)/\lambda$, i.e.

$$\alpha < \lambda.$$

Of course, if $\alpha \geq 1$ then $\alpha < \lambda$ is never satisfied, and O is always unstable. This result, which is based only on a local stability argument, seems to suggest that extinction is less probable with adaptive expectations than with perfect foresight. However, in the two-dimensional adaptive expectation case, extinction might even occur if the fixed point $(0,0)$ is unstable. Looking at (16), we realize that as long as the fish stock and the expected size of the fish population are positive, the expectation remains positive. On the other hand, due to natural mortality and harvesting, the fish stock may become negative. Even if $(0,0)$ is a fixed point, i.e. an equilibrium of the dynamical system, it might not be a biological equilibrium, because trajectories starting close to it take on negative values for the fish stock. Such a situation has to be interpreted as extinction in finite time (see Clark (1990)). Only a rigorous study of the basins of the equilibria give further information if and when this occurs.

For the study of the basins it is important to notice, first, that the map (16) is a noninvertible map, because given $(X(t+1), X^e(t+1))$ several distinct preimages can be obtained by solving (16) with respect to $(X(t), X^e(t))$. As the map T is continuously differentiable, it is easy to obtain the equation of LC_{-1}, since it is included in the set of points at which the determinant of the Jacobian vanishes, i.e.

$$\det DT\,(X, X^e) = (1-\lambda)\,(1+\alpha-2\beta X) + \lambda H'(X^e) = 0. \qquad (17)$$

It is also easy to obtain the image of LC_{-1} by applying T, i.e. $LC = T\,(LC_{-1})$. These sets constitute the so called critical curves, which separate the phase plane into regions Z_k whose points have k preimages, or, equivalently, where k distinct inverses of T are defined (see e.g. Mira et al., 1996, or chapter 3 of this book). It is interesting to note that for $\lambda < 1$ the curve LC_{-1} can be expressed as

$$X = \frac{1}{2\beta}\left[1+\alpha+\frac{\lambda}{1-\lambda}H'(X^e)\right],$$

whereas if $\lambda = 1$ (i.e. in the case of naive expectations), then $\det DT$ never vanishes, since $\det DT = H'(X^e) = 2Af'(X^e)/\left[1+f(X^e)\right]^2 > 0$ for each X^e. So, for $\lambda = 1$ no critical curves exist. As we shall see in the following, the presence of critical curves may have important consequences on the structure of the basins' boundaries and on the occurrence of global bifurcations which cause qualitative changes in the structure.

We have seen that in a situation where the duopolists have perfect foresight, as harvesting becomes cheaper, the danger of extinction of the resource grows. A similar result seems plausible for the model (16) with adaptive expectations. In the remainder of this section, we give a more rigorous study of the relation between harvesting costs and the dynamics of the resource if fishermen have adaptive expectations. We also compare the results with the benchmark case of perfect foresight. In particular, as before, we are interested in the effects of changes in harvesting costs on the probability of conservation (or extinction). Again, we assume that the players are homogeneous with respect to their cost parameters, i.e. $\gamma_1 = \gamma_2 = \gamma_D$ and, in order to show some numerical simulations, we consider the following set of parameter values: $\alpha = 3$, $\beta = 1$, $A = 5.5$, $B = 2$. Recall that, in the case of a sole owner with perfect foresight, for $\gamma_S = 1$ conservation of the resource occurs independent of the initial size of the fish stock. In the duopoly case, conservation of the resource is achieved if player's cost parameters are such that $\gamma_D > 1.833$, i.e. the value of the cost parameter has to be at least 83.3% higher. Accordingly, if we assume $\gamma_D = 2$, then in the duopoly case under perfect foresight the resource is conserved no matter what the initial size of the fish population is. Moreover, the results on the existence of two positive fixed points X_1^* and X_2^* show that as long as $\gamma_D > 1.812$, conservation is achieved at least from some initial values of the resource stock (to be more precise, for initial stock sizes in the interval $[X_1^*, X_{1(-1)}^*]$). How does the situation change when players have adaptive expectations? The first thing to notice is that for $\gamma_D = 2$, extinction of the resource occurs for all initial values of the actual and the predicted fish stock. Hence, errors in the prediction of the fish stock lead to extinction unless costs are higher than in a situation with perfect foresight. For adaptive expectations with $\lambda = 0.3$, we show the basin of the positive equilibrium (X_2^*, X_2^*) for $\gamma_D = 4$ (fig. 3a), $\gamma_D = 2.6$ (fig. 3b), $\gamma_D = 2.575$ (fig. 3c), $\gamma_D = 2.573$ (fig. 3d) and $\gamma_D = 2.56$ (fig. 3e). Observe that the basin of the equilibrium (X_2^*, X_2^*), which for this set of parameters is the only stable steady state, is represented by the white regions, whereas the grey points denote the set of initial conditions leading to extinction. Also note that the horizontal axis represents the actual level of the fish stock, whereas the vertical axis represents the expected fish stock. Obviously, conservation of the resource depends both on the initial sizes of the fish stock and the expected fish stock. For example, for a high value of the cost parameter, if the player's prediction of the fish stock is roughly accurate (i.e. the initial condition is taken from a neighborhood of the diagonal) then conservation results except in the cases where the (the actual and the predicted) resource stock is quite high (see fig. 3a). On the other hand, if the players initially overestimate the size of the fish stock (i.e. the initial condition is chosen above the diagonal, with $X^e(0) \gg X(0)$), then this might result in extinction of the resource. Roughly speaking, the evolution is as follows. Since the player's expectation of the fish stock is high, harvesting effort is high. This, subsequently, results in a reduction of the actual fish population. Since λ is small, the prediction of the future fish stock remains high, and this leads to a further reduction of the fish stock. This, taken together with the small reproduction of the resource due to its small size, leads to extinction in finite time. Note, however, that the evolution of the trajectories might crucially depend on the initial values of stock and expectation. As an example, consider a situation where both the expected and the actual fish stock are small. Due to low harvesting activities, the fish stock increases and the expected value increases too. Eventually, the actual and the expected fish stock are rather large. Now, due to overpopulation and the increase in harvesting activities, the actual fish stock is severely diminished. The evolution of trajectories is about the same, no matter if extinction results or not. It now depends on the precise initial size of the fish population and the predicted stock if extinction occurs.

Observe that, given the information on the structure of the basins and the critical curves, we can make a qualitative prediction of the evolution of the trajectories, since each grey tongue depicted in figs. 3a-d is the preimage of the corresponding tongue to the right of it. This is due to the folding action of the noninvertible mapping T, see Mira et al. (1996), see also chapter 3 of this book. Indeed, the global bifurcations which change the topological structure of the basin of the positive equilibrium, from simply connected (figs. 3a-b) to multiply connected (fig. 3c) to the union of non connected portions (fig. 3e), can be easily explained on the basis of the theory of critical curves in terms of contacts between the basin boundaries and LC.

For example, the creation of "grey holes" nested inside the white basin of the positive equilibrium are the preimages of the small portion of the grey region, indicated by an arrow in fig. 3c, which entered the region Z_3 after a contact between LC and the basin boundary. Analogously, the transition from a multiply connected into a non connected basin, i.e. from fig. 3c to fig. 3d, is a consequence of a contact between the basin boundary and LC near the point indicated by the arrow in fig. 3d. The splitting of white islands into pairs of smaller ones is due to another contact between LC and the basin boundary at the point indicated by the arrow of fig. 3e. As in the benchmark model, for lower values of the cost parameter, the probability of extinction increases in the case of adaptive expectations (see figs. 3b-d), i.e. the set of stock-expectation-combinations resulting in conservation of the resource shrinks.

Once harvesting costs reach a certain level, the situation changes rapidly and drastically. In order to demonstrate the effect of adaptive expectations with a different degree of inertia, we consider the same set of parameters as above, but with $\lambda = 0.1$ and $\gamma_D = 2.56$ (see fig. 3f). In this case, lower inertia in revising expectations enhance the stability of the positive

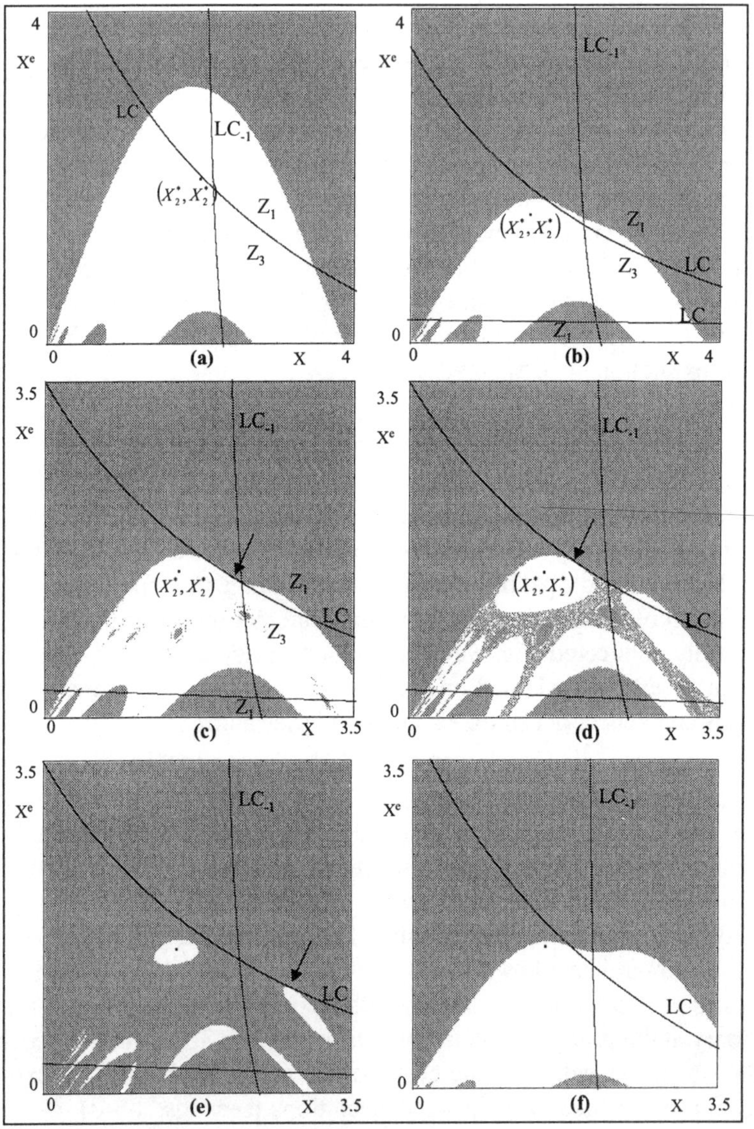

Figure 3: *Some numerical computations of the basins of the positive equilibrium (white regio). The grey region represents the set of initial conditions leading to extinction of the fish stock.*

equilibrium and, consequently, make extinction of the living resource less likely. However, it is worth noticing that the positive equilibrium is very close to the upper boundary of its basin. Despite the fact that the system is close to its equilibrium, a slight overestimation of the fish stock may lead to extinction of the fish population.

6 Concluding remarks

The combination of oligopoly games and resource economics is a challenging task, since the biological law which regulates the natural growth of the resource has to be taken into account. In fact, renewable resources, like a fish population, grow and decline over time. Due to their inherent dynamic nature, their evolution can only be understood using dynamic models. In particular, when the resource is subject to harvest, the problem of conservation is a complex issue. In this chapter we have proposed a discrete-time dynamic model which describes the decisions of duopolists engaged in commercial fishing, where each of the players sells the harvested fish both in a home and in a foreign market. Within this framework we have discussed several important topics, for example, the relation between harvesting costs and conservation of the resource, the influence of the market structure and the changes if agents are not able to predict the future evolution of the fish stock accurately. We studied the duopoly case in order to stress the role of competition and compared this situation with the case of a cooperative venture. Furthermore, adaptive expectations have been proposed to stress the role of expectations formation under bounded rationality, where we have confronted the results with the perfect foresight case.

The assumption of adaptive expectations has led us to a two-dimensional dynamical system in discrete time. Through a global analysis of the resulting noninvertible map, we have shown that, even if we limited our study to a set of parameters which give very simple attractors, namely stable fixed points, another source of complexity arises which is due to the creation of complex topological structures of the basins of attraction. Our analysis has been based on a combination of analytic, geometric and numerical methods. The global (or contact) bifurcations which causes the creation of complex structures of the basins have been studied by the method of critical curves (see Mira et al., 1996, Bischi and Kopel, 2001, see also the chapter 3 of this book).

Our main findings are quite intuitive and can be summarized as follows. Not surprisingly, higher harvesting costs tend to achieve conservation of the resource, independent of the fact if the right to harvest is held by a sole owner or if several players compete. Competition increases the probability of extinction. For example, if we take the case of a sole owner as a reference point and assume that harvesting costs of the sole owner are such that conservation is achieved, then with only two players, the harvesting costs of the competitors have to be twice as high to achieve the same result. Moreover, the numerical example given in this paper indicates that the long run fish stock in the duopoly case is smaller than in the cooperative venture case and the permanent catch in the latter situation is higher. If we, additionally (and realistically), assume that agents are not able to predict the future fish stock accurately and instead use an adaptive expectation scheme, extinction becomes more likely with respect to the case of perfect foresight, especially when fishermen overestimate the fish stock.

Acknowledgments. This work has been performed under the auspices of CNR, Italy, and under the activity of the national research project "Nonlinear Dynamics and Stochastic Models in Economics and finance", MURST, Italy.

References

Bischi, G.I. and Kopel, M., 2001, "Equilibrium Selection in a Nonlinear Duopoly Game with Adaptive Expectations" *Journal of Economic Behavior and Organization*, vol. 46/1, 73-100.

Clark, C.W., 1990, *Mathematical Bioeconomics*, 2nd edition, Wiley Interscience

Conrad, J.M. and Clark, C.W., 1987, *Natural Resource Economics. Notes and problems.* Cambridge University Press.

Conrad, J.M., 1995, "Bioeconomic Models of the fishery", in: *Handbook of Environmental Management*, D.W. Bromley, Ed., Ch. 18, 405-432.

Conrad, J.M., 1999, *Resource Economics*, Cambridge University Press.

Dawid, H. and Kopel, M., 1997, "On the Economically Optimal Exploitation of a Renewable Resource: The Case of a Convex Environment and a Convex Return Function", *Journal of Economic Theory*, 76, 272-297.

Dutta, P.K. and Sundaram, R.K., 1993, "The Tragedy of the Commons?" *Economic Theory*, 3, 413-426.

Levhari, D. and Mirman, L.J., 1982, "The Great fish War: An Example Using a Dynamic Cournot-Nash Solution", in: *Essays in the Economics of Renewable Resources*, L.J. Mirman and D.F. Spulber, Eds., North-Holland, 243-258.

May, R.M., 1987, "Chaos and the Dynamics of Biological Populations", *Proc. R. Soc. London*, A 413, 27-44.

May, R.M., 1976, "Simple Mathematical Models With Very Complicated Dynamics", *Nature*, 261, 459-467.

May, R.M. and Oster, G.F., 1976, "Bifurcations and Dynamic Complexity in Simple Ecological Models", *The American Naturalist*, 110, 573-599.

Mira, C., Fournier-Prunaret, D., Gardini, L., Kawakami, H., and Cathala, J.C., 1994, "Basin bifurcations of two-dimensional noninvertible maps: fractalization of basins", *International Journal of Bifurcations and Chaos*, 4, 343-381.

Mira, C., Gardini, L., Barugola, A., and Cathala, J.C., 1996, *Chaotic Dynamics in Two-Dimensional Noninvertible Maps*, World Scientific.

Neher, P.A., 1990, *Natural Resource Economics. Conservation and Exploitation*, Cambridge University Press.

Okuguchi, K., 1998, "Long-run fish stock and imperfectly competitive international commercial fishing", *Keio Economic Studies*, 35, 9-17.

Pearce, D.W. and Turner, R.K., 1990, *Economics of Natural Resources and the Environment*, John Hopkins University Press.

Rosser, J.B., Jr., 2001, "Complex ecologic-economic dynamics and environmental policy" *Ecological Economics*, 23-37.

Szidarovszki, F. and Okuguchi, K., 1998, "An Oligopoly model of commercial fishing", *Seoul Journal of Economics*, Vol. 11, 321-330.

5 Cournot Duopoly with Kinked Demand According to Palander and Wald

Tönu Puu[1], *Laura Gardini*[2], *and Irina Sushko*[3]

[1] Centre for Regional Science, University of Umeå, Sweden
[2] Department of Economics, University of Urbino, Italy
[3] Institute of Mathematics, National Academy of Sciences, Kiev, Ukraine

1 Introduction

In a short printed abstract of contributions to a Cowles Commission conference at Colorado Springs in 1936, and an extensive follow up, in Swedish, dating from 1939, Tord Palander focused some interesting dynamics problems in Cournot duopoly when the demand curve was kinked linear and the marginal revenue curve hence jumped up, producing two different local profit maximising intersections with the curve of marginal cost, which was assumed constant or even zero. The Cournot reaction functions, as well, then became piecewise linear, including a jump, and they could produce several, coexistent and locally stable equilibria. Palander specified the basins for these, though nor completely, and also the basin for initial conditions from which the system in stead went to an attractive 2-periodic oscillation.

For another case, where the reaction functions did not intersect, and there hence did not exist any Nash equilibrium, Palander recognised the existence of an attractive 3-periodic cycle. Given the numerical tools at that time, the accuracy of his calculations is in fact amazing, though he missed the coexistence of a 6-period cycle and again the complete characterisation of the basins.

Palander gave his argument in terms of two numerical cases, one with identical firms, another with a slight asymmetry between them.

It is noteworthy that, also in 1936, Abraham Wald considered the same type of problems, in an article which later became celebrated as first rigorous statement of existence problems for multi-market equilibria, left open by Walras and later elaborated by Arrow and Debreu.

Like Palander, Wald gave his argument in terms of two numerical examples, and also assumed the demand curve to start and end with straight line segments, but, unlike Palander, he did not assume the curves to meet under a sharp angle. Rather the two line segments were smoothly joined by either a circular segment, or by a demand curve where price was reciprocal to supply squared. The lines and curves were hence joined at tangency, so the demand function was not only continuous but differentiable. The first example, with a circular segment, seems miscalculated, but the second, hyperbola like, is most interesting, as it results in a nondenumerable infinity of coexistent Nash equilibria.

Given its context, Wald was not interested in the dynamics of the reaction functions. Had he cared to elaborate the dynamics, he would have found that none of the coexistent equilibria was even locally attractive. Rather the whole state space, except the sparse subset of the diagonal, would provide basins for oscillatory motions, quite as in Palander's cases, though all different and nondenumerably infinite in number.

It seems sufficiently interesting in itself to lift forward this neglected material, and completing the analysis. However, there is more to it. Sectional linearity of the demand function, combined with constant marginal costs, makes all falling segments of the reaction function have slope -1/2, and this makes all Nash equilibria stable. There may be cycles, even coexistent, but no such things as chaos. However, considering falling marginal costs, the slope becomes steeper, and we may in fact obtain much more interesting dynamics. Globally decreasing marginal costs, of course, do not make sense, but locally, as an approximation to the falling section of a U-shaped marginal cost curve, they indeed do.

We should note that, in all these cases, demand sensitivity to price goes up as price goes down. Joan Robinson in 1933 gave factual arguments for this, in terms of new and more numerous groups of consumers entering the market once they could afford the good, in fact the same type of assumptions as used in all textbook discussions of price discrimination.

The so called "kinked demand case" for duopoly is something quite different. There demand was assumed to become less, not more, elastic as price went down. The explanation for this was not in terms of behaviour of the consumers, but of the competitors, who were supposed to retaliate to increased demand, aimed at bringing price down, but not to reduced demand to bring about the reverse. This type of model was used to the purpose of explaining an allegedly observed extreme stability in duopoly pricing, and so had a strong flavour of ad hoc explanation. See Sweezy (1939). In this latter case the jump in marginal revenue was down, not up, but this is not the issue Palander and Wald dealt with, which no doubt is the more interesting.

2 Palander's First Case

In his first case Palander assumed two identical duopoly firms (both having zero average = marginal cost), facing the kinked (inverse) demand function:

$$p = f(q) = \begin{cases} 100 - \dfrac{q}{20} & q < 1800 \\ \dfrac{100}{7} - \dfrac{q}{420} & q > 1800 \end{cases} \tag{1}$$

shown as the broken black curve in Fig. 1. If we had dealt with a case of simple monopoly, the marginal revenue would have been:

$$MR = \begin{cases} 100 - \dfrac{q}{10} & q < 1800 \\ \dfrac{100}{7} - \dfrac{q}{210} & q > 1800 \end{cases} \tag{2}$$

as usual with the same intercepts as the bits of the demand function, but with doubled slope for each. This produces a jump up from -80 to 40/7 in the marginal revenue function at the kink point, i.e. for $q = 1800$, as we can calculate from (2).

With zero marginal cost we obtain two *local* profit maxima for the monopolist from $MR = MC = 0$, $q = 1000$ and $q = 3000$ respectively. The corresponding monopoly prices from (1) then become $p = 50$, and $p = 100/21$, and the profits (in absence of production costs) become $pq = 50000$ and $pq = 150000/7$, so it is obvious that the monopolist would select the first of the local maxima.

We can see the marginal revenue curve for the monopolist in Fig. 1 as the curve supported by the darkest shade area. Fig. 1 was drawn for a duopolist, but when one competitor supplies zero the other duopolist becomes a monopolist. The horizontal axis in the present case also is the marginal cost curve. It intersects the discontinuous marginal revenue curve in two points, the local profit maxima. Economists are used to establish the global maximum by comparing the triangular areas formed by the marginal revenue curve above and below the marginal cost curve (at present the zero line). It is obvi-

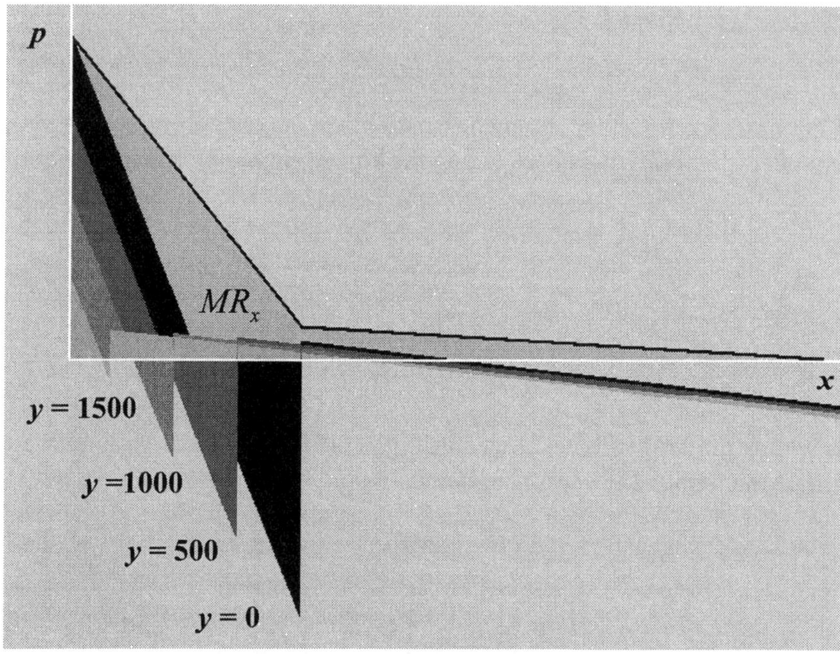

Fig. 1. Demand curve in Palander's first case, and marginal revenue curves for the first firm. Range for x: [0, 10000], *range for p:* [0, 100].

ous that the loss area, below the axis, outweighs the gain area, above the axis, by several times, so that the lower intersection with the axis establishes the global optimum (as already demonstrated numerically). At least this is true with the darkest marginal revenue curve, for the case of monopoly, or zero supply by the competitor. We can already note that this is not always the case in duopoly with positive supplies by the competitor. As shown by the marginal revenue curves with support areas in brighter shade, it may be the higher intersection point that represents the global optimum provided the competitor supplies sufficiently much.

In duopoly, denote the supplies of the competitors, x and y, so $q = x + y$. Now the marginal revenue of the first firm is:

$$MR_x = \begin{cases} 100 - \dfrac{x}{10} - \dfrac{y}{20} & x < 1800 - y \\ \dfrac{100}{7} - \dfrac{x}{210} - \dfrac{y}{420} & x > 1800 - y \end{cases} \tag{3}$$

and likewise for the second firm:

$$MR_y = \begin{cases} 100 - \dfrac{x}{20} - \dfrac{y}{10} & y < 1800 - x \\ \dfrac{100}{7} - \dfrac{x}{420} - \dfrac{y}{210} & y > 1800 - x \end{cases} \tag{4}$$

As we see, the supply by the competitor subtracts from the intercept of the marginal revenue curve, and also pushes the kink point to the left. See Fig. 1. Note however that in equations (3)-(4) the competitor's supply enters with only half the coefficient of the duopolist's own supply. Hence, the presence of a positive supply by the competitor not only pushes the whole marginal revenue curve to the left, but it also decreases the size of the dip, thereby changing the loss and gain areas. As already noted it may be that the global optimum shifts from small supply to large when the competitor increases his supply.

To establish the local optimum, given zero marginal costs, the first firm solves the equation: $MR_x = MC_x = 0$ with respect to x, and the second firm solves the equation $MR_y = MC_y = 0$ with respect to y. This results in the reaction functions $x' = \phi(y)$ and $y' = \varphi(x)$, which we will expect to become identical in view of the symmetry of the present case. Equating (3) to zero gives us two solutions for $x' = \phi(y)$:

$$x' = \phi(y) = 1000 - \frac{y}{2} \tag{5}$$

$$x' = \phi(y) = 3000 - \frac{y}{2} \tag{6}$$

so we have to establish a criterion for the choice. (Note that the dash denotes advancing the map one step in a dynamical sense from t to $t+1$.) There are now two different things to consider.

First, (5) and (6) are not applicable everywhere. We have to consider that (5) applies if $q < 1800$, (6) if $q > 1800$. Given $q = x + y$ we find by substitution from (5) in $q < 1800$ the condition $y < 1600$ for the validity of (5). Likewise, substituting from (6) in $q > 1800$, we find the condition $y > -2400$ for the validity of (6), which is not restrictive, as the supply y anyhow has to be nonnegative. So, we rephrase (5)-(6):

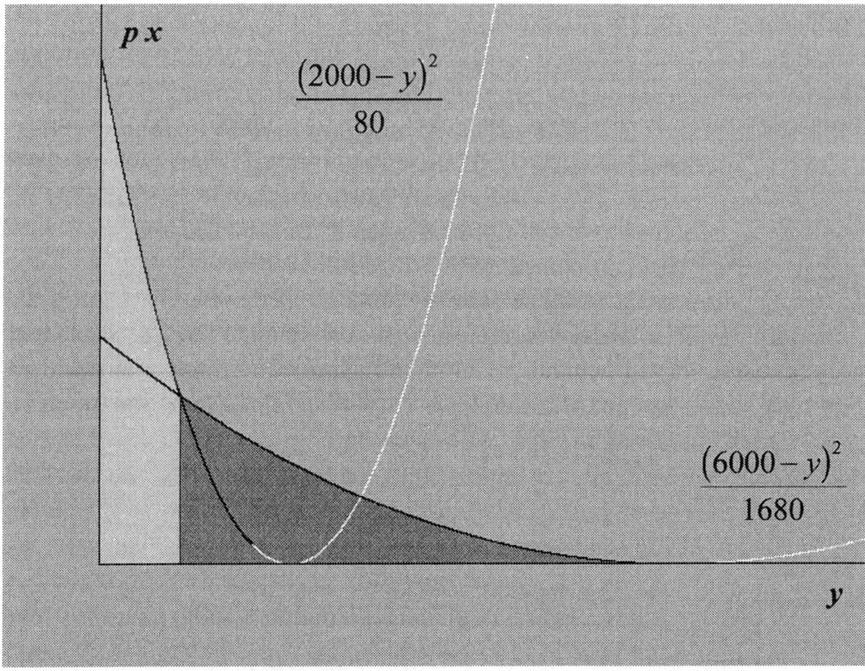

Fig. 2. Local profit maxima for the first firm as dependent on the competitor's supply. Black sections of the profit parabolas are the relevant ones. Range for y: [0, 7000], *range for px (profits):* [0, 55000].

$$x' = \phi(y) = 1000 - \frac{y}{2} \qquad y < 1600 \tag{7}$$

$$x' = \phi(y) = 3000 - \frac{y}{2} \qquad -2400 < y \tag{8}$$

Further, not only the supply of the competitor, but the supply of the firm itself has to be positive in order to make sense. From (5) we therefore find $y < 2000$, which is not operative because it is already implied by $y < 1600$. Likewise from (6) we find $y < 6000$, which, however, is. Consequently, considering all constraints, we can now write (7)-(8):

$$x' = \phi(y) = 1000 - \frac{y}{2} \qquad 0 < y < 1600 \tag{9}$$

$$x' = \phi(y) = 3000 - \frac{y}{2} \qquad 0 < y < 6000 \tag{10}$$

We are not quite done yet, because the expressions (9)-(10) have a common range of validity, $0 < y < 1600$, and so do not yield a unique solution in this interval. To achieve this we have to consider the following fact. Both solutions provide local profit maxima, but we have to select the global one.

As production costs are zero, profits equal revenues, so we obtain by substitution from (9) in (1):

$$px' = \frac{(2000 - y)^2}{80} \qquad 0 < y < 1600 \tag{11}$$

Similarly, substituting from (10) in (1):

$$px' = \frac{(6000 - y)^2}{1680} \qquad 0 < y < 6000 \tag{12}$$

Equating (11) to (12), we get two solutions:

$$y = 1800 \pm 200\sqrt{21} \tag{13}$$

but only the smaller root makes sense, as only this is in the common validity range $0 < y < 1600$. This always is the case when we compare profits. Even if we solve a second order equation only one root counts. The geometry of the case is illustrated in Fig. 2, where the dark curves are the relevant ones.

So, assembling the pieces we have:

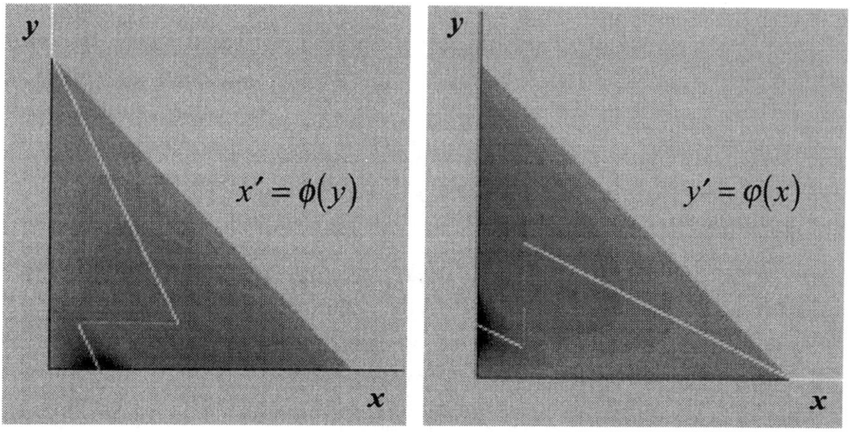

Fig. 3. Profit surfaces for the two firms in Palander's first case represented in terms of shading and the raction curves for the firms. Ranges for x, y: [0, 7000].

$$x' = \phi(y) = \begin{cases} 1000 - y/2 & 0 < y < 1800 - 200\sqrt{21} \\ 3000 - y/2 & 1800 - 200\sqrt{21} < y < 6000 \\ 0 & 6000 < y \end{cases} \tag{14}$$

for the first firm, and likewise for the second firm:

$$y' = \varphi(x) = \begin{cases} 1000 - x/2 & 0 < x < 1800 - 200\sqrt{21} \\ 3000 - x/2 & 1800 - 200\sqrt{21} < x < 6000 \\ 0 & 6000 < x \end{cases} \tag{15}$$

These piecewise linear reaction functions are displayed in Fig. 3 in white against a background of the profit surfaces for the first and second firms respectively, represented in terms of shading, the darker shade indicating higher profits. Due to the symmetry of the cases the pictures are rotations of each other with the diagonal as rotation axis.

In Fig. 4 we superpose the two reaction functions quite as Palander did. Equations (14)-(15) have two intersections, $x = y = 2000/3$ and $x = y = 2000$, obtained by either equating x' to y in (14) or y' to x in (15). As the slopes of the reaction functions (both pieces) are -1/2, both equilibria, the Cournot points, are locally stable. They hence coexist with each its proper basin of attraction in the space of initial conditions.

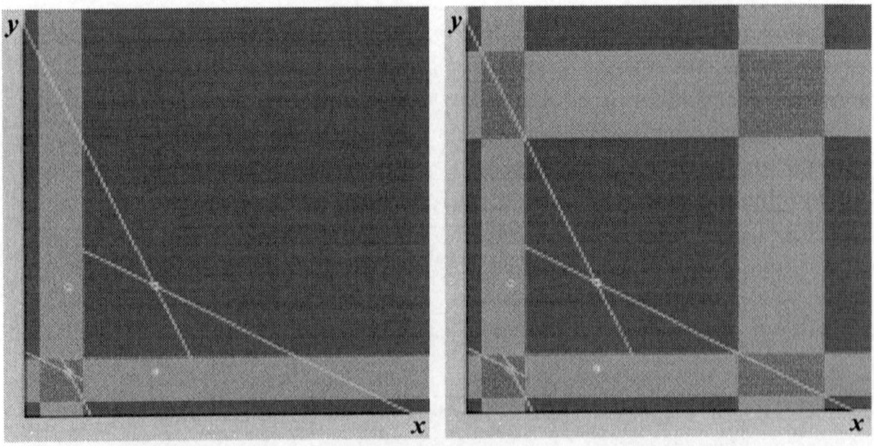

Fig. 4. Reaction functions and attraction basins according to Palander (left), and complete description (right). Ranges displayed for x and y: [0, 6500].

Note that the slopes of the reaction function always have this value, so all Cournot equilibria are stable. This is due to the linearity of the demand function. In all profit expressions the output of the optimising firms enters by its square, while that of the competitor enters linearly. By differentiation then, to get marginal revenue, as we see from (3)-(4), the competitor's sales enters with half the coefficient of the firm's own sales. So, this constantly low slope makes all the intersections of reaction functions into stable equilibrium points.

At least this is so with two competitors. We can note in passing that in the general case of n competitors, the Jacobian matrix has $(n - 1)$ eigenvalues equal to 1/2 and 1 eigenvalue equal to $-(n - 1)/2$. Accordingly, with $n > 3$, the equilibrium points are no longer stable. For instance first explorative simulations indicate that with four competitors the first Palander case has an attractive 13-period cycle.

The left picture in Fig. 4 displays the attraction basins for the two coexistent fixed points exactly as described by Palander, and a basin from which the process was supposed to go to a 2-period oscillation.

To be quite exact, in discussing the dynamics of the iterated map $(x',y') = (\phi(y), \varphi(x))$ according to (14)-(15), Palander, like most authors of his day, was very keen on distinguishing between two different systems, respectively called "*simultaneous adjustment*", and "*alternative adjustment*". In the first case both competitors react on the actual supplies recorded from the previous period, in the latter they take turns in reacting. In terms of dynamics there is actually no difference. We can always compose two *inde-*

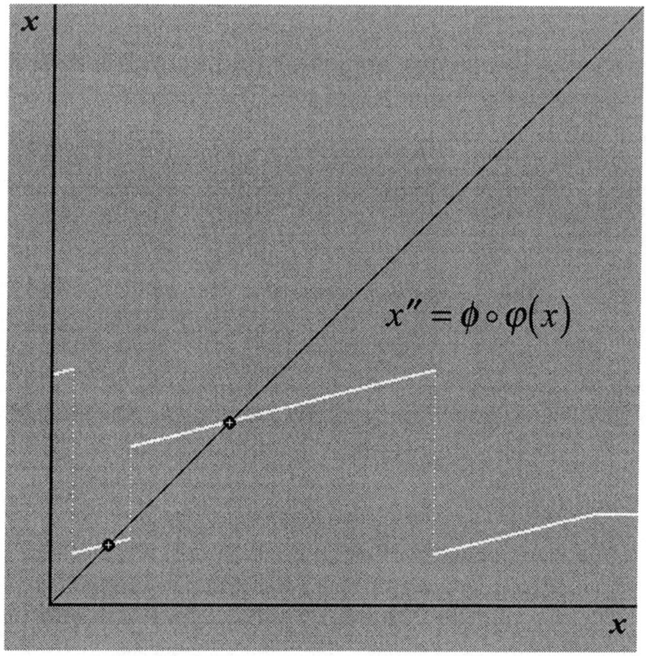

Fig. 5. Composite of reaction functions and diagonal. Range shown [0, 6500].

pendent iterations $x'' = \phi \circ \varphi(x)$ and $y'' = \varphi \circ \phi(y)$ *in each variable alone*, only note that the processes then run with a basic period twice as long as the one originally introduced. The dynamics is exactly *the same* in *both* of Palander's adjustment schemes. The only difference lies in the fact that, for the "alternative" case, the process cannot start out from every initial condition in x, y -space, the initial y-coordinate being a first iterate of the first x-coordinate, and vice versa. So, alternative adjustment just does not recover the full dynamics possibilities of Palander's model.

The composite iteration, which can be seen in Fig. 5, is:

$$x'' = \phi \circ \varphi(x) = \begin{cases} 2500 + x/4 & 0 < x < 400\sqrt{21} - 1600 \\ 500 + x/4 & 400\sqrt{21} - 1600 < x < 1800 - 200\sqrt{21} \\ 1500 + x/4 & 1800 - 200\sqrt{21} < x < 400\sqrt{21} + 2400 \\ -500 + x/4 & 400\sqrt{21} + 2400 < x < 6000 \\ 0 & 6000 < x \end{cases} \quad (16)$$

and, as mentioned, there is an identical expression for the composite $y'' = \varphi \circ \phi(y)$. The Cournot equilibria can also be recovered from (16) by equating x'' to x, though only from the second and third pieces because the ones obtained from the first and fourth are not in the intervals of relevance specified in the composite function. Put in other words, in Fig. 5 the relevant pieces of (16) do not intersect the diagonal.

To describe the basins, we define the two strips:

$$S_x = \left\{(x,y) \middle| 400\sqrt{21} - 1600 < x < 1800 - 200\sqrt{21}\right\} \tag{17}$$

$$S_y = \left\{(x,y) \middle| 400\sqrt{21} - 1600 < y < 1800 - 200\sqrt{21}\right\} \tag{18}$$

They are shown in the brightest shade in Fig. 4. The extrema defining the strips S_x and S_y are the first two discontinuity points of the composite reaction function (16) displayed in Fig. 5. For initial conditions $(x_0, y_0) \in S_x \cap S_y$, i.e. in the intersection of the two strips, the process goes to the lower Cournot equilibrium point. However, in a complete analysis, there are also initial points to the right of the third discontinuity point in (16), which also belong to the basin of the lower fixed point, as we will see below. For the moment, let us follow Palander's reasoning: For initial conditions in the union of the strips, net of the intersection, $(x_0, y_0) \in (S_x \cap S_y') \cup (S_x' \cap S_y)$, the process goes to a 2-period oscillation, and for all other initial conditions it goes to the higher Cournot equilibrium. In all Palander described one quadratic basin for the lower equilibrium, four rectangular basin pieces for the higher equilibrium, and four rectangular basin pieces for oscillation, nine pieces in all.

Palander missed the full description of the basins, because he did not note that the various areas in his account also had preimages under the mapping. So, suppose we define two more strips:

$$T_x = \left\{(x,y) \middle| 400\sqrt{21} + 2400 < x < 9200 - 800\sqrt{21}\right\} \tag{19}$$

$$T_y = \left\{(x,y) \middle| 400\sqrt{21} + 2400 < y < 9200 - 800\sqrt{21}\right\} \tag{20}$$

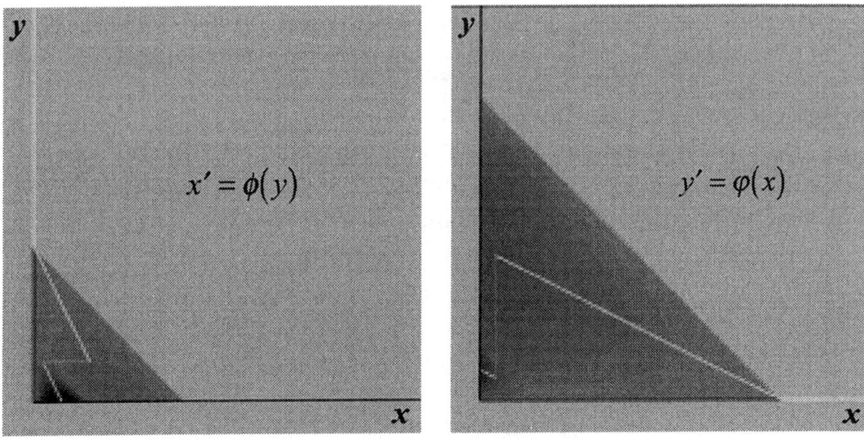

Fig. 6. Profit surfaces for the two firms in Palander's second case represented in terms of shading and the raction curves for the firms. Ranges for x,y: [0, 6500].

These strips are the new bright shade ones in the right picture of Fig. 4. Their intersections with the previous strips give rise to three new basin pieces for the lower Cournot equilibrium, further adding to and fragmenting the basins for the upper equilibrium and for oscillation so that the total number of pieces increases from 9 to 25. If it were not for the horizontal sections of the reaction functions, we could continue this process ad infinitum, but, in fact, it stops here. Note from Fig. 4 how the new strips (19)-(20) form by the intersections of the reaction functions with the old strips (17)-(18).

3 Palander's Second Case

As an alternative Palander, in his Swedish article, but not in the abstract, suggested:

$$p = f(q) = \begin{cases} 100 - \dfrac{q}{10} & q < 900 \\ \dfrac{500}{41} - \dfrac{q}{410} & q > 900 \end{cases} \tag{21}$$

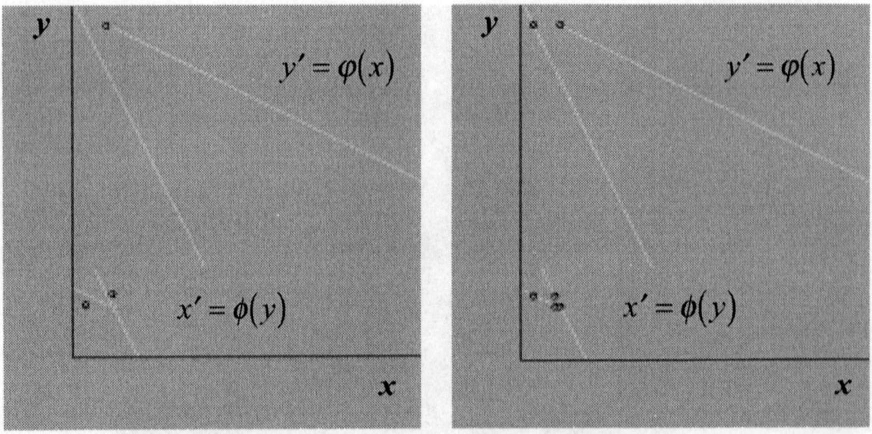

Fig. 7. Three-cycle for Palander's second case (left) and coexistent six-cycle (right). Ranges for x, y: [0, 3000].

where the first firm has a constant marginal = average cost equal to 6, whereas the second firm still has zero production cost. By exactly the same procedure as before, this time not forgetting to subtract costs from the revenues for the first firm, we obtain the reaction functions:

$$x' = \phi(y) = \begin{cases} 470 - \dfrac{y}{2} & 0 < y < 900 - 40\sqrt{41} \\ 1270 - \dfrac{y}{2} & 900 - 40\sqrt{41} < y < 2540 \\ 0 & 2540 < y \end{cases} \tag{22}$$

$$y' = \varphi(x) = \begin{cases} 500 - \dfrac{x}{2} & 0 < x < 900 - 100\sqrt{41} \\ 2500 - \dfrac{x}{2} & 900 - 100\sqrt{41} < y < 5000 \\ 0 & 5000 < y \end{cases} \tag{23}$$

They are displayed in Fig. 6, again against the background of the profit surfaces indicated by shading. It is easy to check that the reaction functions do

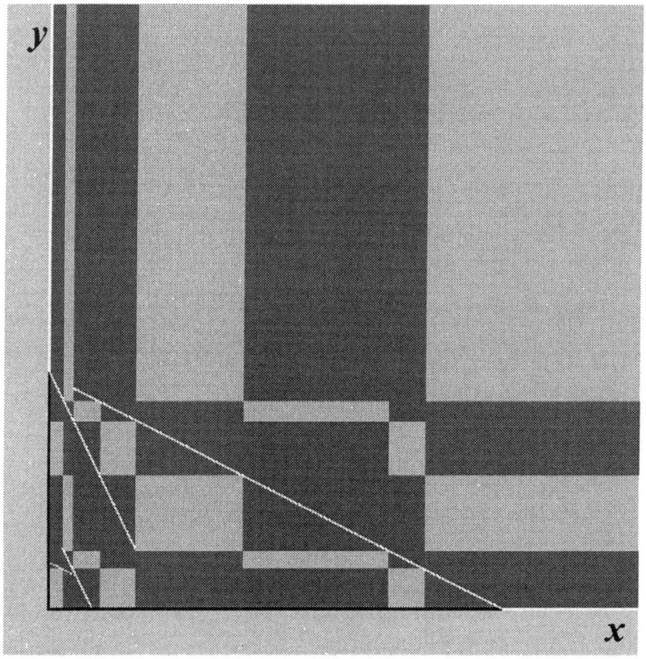

Fig. 8. The basins for the 3-cycle (bright) and the 6-cycle (dark). Range for x and y: [0, 6500]

not intersect, and so there exist no Cournot equilibria at all. Palander noted the existence of a stable 3-period cycle. See Fig. 7, the left picture. However, there is a coexistent 6-cycle, as we see in the right picture of Fig. 7, in a somewhat increased resolution in order to see all the points as distinct. The basin, shown in Fig. 8 is quite complex, as we see. Again the pattern would be repeated ad infinitum if we did not have the (white) vertical and horizontal segments of the reaction function, which are again due to the fact that supply cannot become negative.

In his picture for the second case, Palander showed a shaded area, without commenting its significance at all. He definitely did not designate it as a basin for the 3-period oscillation. In view of Fig. 8 it is, in fact, a mixture of the basins for the two coexistent oscillations. Probably he checked this area and noted that the system went to oscillatory motion, whereas he left the rest of the plane unchecked. The coexistence of the different cycles for a more general class of models has recently been established by Bischi, Mammana, and Gardini (2000).

It is interesting to note, in numerical simulations, though difficult to see in Fig. 7, that both in the 3-cycle and in the 6-cycle it is always the same three values for the x-coordinate and the y-coordinate, though different for the coordinates, which are visited. The way the 3-cycle becomes a 6-cycle is by each competitor lingering two periods at the same supply. As the time series for x and y are displaced, it then takes six periods to arrive at the same combination of coordinate values. It is also noteworthy that, whereas in the 3-cycle the points are not on the reaction curves, in the 6-cycle they in fact are. So, though assuming Palander's "simultaneous" adjustment, the process itself settles at "alternative" adjustment. See Table 1.

Table 1. Coordinates for 3-cycle and 6-cycle.

3-period cycle		6-period cycle	
x-coordinate	*y*-coordinate	*x*-coordinate	*y*-coordinate
280.6349	454.9206	280.6349	378.7302
242.5396	2359.6825	280.6349	2359.6825
90.1587	378.7302	90.1587	2359.6825
280.6349	454.9206	90.1587	454.9206
242.5396	2359.6825	242.5396	454.9206
90.1587	378.7302	242.5396	378.7302
280.6349	454.9206	280.6349	378.7302

4 Cubic Demand Functions

It is also interesting in this context to consider the case of a cubic demand curve, as considered by for instance Joan Robinson (1933), and hence also a cubic marginal revenue curve, especially as combined with the standard textbook case of a quadratic (U-shaped) marginal cost curve. We now only need one single expression to state price:

$$p = A - B(x+y) + C(x+y)^2 - D(x+y)^3 \tag{24}$$

Provided we choose all the coefficients positive and such that

$$8BD < 3C^2 < 9BD \tag{25}$$

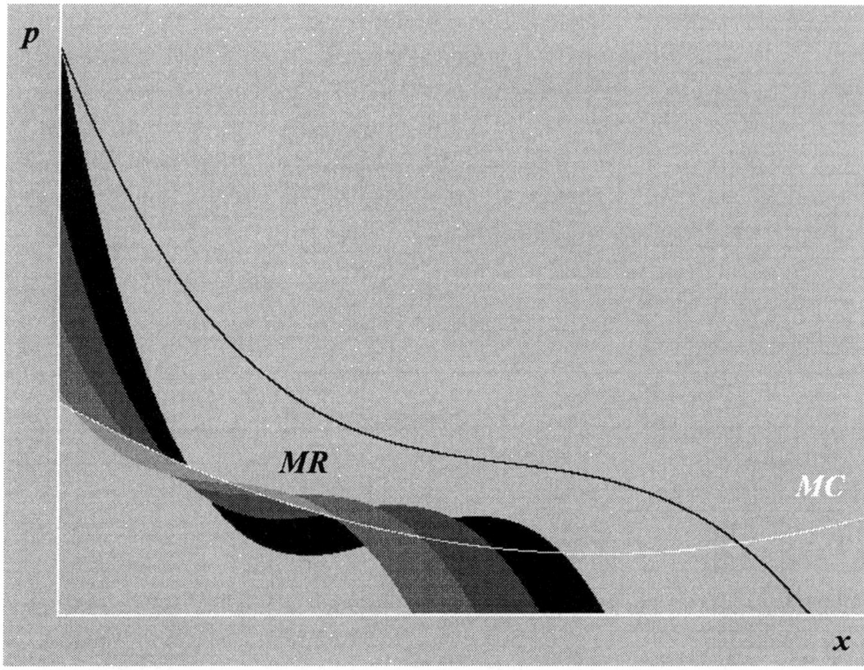

Fig. 9. Cubic demand, marginal cost, and marginal reveue curves for different supplies of the competitor. Range for x: [0, 7.5], *range for p:* [0, 6].

then the marginal revenue curve has a minimum followed by a maximum, while the demand curve is monotonically decreasing. Marginal revenues are obtained by multiplying (24) through by x for the first firm, by y for the second, and then differentiating with respect to that variable:

$$MR_x = \left(A - By + Cy^2 - Dy^3\right) - 2\left(B - 2Cy + 3Dy^2\right)x + 3(C - 3Dy)x^2 - 4Dx^3 \qquad (26)$$

$$MR_y = \left(A - Bx + Cx^2 - Dx^3\right) - 2\left(B - 2Cx + 3Dx^2\right)y + 3(C - 3Dx)y^2 - 4Dy^3 \qquad (27)$$

Supposing for simplicity the symmetrical case, so that the firms have identical, U-shaped marginal cost curves, we now write the marginal cost as:

$$MC_x = E - 2Fx + 3Gx^2 \tag{28}$$

$$MC_y = E - 2Fy + 3Gy^2 \tag{29}$$

Again all coefficients are positive, which gives the right kind of curvature, and we now require:

$$F^2 < 3EG \tag{30}$$

in order that in its minimum point marginal cost be still positive.

An example of suitable numerical coefficients are $A = 5.6, B = 2.7, C = 0.62, D = 0.05, E = 2.1, F = 0.3$, and $G = 0.02$. With these we have two local profit maxima, and now a profit minimum in between.

In Fig. 9 we display the situation quite as above in Fig. 1. Again, the darkest shade supports the marginal revenue function when the competitor supplies nothing, i.e. the firm is a monopolist. The more the competitor supplies the brighter the support shade. We can again compare the loss areas below the marginal cost curve with the gain areas above the marginal cost curve. It so seems that for the brightest shade, in the foreground, the loss outweighs the gain, and it also seems that the same holds true for the darkest shade, in the absolute background. Intermediate cases yield a larger gain area, which causes switching and reswitching. Fig. 10 displays the reaction functions $x' = \phi(y)$ and $y' = \varphi(x)$ for the two firms against a background of the profit surfaces, again in terms of shading. We see how they switch up from the lower local maximum to the upper, and reswitch back to the lower.

This classical textbook case would merit closer study, but it is computationally awkward. To get $x' = \phi(y)$ and $y' = \varphi(x)$ we have to solve the following cubic equations for local profit maxima, obtained through equating (26) to (28) and (27) to (29).

$$\begin{aligned}\left(A - E - By + Cy^2 - Dy^3\right) - 2\left(B - F - 2Cy + 3Dy^2\right)x' \\ + 3\left(C - G - 3Dy\right)x'^2 - 4Dx'^3 = 0\end{aligned} \tag{31}$$

$$\begin{aligned}\left(A - E - Bx + Cx^2 - Dx^3\right) - 2\left(B - F - 2Cx + 3Dx^2\right)y' \\ + 3\left(C - G - 3Dx\right)y'^2 - 4Dy'^3 = 0\end{aligned} \tag{32}$$

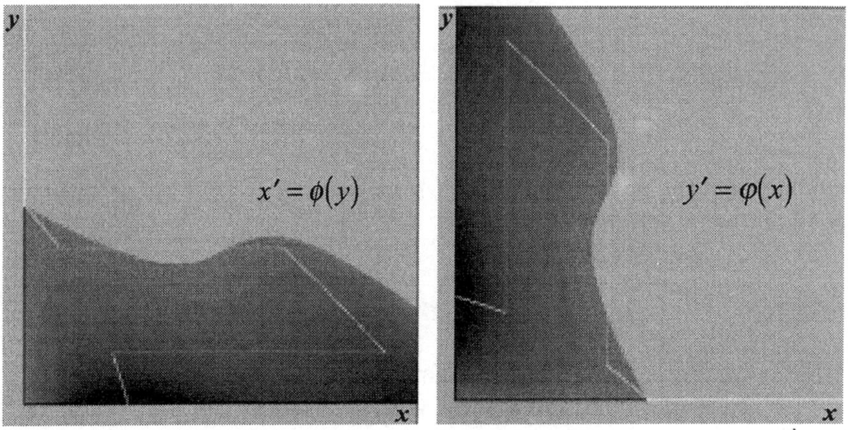

Fig. 10. Profit surfaces and reaction curves for the cubic demand case. Range shown for x and y: [0, 4.5].

Being third order equations, (31) and (32) may have three real roots for x' and y' respectively, which is the really interesting case. Then the intermediate root is a profit minimum, whereas the smallest and the largest provide local profit maxima. To decide between them it must be determined which is the global maximum, by evaluating the relevant quartic profit function:

$$\Pi_x(x',y)=\left(A-B(x'+y)+C(x'+y)^2-D(x'+y)^3-E+Fx'-Gx'^2\right)x' \quad (33)$$

$$\Pi_y(x,y')=\left(A-B(x+y')+C(x+y')^2-D(x+y')^3-E+Fy'-Gy'^2\right)y' \quad (34)$$

As we see in Fig. 10, we get nonlinear decreasing reaction curves with one switch up and another reswitch down. The lower segments in each picture are part of one single curve, representing the lower local profit maximum. Between this and the upper curve segment there is a locus for the minima, not drawn in the pictures, but with a position which can be inferred from the shading.

Equating $x=x'$ and $y=y'$, subtracting (31) from (32), and factoring, we get the equation:

$$(x-y)\left(3D(x+y)^2-(2C-3G)(x+y)+B-2F\right)=0 \quad (35)$$

for the location of possible Cournot equilibria. The first factor yields $x=y$, which substituted into (31) and (32), yields:

$$(A-E)-(3B-2F)x+(8C-3G)x^2-20Dx^3=0 \tag{36}$$

$$(A-E)-(3B-2F)y+(8C-3G)y^2-20Dy^3=0 \tag{37}$$

The three roots, possible Cournot equilibria, are two intersections between the profit maximising branches or the reaction functions, and one between the minimising, though one or both of the former may, but need not in general, be excluded by the jump conditions.

From (31) and (32) we can also (by implicit differentiation) obtain the slopes of the reaction functions in the Cournot points (where as we remember $x=y$):

$$\frac{dx'}{dy}=-\frac{B-6Cy+24Dy^2}{2B-2F-(10C-6G)y+36Dy^2} \tag{38}$$

$$\frac{dy'}{dx}=-\frac{B-6Cx+24Dx^2}{2B-2F-(10C-6G)x+36Dx^2} \tag{39}$$

If they multiply to a value larger than unity, then the Cournot equilibria lose stability and there is a possibility for complex dynamics. To judge from our numerical case, illustrated in Fig. 10, the slopes might be sufficiently high. However, the reaction curves do not intersect at all, so we are in a case similar to Palander's second. Provided we skip the simplifying assumption of identical firms, things become even more complicated.

It is better to try things out through linear approximation, replacing the falling branch of the marginal cost curve with a straight line and letting the kinked linear demand curve replace the qualitatively similar cubic. As we will find things become interesting enough.

5 Decreasing Marginal Cost

So, let us revert to the kinked demand curve case as suggested by Palander, and assume:

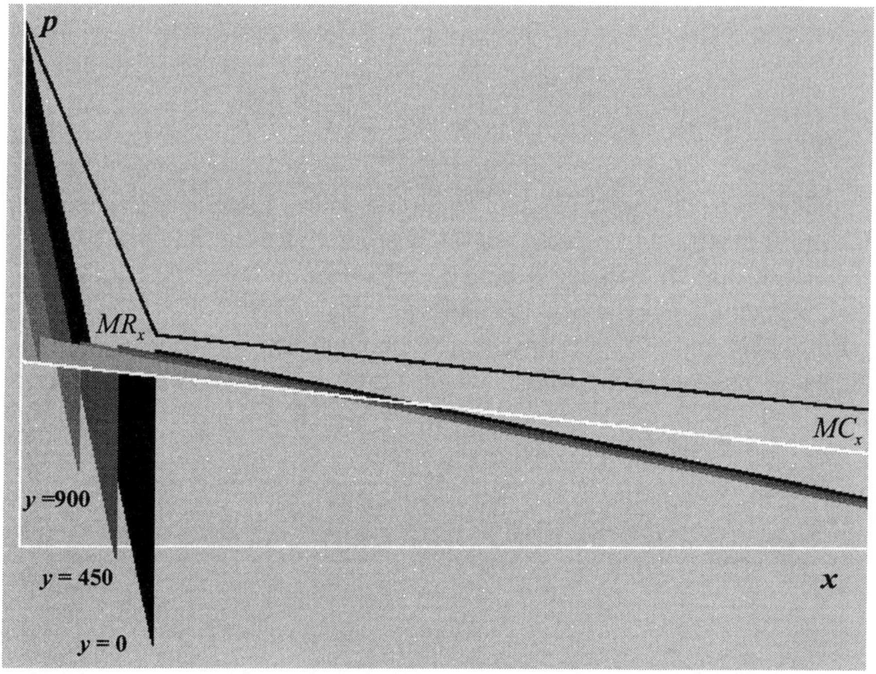

Fig. 11. Demand, falling marginal cost, and marginal reveue curves for different supplies of the competitor. Range for x: [0, 10000], *range for p:* [0, 155].

$$p = f(x+y) = \begin{cases} \alpha_1 - \beta_1(x+y) & x+y \le \overline{q} \\ \alpha_2 - \beta_2(x+y) & x+y > \overline{q} \end{cases} \tag{40}$$

where

$$\overline{q} = \frac{\alpha_1 - \alpha_2}{\beta_1 - \beta_2} \tag{41}$$

is the quantity coordinate at the kink point. Note that to have the right kind of kink we must have $\alpha_1 > \alpha_2$, $\beta_1 > \beta_2$. Given we take $\alpha_1 = 150$, $\alpha_2 = 450/7$, $\beta_1 = 0.0565$ and $\beta_2 = 0.00249$, this indeed is the case. The facts are illustrated in Fig. 11, We see the kinked demand curve, familiar by now from the original Palander cases, and further a selection of marginal revenue curves, drawn at regular intervals of the supply for the competitor 0, 450, 900, and 1350. As usual the darker shade represents a smaller supply by the competitor, the darkest being the case of a monopolist.

Further suppose we have quadratic total cost functions, i.e. linear marginal cost functions:

$$C_x = g_x(x) = a_x x - b_x x^2 \tag{42}$$

$$C_y = g_y(y) = a_y y - b_y y^2 \tag{43}$$

In general we take a_x, b_x, a_y, b_y positive to have down sloping marginal cost curves. In particular take the numerical coefficients $a_x = a_y = 53$ and $b_x = 1/780$ and $b_y = 1/800$. The marginal cost curves hence have the same intercept, 53, whereas the slopes, 1/390 and 1/400, are very slightly different. Fig. 11 was drawn for the first firm, but the difference is so slight that it would almost be imperceptible in the picture. As we see, though the marginal cost curves are down sloping, we took care that the firms never operate in the region where there is in any danger for marginal costs becoming negative. Again we can, by comparing loss and gain areas, see that in the case of virtual monopoly, $y = 0$, the first firm would choose the lower intersection of the marginal cost and revenue curves, whereas for $y = 1350$ it would definitely choose the higher one, as there is hardly any loss area at all.

From (40) and (42)-(43) we now get profits:

$$\Pi_x = f(x+y)x - g_x(x) \tag{44}$$

$$\Pi_y = f(x+y)y - g_y(y) \tag{45}$$

The pieces of the reaction functions for the first firm are obtained by differentiating profits (44) with respect to x, equating to zero, and solving, equivalent to the procedure of equating marginal revenue to marginal cost. We get two solutions:

$$x' = \frac{1}{2}\frac{\alpha_1 - a_x}{\beta_1 - b_x} - \frac{y}{2}\frac{\beta_1}{\beta_1 - b_x} \qquad x' = \frac{1}{2}\frac{\alpha_2 - a_x}{\beta_2 - b_x} - \frac{y}{2}\frac{\beta_2}{\beta_2 - b_x} \tag{46}$$

As usual these are results of local optimization. The choice of branch is due to globally maximal profits. Substituting from the reaction curve branches in

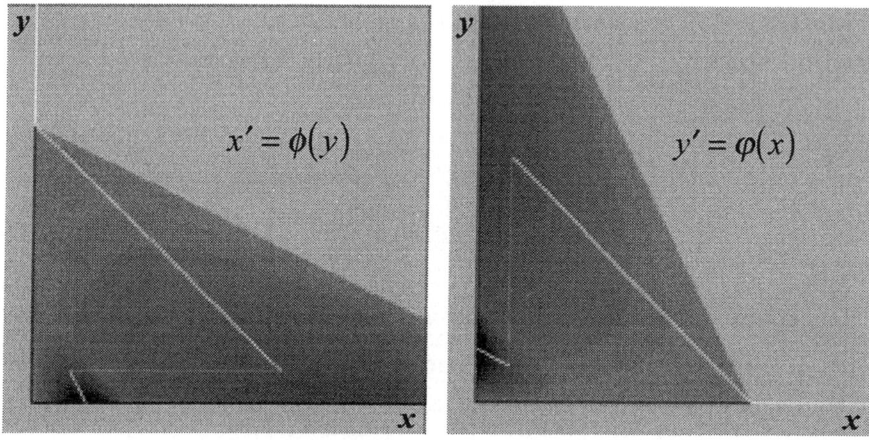

Fig. 12. Profit surfaces and reaction curves for the decreasing linear marginal cost case. Ranges shown for x and y: [0, 6500].

the profit expression (44) we get local maximum profits for the first firm as dependent on the supply of the competitor alone:

$$\Pi_x^1 = \frac{(\alpha_1 - a_x - \beta_1 y)^2}{4(\beta_1 - b_x)} \qquad \Pi_x^2 = \frac{(\alpha_2 - a_x - \beta_2 y)^2}{4(\beta_2 - b_x)} \tag{47}$$

The procedure now runs as in Section 2, by establishing the ranges of validity for the branches, by substituting the reaction function pieces (46) in the condition for the choice of branch, according to (40) and (41), and considering that no supply quantity can be negative. In this way we find, this time skipping details, that again only one of the roots in the quadratic for y, obtained from equating the two expressions (47), is in the range where both branches are relevant. This root is:

$$\bar{y} = \frac{\alpha_1\sqrt{\beta_2 - b_x} - \alpha_2\sqrt{\beta_1 - b_x} - a_x\left(\sqrt{\beta_2 - b_x} - \sqrt{\beta_1 - b_x}\right)}{\beta_1\sqrt{\beta_2 - b_x} - \beta_2\sqrt{\beta_1 - b_x}} \tag{48}$$

and we can state the reaction function for the first firm as:

$$x' = \phi(y) = \begin{cases} \frac{1}{2}\frac{\alpha_1 - a_x}{\beta_1 - b_x} - \frac{y}{2}\frac{\beta_1}{\beta_1 - b_x} & 0 < y < \overline{y} \\ \frac{1}{2}\frac{\alpha_2 - a_x}{\beta_2 - b_x} - \frac{y}{2}\frac{\beta_2}{\beta_2 - b_x} & \overline{y} < y < \frac{\alpha_2 - a_x}{\beta_2} \\ 0 & \frac{\alpha_2 - a_x}{\beta_2} < y \end{cases} \tag{49}$$

Likewise, we obtain for the second firm:

$$y' = \varphi(x) = \begin{cases} \frac{1}{2}\frac{\alpha_1 - a_y}{\beta_1 - b_y} - \frac{x}{2}\frac{\beta_1}{\beta_1 - b_y} & 0 < x < \overline{x} \\ \frac{1}{2}\frac{\alpha_2 - a_y}{\beta_2 - b_y} - \frac{x}{2}\frac{\beta_2}{\beta_2 - b_y} & \overline{x} < x < \frac{\alpha_2 - a_y}{\beta_2} \\ 0 & \frac{\alpha_2 - a_y}{\beta_2} < x \end{cases} \tag{50}$$

with

$$\overline{x} = \frac{\alpha_1\sqrt{\beta_2 - b_y} - \alpha_2\sqrt{\beta_1 - b_y} - a_y\left(\sqrt{\beta_2 - b_y} - \sqrt{\beta_1 - b_y}\right)}{\beta_1\sqrt{\beta_2 - b_y} - \beta_2\sqrt{\beta_1 - b_y}} \tag{51}$$

We do not replicate the derivation, all the arguments run as before, just interchange x and y everywhere in the formulas (41)-(43).

The facts are illustrated in Fig. 12, where we as usual display the profit surfaces in terms of shading, according to (44) and (45). We also see the reaction functions.

The big difference now is that the slopes of some pieces of the reaction functions exceed unity and hence destabilise any Cournot equilibrium, provided it exists. Actually, as we may see in Fig. 13, there is no such equilibrium in the example, but the case nevertheless involves more complex dynamics than encountered up to now. It is a bit difficult, in this resolution, to see that the long almost concurrent reaction curve segments do not intersect, but they do not.

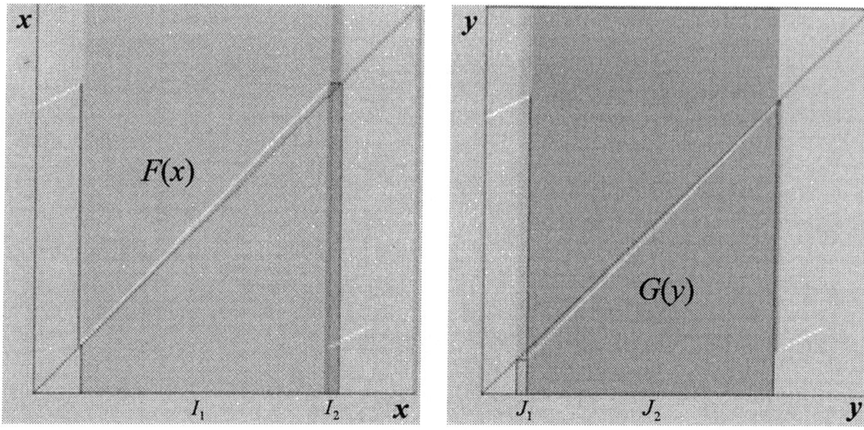

Fig. 13. Composite reaction functions. Range for x, y: [0, 5000]

Using the numerical specifications, we find the two slopes in equation (49) equal to $-\beta_1/2/(\beta_1-b_x)\approx-0.5116$ and $-\beta_2/2/(\beta_2-b_x)\approx-1.0307$, those in (50) to $-\beta_1/2/(\beta_1-b_y)\approx-0.5113$ and $-\beta_2/2/(\beta_2-b_y)\approx-1.004$. As already seen in the previous examples, the dynamics of the duopoly are strictly related to the dynamics of the one dimensional maps obtained as composite functions.

The graph of $x''=\phi\circ\varphi(x)=F(x)$ is shown in Fig. 13 (left) and we see that the map is absorbing in the interval $I=I_1\cup I_2$ because any initial condition is mapped by *F* into *I*, and *I* is invariant, *F*(*I*) = *I*. Moreover, it is easy to see that the dynamics of *F* are completely chaotic in the interval *I*. In fact, the function *F* in *I* is made up of two linear pieces: an expansion in I_1 with slope (-1.004)(-1.0307) = 1.034828, and a contraction in I_2, with slope (-1.004)(-0.5116) = 0.5136464. Then, as *F* is an expansion in I_1, any cycle of *F* in *I* must have at least one point in the interval I_2, but then any point $x_0\in I_2$ is mapped by *F* in I_1: $x_1=F(x_0)\in I_1$, and the expanding process starts. The number of iterations which are necessary in order to again get a point in the interval I_2 is higher than 21 (actually more than 50). As $(0.5136464)(1.034828)^{21}>1$, we can conclude that the multiplier associated with any cycle of *F* in *I* is higher than 1, and all the cycles are thus unstable.

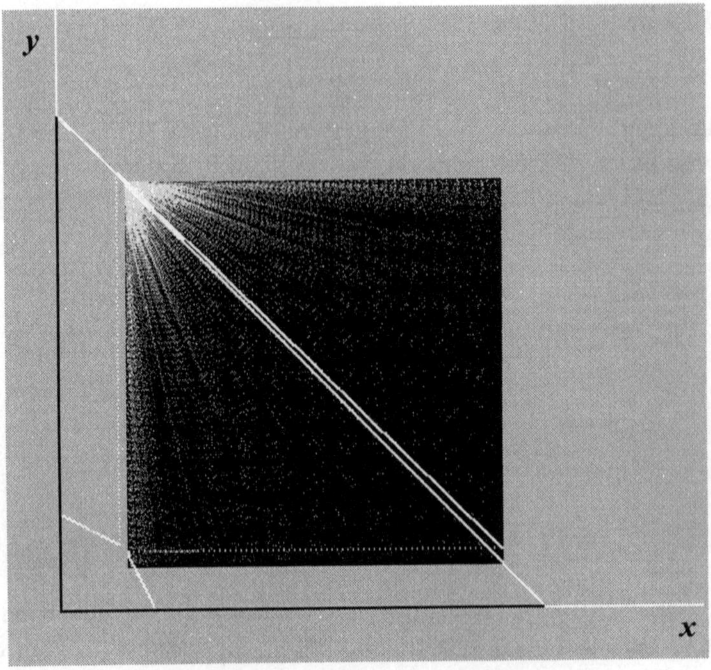

Fig. 14. Reaction functions in case of decreasing marginal costs and chaotic attractor. Range for x and y: [0, 6000].

As F is invariant in I we conclude that any initial condition is either asymptotically periodic to a repelling cycle or aperiodic, and thus F is chaotic in I. Similarly, $y'' = \varphi \circ \phi(y) = G(y)$, the graph of which is shown in Fig. 13 (right), is absorbing and invariant in the interval $J = J_1 \cup J_2$, and it too is chaotic in that interval.

Summing up, for the dynamics of this duopoly game, whose double iterate is governed by the two maps F and G, we find that the rectangle $I \times J$ (the cartesian product) is absorbing and invariant in the phase space $\Re^2$ and that the dynamics are chaotic in $I \times J$. An example is shown in Fig. 14. Note that from Fig. 13 we can deduce that all the trajectories spend more time (more iterations) in the lower part of the interval I and the upper part of interval J where the graphs of F and G are close to the diagonal. As a consequence the dynamics of the duopoly game visit the upper left corner of the rectangle $I \times J$ most frequently. This is shown in Fig. 14 where we used a colouring process: the brightness of shade represents the frequency of visiting in the iteration process. More on this kind of dynamics can be found in Chapter 6.

6 Wald's Second Model

As mentioned Abraham Wald suggested two cases where the kink in a piecewise linear demand function was replaced by a curve segment which smoothly and tangentially joined the line segments. The most interesting is the second example with the demand function:

$$p = g(x+y) = \begin{cases} 3-2(x+y) & x+y \leq 1 \\ \dfrac{1}{(x+y)^2} & 1 \leq x+y \leq 4 \\ \dfrac{3}{16} - \dfrac{x+y}{32} & x+y \geq 4 \end{cases} \tag{52}$$

and zero marginal cost for both firms. To construct the reaction functions, profits (=revenues) according to the various expressions (52) are maximised. For the first firm we have the profits: $(3-2(x+y))x$, $x/(x+y)^2$, and $(6-(x+y))x/32$ respectively, and maximise with respect to x. In this way the reaction function branches:

$$x = \frac{3}{4} - \frac{y}{2} \tag{53}$$

$$x = y \tag{54}$$

$$x = 3 - \frac{y}{2} \tag{55}$$

are obtained. Wald does not need to compare profits for the various branches, it just suffices to consider the validity due to the inequality constraints in (52). Take the constraint $x+y \leq 1$, substitute from (53), and solve for $y \leq 1/2$. Likewise, take $1 \leq x+y \leq 4$, substitute from (54) and solve for $1/2 \leq y \leq 2$. Finally, take $x+y \geq 4$, substitute from (55), and solve for $y \geq 2$. As we see the three reaction curve pieces have exclusive ranges of relevance, so there are only two points, $y = 1/2$ and $y = 2$, in which the ranges meet.

The three local maximum profit expressions are easily calculated as $(3-2y)^2/6$, $1/(4y)$, and $(6-y)^2/128$ respectively, so in the points $y=1/2$ and $y=2$ profits are pairwise the same for the expressions, 1/2 and 1/8 respectively. So, unlike Palander, Wald does not need to compare profits for selecting a global optimum. The ranges of relevance are sufficient.

Considering also that no supply quantity can be negative, we add a lower bound for the first section and an upper bound in the third, and can so piece the reaction function for the first firm together:

$$x' = \phi(y) = \begin{cases} 3/4 - y/2 & 0 \le y \le 1/2 \\ y & 1/2 \le y \le 2 \\ 3 - y/2 & 2 \le y \le 6 \\ 0 & 6 \le y \end{cases} \tag{56}$$

and a similar one for the second firm:

$$y' = \varphi(x) = \begin{cases} 3/4 - x/2 & 0 \le x \le 1/2 \\ x & 1/2 \le x \le 2 \\ 3 - x/2 & 2 \le x \le 6 \\ 0 & 6 \le x \end{cases} \tag{57}$$

As we see the reaction functions coincide along the entire middle segment, so, as Wald noted, there is a continuum of Nash equilibria $x=y$ along the entire line segment.

Considering the dynamics, which Wald did not, it is easy to see from the definition of the reaction functions $y' = \varphi(x)$ and $x' = \phi(y)$ that, given any initial condition (x_0, y_0) with $x_0 > 0$ and $y_0 > 0$, in one or at most two iterations $y_1 = \varphi(x_0)$ or $x_2 = \phi(y_1)$ jumps into the interval [0.5, 2], and $x_1 = \phi(y_0)$ or $y_2 = \varphi(x_1)$ as well into the interval [0.5, 2]. Thus any further iterations remain confined inside the square $S = [0.5, 2] \times [0.5, 2]$ of the phase plane. In fact, either $x_2 = y_2$, so we arrive at a fixed point. Or $x_2 \ne y_2$, so we are in a cycle of period two: (x_0, y_0), (x_1, y_1), (x_2, y_2), (y_2, x_2), (x_2, y_2)...

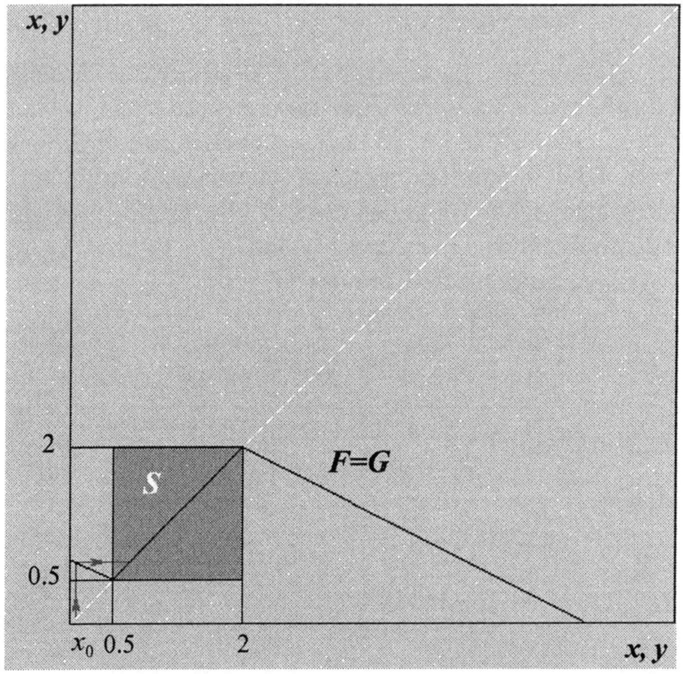

Fig. 15. Composite Wald reaction functions F(x) and G(y), (identical). Variable ranges [0, 7].

Thus all the points on the diagonal of S, (x, x) with $0.5 \le x \le 2$, are infinitely many (uncountable) fixed points of the game, while all the points in S off the diagonal, say (x, y), with $x \neq y$ and $x, y \in [0.5, 2]$, belong to infinitely many (uncountable) cycles of period two $((x, y), (y, x), (x, y), \ldots)$. Note also that all the fixed points are stable but not asymptotically stable, as any initial condition in a neighbourhood of a fixed point (x, x) will not give rise to a sequence converging to that fixed point, but either to a two-cycle or to a different fixed point in a neighbourhood of (x, x). Similarly, for the same arguments, all the two-cycles are stable but not asymptotically stable. As there are infinitely many such oscillations and basins for them, there is no point in drawing basin pictures. This type of coexistence has a strong flavour of structural instability.

Unfortunately, we cannot remove the symmetry of the firms from Wald's model, because any nonzero cost (the only characterising difference between the competitors) would cause us to have to solve a third degree equation for getting the reaction function for the middle piece. If so, we could as well go back to the cubic case treated above.

Accordingly the Palander examples are much more useful to elaborate than the Wald cases.

7 A Modified Wald Model

Departing from Wald's second example we may construct a simpler demand function, where the middle segment is just a hyperbola replaced by tangent straight lines at the ends, say:

$$p = \begin{cases} 4-4q & q \leq 0.5 \\ 1/q & 0.5 \leq q \leq 2 \\ 1-q/4 & q \geq 2 \end{cases} \tag{58}$$

The original Wald cases must be regarded as mere examples, without any subject matter significance. The hyperbola, $p = 1/q$, however, arises in a natural way in basic economic models. For instance, whenever the utility functions of the consumers have Cobb-Douglas shape, then utility maximisation results in constant budget shares for each commodity. Constant budget shares, however, mean that demand of each commodity is reciprocal to its price. As this holds for all consumers, this is one of the rare cases where the aggregation problems are easily solved, and aggregate demand as well becomes reciprocal to price.

There is a problem with this type of demand function. When used as demand function for a monopolist, the result is no definite profit maximum. This is because price times quantity sold, i.e. total revenue, is independent of price. By letting price become infinite and the quantity go to zero the monopolist can retain the same constant revenue, but cut any production costs down to zero. This, of course, is a symptom of something basically wrong in the assumptions. In no consumption theory does it make sense to buy zero fractions of units of a commodity at infinite prices. The function (58) is one way of dealing with this problem. (In another chapter we deal with this by adding a positive constant in the denominator instead.)

Now, dealing with duopolists having constant marginal costs c_x, c_y, we obtain the reaction functions:

$$x' = \phi(y) = \begin{cases} (4-c_x)/8 - y/2 & 0 < y < c_x/4 \\ \sqrt{y/c_x} - y & c_x/4 < y < 4c_x \\ 2(1-c_x) - y/2 & 4c_x < y < 4(1-c_x) \\ 0 & 4(1-c_x) < y \end{cases} \tag{59}$$

$$y' = \varphi(x) = \begin{cases} (4-c_y)/8 - x/2 & 0 < x < c_y/4 \\ \sqrt{y/c_y} - x & c_y/4 < x < 4c_y \\ 2(1-c_y) - x/2 & 4c_y < x < 4(1-c_y) \\ 0 & 4(1-c_y) < x \end{cases} \tag{60}$$

These are similar to Wald's original reaction functions, but in the middle section the straight line is replaced by a curve segment. Further, we may now introduce an asymmetry between the firms (different c_x, c_y), as it is possible to solve for the middle sections of the reaction functions in closed form even with nonzero costs.

Let us fix the parameter $c_y = 0.53$ and study what happens as we successively decrease the parameter c_x, from $c_x = 0.1$. At $c_x = 0.1$, the two composite functions $x'' = \phi \circ \varphi(x) = F(x)$ and $y'' = \varphi \circ \phi(y) = G(y)$ have a globally stable fixed point. Thus also the duopoly game has only one globally stable fixed point. Put $F(x^*) = x^*$. Then, defining $y^* = \varphi(x^*)$, we have $G(y^*) = y^*$. From $F(x^*) = \phi \circ \varphi(x^*) = x^*$ we deduce that $\phi(y^*) = x^*$, so $G(y^*) = \varphi \circ \phi(y^*) = \varphi(x^*) = y^*$. Then (x^*, y^*) is the only fixed point of our map.

However, as c_x decreases, the fixed point becomes repelling and the two composite functions both get an attracting cycle of period 2 (see Fig. 16 drawn for $c_x = 0.05$). Let x_1 and x_2 be the two points of the cycle of F, and define $y_1 = \varphi(x_1)$ and $y_2 = \varphi(x_2)$. Then we see that y_1 and y_2 are the periodic points of G. In fact, as $\phi \circ \varphi(x_1) = F(x_1) = x_2$ we deduce that $\phi(y_1) = x_2$,

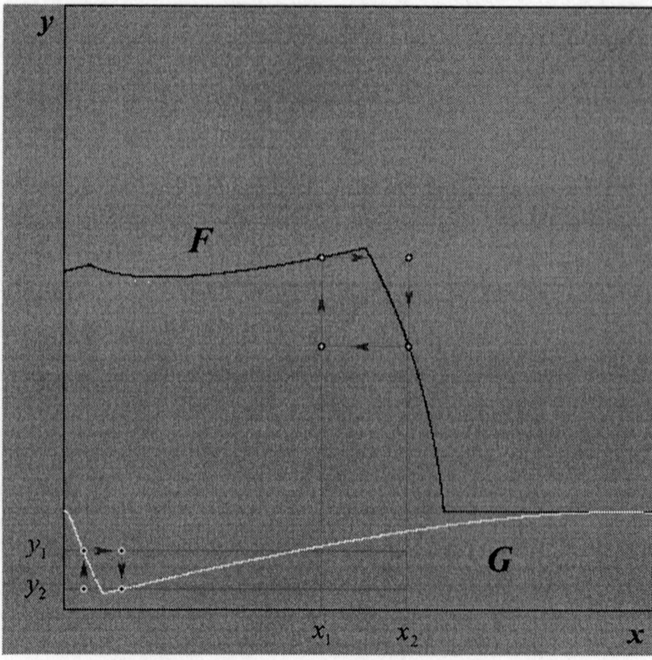

Fig. 16. Composite reaction functions and 4-cycle. Variable range [0, 3].

and as $\phi \circ \varphi(x_2) = F(x_2) = x_1$ we deduce that $\phi(y_2) = x_1$. Then we have $G(y_1) = \varphi \circ \phi(y_1) = \varphi(x_2) = y_2$ and similarly $G(y_2) = \varphi \circ \phi(y_2) = \varphi(x_1) = y_1$. As these two cycles are globally attracting for the composite functions we can deduce that the duopoly game T has a unique attractor which is a cycle of period 4. The point (x_1, y_1) is mapped by T in the following sequence:

$$\begin{aligned} (x_1, y_1) \quad &\rightarrow T \rightarrow (\phi(y_1), \varphi(x_1)) = (x_2, y_1) \\ &\rightarrow T \rightarrow (\phi(y_1), \varphi(x_2)) = (x_2, y_2) \\ &\rightarrow T \rightarrow (\phi(y_2), \varphi(x_2)) = (x_1, y_2) \\ &\rightarrow T \rightarrow (\phi(y_2), \varphi(x_1)) = (x_1, y_1) \end{aligned} \tag{61}$$

Besides the repelling fixed point (x^*, y^*) of the map T, we also have another repelling cycle of period four, in fact, all the periodic points of the map T

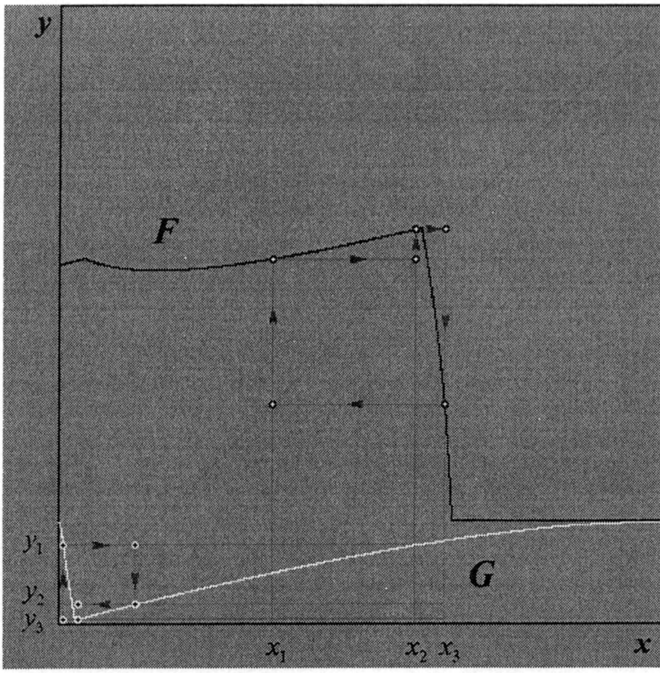

Fig. 17. Composite reaction functions, 3- and 6-cycles. Variable range [0, 3].

belong to the Cartesian product $\{x_1, x_2, x^*\} \times \{y_1, y_2, y^*\}$ and besides the stable 4-cycle and the fixed point (repelling node) we also have a saddle cycle of period 4:

$$\begin{aligned}(x^*, y_1) \quad &\rightarrow T \rightarrow (\phi(y_1), \varphi(x^*)) = (x_2, y^*)\\ &\rightarrow T \rightarrow (\phi(y^*), \varphi(x_2)) = (x^*, y_2)\\ &\rightarrow T \rightarrow (\phi(y_2), \varphi(x^*)) = (x_1, y^*)\\ &\rightarrow T \rightarrow (\phi(y^*), \varphi(x_1)) = (x^*, y_1)\end{aligned} \tag{62}$$

It is easy to predict that as c_x is further decreased, more complex dynamics may occur, including bistability.

As an example we see that at $c_x = 0.02$ the composite functions have an attracting cycle of period three, see Fig. 17 Let $\{x_1, x_2, x_3\}$ be the 3-cycle of F (with $x_{i+1} = F(x_i)$, see Fig. 17), and define $y_i = \varphi(x_i)$ for i= 1, 2, 3. Then

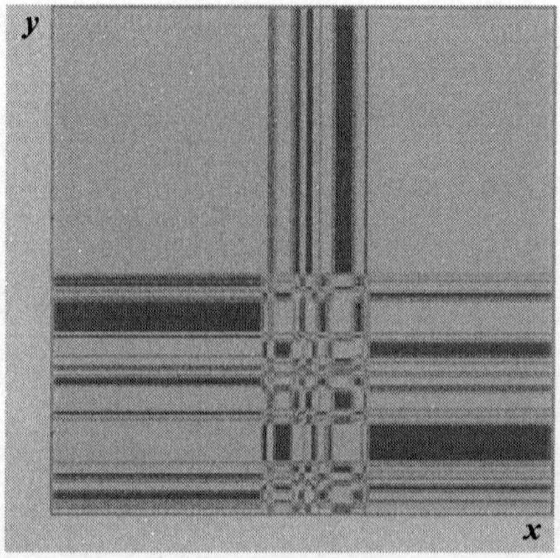

Fig. 18. Basins for coexistent 3- and 6-cycles. Axes: x, y in interval [0, 3].

we have $\phi(y_1)=x_2$, $\phi(y_2)=x_3$, $\phi(y_3)=x_1$, hence $G(y_1)=y_2$, $G(y_2)=y_3$, $G(y_3)=y_1$. It follows that the cartesian product $\{x_1,x_2,x_3\}\times\{y_1,y_2,y_3\}$ includes stable periodic points. We have:

$$
\begin{aligned}
(x_1,y_1) \; &\rightarrow T \rightarrow (\phi(y_1),\varphi(x_1)) \rightarrow (x_2,y_1) \\
&\rightarrow T \rightarrow (\phi(y_1),\varphi(x_2)) \rightarrow (x_2,y_2) \\
&\rightarrow T \rightarrow (\phi(y_2),\varphi(x_2)) \rightarrow (x_3,y_2) \\
&\rightarrow T \rightarrow (\phi(y_2),\varphi(x_3)) \rightarrow (x_3,y_3) \\
&\rightarrow T \rightarrow (\phi(y_3),\varphi(x_3)) \rightarrow (x_1,y_3) \\
&\rightarrow T \rightarrow (\phi(y_3),\varphi(x_1)) \rightarrow (x_1,y_1)
\end{aligned}
\tag{63}
$$

which is a 6-cycle of T, whose points are clearly visible in Fig. 17 But there is also an attracting 3-cycle:

$$
\begin{aligned}
(x_1,y_2) \; &\rightarrow T \rightarrow (\phi(y_2),\varphi(x_1)) \rightarrow (x_3,y_1) \\
&\rightarrow T \rightarrow (\phi(y_1),\varphi(x_3)) \rightarrow (x_2,y_3) \\
&\rightarrow T \rightarrow (\phi(y_3),\varphi(x\ )) \rightarrow (x_1,y_2)
\end{aligned}
\tag{64}
$$

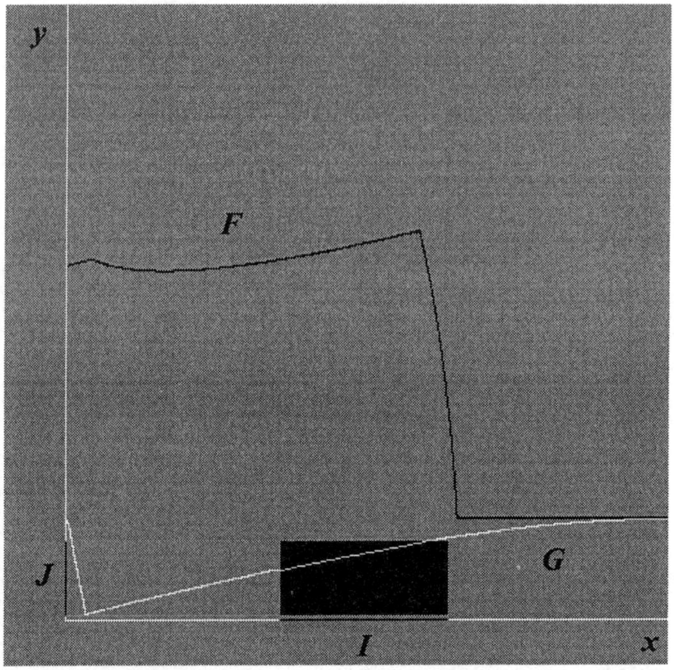

Fig. 19. Chaotic attractor in 1 piece. Axes intervals: x, y: [0, 3].

Further, note that the cartesian product $\{x_1, x_2, x_3, x^*\} \times \{y_1, y_2, y_3, y^*\}$ includes periodic points, and besides the two cycles given above we have the repelling fixed point and the repelling cycle of period 6 of saddle type given by (x^*, y_1)... The stable set of this saddle of period 6, which is made up of vertical and horizontal lines issuing from the periodic points and their preimages, as shown by Bischi, Mammana, and Gardini (2000), separating the two basins of attraction of the map T, as shown in Fig. 18, where the light grey points belong to the basin of the 6-cycle while the dark-grey points constitute the basin of the 3-cycle.

We note however that the basins shown in Fig. 18 are only approximate, and the picture is a consequence of the numerical impossibility to see the fine structure of repelling Cantor sets. For example it includes also the repelling fixed point in some rectangle but it is clear that it is a repellor and that point will not converge to any other cycle. Similarly we have infinitely many other cycles existing in the phase plane, all unstable, and such points, together with their stable sets, cannot be seen in the figures numerically computed, but their existence can be rigorously proved.

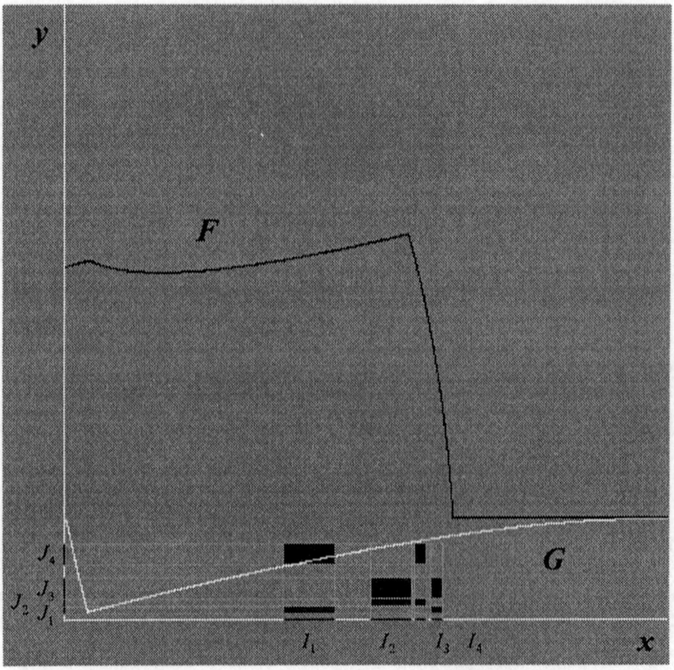

Fig. 20. Chaotic attractor in 8 pieces. Axes intervals: x, y: [0, 3].

Note the great difference in comparison to the case considered in Section 3, where the basins shown in Fig. 8 too were associated with cycles of period 3 and 6 of the map. In that example the composite functions F and G were *discontinuous*, and all pieces constituting F and G had slope 1/4, so that the 3-cycle $\{x_1, x_2, x_3\}$ of F was its only cycle (globally attracting), and similarly for G. The map T can only have those two stable cycles, whose periodic points belong to the cartesian product $\{x_1, x_2, x_3\} \times \{y_1, y_2, y_3\}$, and their basins are separated by the lines in which the discontinuous map changes by definition, and of their preimages.

Differently, in the case shown in Fig. 17, the reaction functions are *continuous*, so that the maps F and G are continuous too, and the Sharkovsky theorem applies, which means that F and G possess infinitely many repelling cycles of any order. It also follows that the map T has infinitely many repelling cycles of any order as the cartesian product of (periodic points of F) $\times$ (periodic points of G) includes all periodic points of the map T. See the proof in Bischi, Mammana, and Gardini (2000).

It is easy to deduce that, as the parameter c_x decreases from 0.05 to 0.02, chaotic dynamics in F and G occur, so that also chaotic dynamics of T exist. An example is shown in Fig. 19 (c_x = 0.025), where F is chaotic on an interval I, G is chaotic on the interval $J = \varphi(I)$, and thus the map T is chaotic in the rectangle $I \times J$. Another example is shown in Fig. 20 (c_x = 0.0295) where F has 4-cyclical chaotic intervals $\{I_1, I_2, I_3, I_4\}$ and as has G, $\{J_1, J_2, J_3, J_4\}$ (where $J_i = \varphi(I_i)$ for i = 1, 2, 3, 4), so that the map T is also chaotic, having 16-cyclical chaotic rectangles belonging to the cartesian product $\{I_1, I_2, I_3, I_4\} \times \{J_1, J_2, J_3, J_4\}$.

References

Arrow, K. J. and Debreu, G., 1954, "Existence of equilibrium for a competitive economy", *Econometrica* 22:265-290

Bischi, G. I., Mammana, C., and Gardini, L., 2000, "Multistability and cyclic attractors in duopoly games", *Chaos, Solitons & Fractals* 11:543-564

Palander, T., 1936, "Instability in competition between two sellers", in Abstracts of Papers Presented at the Research Conference on Economics and Statistics held by the Cowles Commission at Colorado College, *Colorado College Publications*, General Series No. 208, Studies Series No. 21.

Palander, T, 1939, "Konkurrens och marknadsjämvikt vid duopol och oligopol" ("Competition and market equilibrium in duopoly and oligopoly"), *The Scandinavian Journal of Economics* (then Ekonomisk Tidskrift) 41:124-145, and 222-250

Robinson, J., 1933, *The Economics of Imperfect Competition*, MacMillan, London

Sweezy, P. M., 1939, "Demand under conditions of oligopoly", *Journal of Political Economy* 47:568-573

Wald, A., 1936, "Über einige Gleichungssysteme der mathematischen Ökonomie", *Zeitschrift für Nationalökonomie* (Dezember) 637-670

6 Duopoly with Piecewise Linear Discontinuous Reaction Functions

Irina Sushko[1], *Laura Gardini*[2], *and Tönu Puu*[3]

[1]Institute of Mathematics, National Academy of Sciences of Ukraine, Kiev, Ukraine
[2]Department of Economics, University of Urbino, Italy
[3]Centre for Regional Science, University of Umeå, Sweden

1 Introduction

In the previous Chapter (Section 5) we have considered a variant of Palander's model of duopoly with kinked demand, and decreasing marginal costs. In the original model, with constant, or even zero, marginal costs no more interesting dynamics than periodic orbits can occur. This is because the resulting slopes of the reaction functions are too low. A way to raise the slopes is assuming the marginal cost line to be downsloping.

Of course, a globally downsloping marginal cost line does not make any sense at all, because it will eventually (once it reaches negative values) result in production costs decreasing with increasing production, and further even in negative production costs.

However, a piecewise linear marginal cost function could approximate the traditional U-shaped textbook case, which itself is awkward to work with as even in the simplest case we have to consider roots of cubic equations (see Section 4 of the previous Chapter). We could then focus the study on just what happens in one downsloping section of the approximating train of straight line segments, and take care to check that we only study processes which are confined to this section. We already found indications of chaos, and will now proceed to a more detailed and more formal study of these cases. In particular we focus on the changes in the dynamics as we change two of the parameters, more precisely the slopes of the two pieces in the kinked demand function.

2 Description of the General Model

Consider the general duopoly model with a piecewise linear continuous demand function of the form

$$p = f(x+y) = \begin{cases} \alpha_1 - \beta_1(x+y), & 0 \le x+y \le q; \\ \alpha_2 - \beta_2(x+y), & q < x+y \le q_1; \\ 0, & x+y > q_1; \end{cases} \quad (1)$$

where

$$q = \frac{\alpha_1 - \alpha_2}{\beta_1 - \beta_2}, \quad q_1 = \frac{\alpha_2}{\beta_2}$$

are kink points; $\alpha_i > 0, \beta_i > 0,\ i = 1, 2$, (see fig.1).

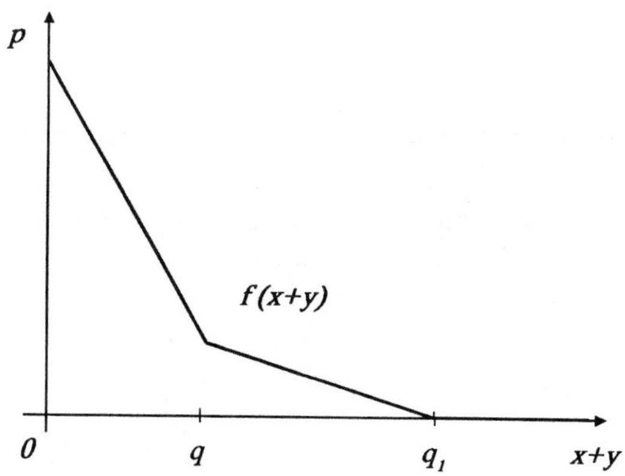

Figure 1: *Demand function.*

Let $\alpha_1 > \alpha_2$, then $\beta_1 > \beta_2$, $q > 0$, $f(q) = (\alpha_2\beta_1 - \alpha_1\beta_2)/(\beta_1 - \beta_2) > 0$.

Let the cost function of the first competitor be

$$C_x(x) = a_x x - b_x x^2, \quad 0 < x < \frac{a_x}{2b_x}; \quad (2)$$

and of the second

$$C_y(y) = a_y y - b_y y^2, \quad 0 < y < \frac{a_y}{2b_y}. \quad (3)$$

We define the function $C_x(x)$ in the interval $0 < x < a_x/2b_x$ because only when the marginal cost is positive does the model make sense, and positive marginal cost corresponds to the increasing branch of the cost function. For the same reason the function $C_y(y)$ is defined for $0 < y < a_y/2b_y$.

Note that if $a_x = 0$ and $b_x > 0$, then $C_x(x) < 0$, so that this case is not allowed. We shall consider $a_x > 0$, $b_x \geq 0$ as well as $a_y > 0$, $b_y \geq 0$.

Let $0 \leq x+y \leq q_1$. The marginal revenue $MR_x(x,y)$ and the marginal cost $MC_x(x)$ of the first firm are:

$$\begin{aligned} MR_x(x,y) &= (f(x+y)x)'_x = \begin{cases} \alpha_1 - 2\beta_1 x - \beta_1 y, & 0 \leq x+y \leq q; \\ \alpha_2 - 2\beta_2 x - \beta_2 y, & q < x+y < q_1; \end{cases} \\ MC_x(x) &= a_x - 2b_x x; \end{aligned}$$

and of the second one are:

$$\begin{aligned} MR_y(x,y) &= (f(x+y)y)'_y = \begin{cases} \alpha_1 - 2\beta_1 y - \beta_1 x, & 0 \leq x+y \leq q; \\ \alpha_2 - 2\beta_2 y - \beta_2 x, & q < x+y < q_1; \end{cases} \\ MC_y(y) &= a_y - 2b_y y. \end{aligned}$$

The profit functions of the two competitors are respectively

$$\begin{aligned} \Pi_x(x,y) &= f(x+y)x - C_x(x) \\ &= \begin{cases} (\alpha_1 - a_x - \beta_1 y)x - (\beta_1 - b_x)x^2, & 0 \leq x+y \leq q; \\ (\alpha_2 - a_x - \beta_2 y)x - (\beta_2 - b_x)x^2, & q < x+y < q_1; \end{cases} \end{aligned} \tag{4}$$

and

$$\begin{aligned} \Pi_y(x,y) &= f(x+y)y - C_y(y) \\ &= \begin{cases} (\alpha_1 - a_y - \beta_1 x)y - (\beta_1 - b_y)y^2, & 0 \leq x+y \leq q; \\ (\alpha_2 - a_x - \beta_2 x)y - (\beta_2 - b_y)y^2, & q < x+y < q_1. \end{cases} \end{aligned}$$

In order to obtain local maximal profits of the first competitor we have to solve the equation

$$\frac{\partial}{\partial x}(\Pi_x(x,y)) = 0 \tag{5}$$

with respect to x (or, in other words, we solve the equation $MR_x(x,y) = MC_x(x)$ with respect to x). Note that the solution of (5) is indeed a profit maximum if the second derivative of $\Pi_x(x,y)$ is negative, i.e.

$$\frac{\partial^2}{\partial x^2}(\Pi_x(x,y)) = -2(\beta_i - b_x) < 0,$$

which is true if

$$\beta_i > b_x, \ \ i = 1, 2.$$

Using the same arguments for the second competitor we get the following conditions:

$$\beta_i > b_y, \ i = 1, 2.$$

Solving (5) with respect to x we get the following two functions denoted by $\varphi_1(y)$ and $\varphi_2(y)$ (associated with the two functions defining $\Pi_x(x,y)$ in (4)):

$$x = \varphi_1(y) = \frac{(\alpha_1 - \beta_1 y - a_x)}{2(\beta_1 - b_x)}; \tag{6}$$

$$x = \varphi_2(y) = \frac{(\alpha_2 - \beta_2 y - a_x)}{2(\beta_2 - b_x)}. \tag{7}$$

The function $\varphi_1(y)$ is defined for $0 \leq x + y \leq q$, i.e. $0 \leq \varphi_1(y) + y \leq q$, from which it follows that if $\beta_1 > 2b_x$ then $\varphi_1(y)$ is given for

$$0 \leq y \leq y_1 \ ,$$

where

$$y_1 = \frac{2q(b_x - \beta_1) + \alpha_1 - a_x}{2b_x - \beta_1} = \frac{f(q) - a_x}{2b_x - \beta_1} + q \ ;$$

while if $\beta_1 < 2b_x$ then $\varphi_1(y)$ is defined for

$$y_1 \leq y \leq y_2 \ ,$$

where

$$y_2 = \frac{\alpha_1 - a_x}{2b_x - \beta_1}.$$

The condition $x = \varphi_1(y) \geq 0$ also has to be fulfilled and it holds for

$$y \leq y_{0,1} = \frac{\alpha_1 - a_x}{\beta_1}.$$

Note that from $y \geq 0$ it follows that $y_{0,1} \geq 0$, i.e. the following condition has to be satisfied:

$$\alpha_1 \geq a_x.$$

The function $\varphi_2(y)$ is defined for $q < x + y < q_1$, i.e. $q < \varphi_2(y) + y < q_1$, from which it follows that if $\beta_2 > 2b_x$ then $\varphi_2(y)$ is given for

$$y_3 < y < y_4;$$

where

$$y_3 = \frac{2q(b_x - \beta_2) + \alpha_2 - a_x}{2b_x - \beta_2} = \frac{f(q) - a_x}{2b_x - \beta_2} + q;$$

$$y_4 = q_1 - \frac{a_x}{2b_x - \beta_2};$$

while if $\beta_2 < 2b_x$ then $\varphi_2(y)$ is defined for

$$y_4 < y < y_3.$$

From $x = \varphi_2(y) \geq 0$ we get the inequality

$$y \leq y_{0,2} = \frac{\alpha_2 - a_x}{\beta_2}$$

which together with the condition $y \geq 0$ gives the following condition

$$\alpha_2 \geq a_x.$$

In order to have the cost function defined in the suitable range (given in (2)) we substitute $x = \varphi_1(y)$ and $x = \varphi_2(y)$ into the marginal cost expression and ask for positivity:

$$MC_x(\varphi_i(y)) = a_x - 2b_x\varphi_i(y) > 0,$$

which, when $b_x \neq 0$, holds if

$$y > \frac{\alpha_i}{\beta_i} - \frac{a_x}{b_x}, \quad i = 1, 2.$$

A sufficient condition for the inequality above to be fulfilled is

$$\frac{\alpha_i}{\beta_i} < \frac{a_x}{b_x}, \quad i = 1, 2.$$

Summarizing, it can be shown that two qualitatively different cases occur for the range of definition of the functions $\varphi_1(y)$ and $\varphi_2(y)$ depending on $a_x \lesseqgtr f(q)$.

1) If $a_x \leq f(q)$, then $y_{0,1} \leq y_{0,2}$. Moreover, the function $\varphi_1(y)$ is defined only for $\beta_1 > 2b_x$ (as $y_1 > y_{0,1}$ for $\beta_1 < 2b_x$) :

$$x = \varphi_1(y) \text{ for } 0 \leq y \leq y_1, \ \beta_1 > 2b_x.$$

The function $\varphi_2(y)$ is given as follows:

$$x = \varphi_2(y) \begin{cases} \text{for } y_3 < y < y_{0,2} \text{ if } \beta_2 > 2b_x; \\ \text{for } 0 < y < y_{0,2} \text{ if } \beta_2 < 2b_x. \end{cases}$$

In the last inequality the function $\varphi_2(y)$ is defined for $0 < y < y_{0,2}$ because when $\beta_2 < 2b_x$ then $y_4 < 0$.

2) If $a_x > f(q)$, then $y_{0,1} > y_{0,2}$. The functions $\varphi_1(y)$ and $\varphi_2(y)$ are defined as follows:

$$x = \varphi_1(y) \begin{cases} \text{for } 0 \leq y \leq y_{0,1} \text{ if } \beta_1 > 2b_x; \\ \text{for } y_1 \leq y < y_{0,1} \text{ if } \beta_1 < 2b_x; \end{cases}$$

$$x = \varphi_2(y) \quad \text{for } 0 < y < y_3, \ \beta_2 < 2b_x;$$

(as $y_3 > y_{0,2}$ for $\beta_2 > 2b_x$).

Using the same arguments for the second competitor we can repeat all the computations performed above putting x instead of y and vice versa, denoting the two solutions of the equation

$$\frac{\partial}{\partial y}(\Pi_y(x, y)) = 0$$

by $\psi_i(x)$ (instead of $\varphi_i(y)$), $i = 1, 2$.

Let us collect all the parameter conditions which are to be fulfilled in order to have a meaningful duopoly model with the demand function (1) and the cost functions (2) and (3):

$$\begin{aligned}
&\alpha_i > 0,\ \ \beta_i > 0;\\
&\alpha_1 > \alpha_2,\ \ \beta_1 > \beta_2,\ (\alpha_2\beta_1 - \alpha_1\beta_2) > 0;\\
&a_x > 0,\ b_x \geq 0;\ \ a_y > 0,\ b_y \geq 0;\\
&a_x \leq \alpha_i,\ \ b_x < \beta_i,\ \ a_y \leq \alpha_i,\ \ b_y < \beta_i;\\
&\alpha_i/\beta_i < a_x/b_x \ \text{ for } b_x \neq 0;\\
&\alpha_i/\beta_i < a_y/b_y \ \text{ for } b_y \neq 0.
\end{aligned} \tag{8}$$

The functions $\varphi_1(y)$ and $\varphi_2(y)$ are the solutions of the local optimization problem for the first competitor. In order to get a global maximal profit (denote it by $\Pi_1(y)$) we have to substitute the functions $\varphi_1(y)$ and $\varphi_2(y)$ in the profit expression (4), compare two profits and choose the function that corresponds to the maximum. In this way we obtain a reaction function of the first competitor x defined (at each y) by the function $\varphi_i(y)$, $i = 1, 2$, which corresponds to the global maximal profit.

Denote $\Pi_x(\varphi_1(y), y)$ by $\Pi_{x,1}(y)$ and $\Pi_x(\varphi_2(y), y)$ by $\Pi_{x,2}(y)$:

$$\Pi_{x,1}(y) = \frac{(\alpha_1 - \beta_1 y - a_x)^2}{4(\beta_1 - b_x)},\ \Pi_{x,2}(y) = \frac{(\alpha_2 - \beta_2 y - a_x)^2}{4(\beta_2 - b_x)};$$

the graphs of these functions are convex parabolas which intersect the vertical axis in

$$\Pi_{x,1}(0) = \frac{(\alpha_1 - a_x)^2}{4(\beta_1 - b_x)},\ \Pi_{x,2}(0) = \frac{(\alpha_2 - a_x)^2}{4(\beta_2 - b_x)},$$

and have minimum value (zero) at the following points:

$$\Pi_{x,1}(y) = 0 \text{ at } y = y_{0,1};\ \ \Pi_{x,2}(y) = 0 \text{ at } y = y_{0,2}.$$

It follows from the conditions $\varphi_1(y) \geq 0$ that the function $\Pi_{x,1}(y)$ is defined for $0 \leq y \leq y_{0,1}$; from the condition $\varphi_2(y) \geq 0$ we get that $\Pi_{x,2}(y)$ is defined for $0 \leq y \leq y_{0,2}$. In order to have the function $\Pi_1(y)$ defined for all positive y, let $\Pi_1(y)$ be given by

$$\Pi_1(y) = \begin{cases} \max\{\Pi_{x,1}(y), \Pi_{x,2}(y)\}, & 0 \leq y \leq \max\{y_{0,1}, y_{0,2}\};\\ 0, & y > \max\{y_{0,1}, y_{0,2}\}. \end{cases}$$

It is easy to get that $\Pi_{x,1}(y) \cap \Pi_{x,2}(y) = \{\overline{y}, \overline{y}_1\}$, where

$$\overline{y} = \frac{(\alpha_1 - a_x)\sqrt{(\beta_2 - b_x)} - (\alpha_2 - a_x)\sqrt{(\beta_1 - b_x)}}{\beta_1\sqrt{(\beta_2 - b_x)} - \beta_2\sqrt{(\beta_1 - b_x)}};$$

$$\overline{y}_1 = \frac{(\alpha_1 - a_x)\sqrt{(\beta_2 - b_x)} + (\alpha_2 - a_x)\sqrt{(\beta_1 - b_x)}}{\beta_1\sqrt{(\beta_2 - b_x)} + \beta_2\sqrt{(\beta_1 - b_x)}}.$$

Note that in order to avoid division by zero in the expression for $\overline{y}$, the following condition should be checked:

$$b_x \neq \frac{\beta_1\beta_2}{\beta_1 + \beta_2}.$$

Proposition 1 *Let* $a_x < f(q)$, $\beta_1 > 2b_x$, $\overline{y} > 0$ *and*

$$b_x < \frac{\beta_1\beta_2}{\beta_1 + \beta_2}. \tag{9}$$

Then the reaction function of the first competitor is

$$x = \begin{cases} \varphi_1(y), & 0 \leq y < \overline{y}; \\ \varphi_2(y), & \overline{y} \leq y \leq y_{0,2}; \\ 0, & y > y_{0,2}. \end{cases} \tag{10}$$

Proof. In fact, it follows from the assumptions on the parameters that $y_{0,1} < y_{0,2}$, $\overline{y} < \overline{y}_1$, $\overline{y} < y_{0,1}$, $y_{0,1} < \overline{y}_1 < y_{0,2}$, that is the point $\overline{y}_1$ belongs to the interval where only the profit function $\Pi_{x,2}(y)$ is defined, thus only the point $\overline{y}$ is related to a jump from one function to the other.

The function $\varphi_1(y)$ is defined for $0 < y \leq y_1$, $\beta_1 > 2b_x$. It can be shown that $\overline{y} < y_1 < y_{0,1}$.

If $\beta_2 > 2b_x$ then the function $\varphi_2(y)$ is defined for $y_3 < y < y_{0,2}$. It follows from the assumptions that $y_3 < \overline{y}$; if $\beta_2 < 2b_x$ then $\varphi_2(y)$ is defined for $0 < y < y_{0,2}$.

Thus, the global maximal profit for the first competitor is given by

$$\Pi_1(y) = \begin{cases} \Pi_{x,1}(y), & 0 \leq y < \overline{y}; \\ \Pi_{x,2}(y), & \overline{y} \leq y \leq y_{0,2}; \\ 0, & y > y_{0,2}; \end{cases}$$

(see fig.2 a), from which it follows that the reaction function is given by (10).

Proposition 2 *Let* $a_x < f(q), \beta_1 > 2b_x, \overline{y} < 0$ *and*

$$b_x < \frac{\beta_1 \beta_2}{\beta_1 + \beta_2}.$$

Then the reaction function of the first competitor is

$$x = \begin{cases} \varphi_2(y), & 0 \le y \le y_{0,2} \ ; \\ 0, & y > y_{0,2} \ . \end{cases} \tag{11}$$

Proof. It is easy to see that if $\overline{y} < 0$ then the global maximal profit is

$$\Pi_1(y) = \begin{cases} \Pi_{x,2}(y), & 0 \le y \le y_{0,2} \ ; \\ 0, & y > y_{0,2} \ ; \end{cases}$$

and the reaction function of the first competitor in this case is given by (11).

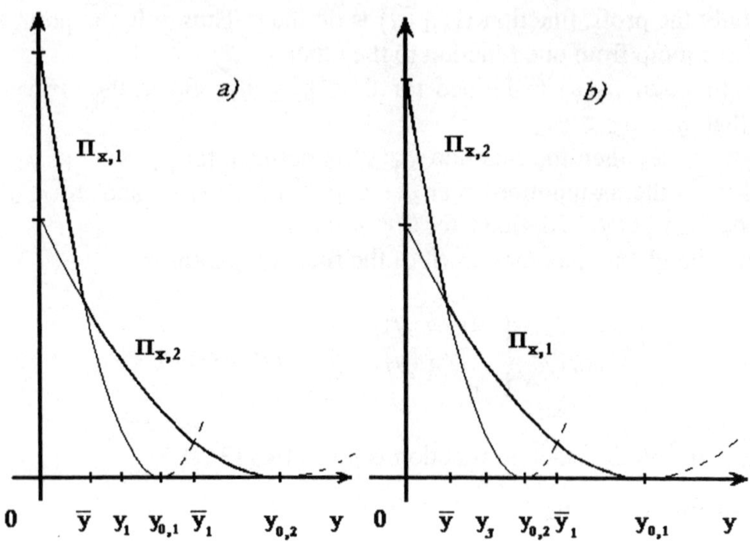

Figure 2: *The profit functions* $\Pi_{x,1}$ *and* $\Pi_{x,2}$ *corresponding to the branches* $\varphi_1(y)$ *and* $\varphi_2(y)$ *of the reaction function of the first competitor in the case of parameter assumptions of Proposition 1 (a) and Proposition 3 (b).*

Corollary 1 *If the assumptions $a_x < f(q)$, $\beta_1 > 2b_x$ are maintained but the inequality (9) is not satisfied (which can occur only if $\beta_2 < 2b_x$) then $\overline{y} > \overline{y}_1$, i.e. $\overline{y}$ belongs to the interval where the profit function is equal to zero. Thus the reaction function of the first competitor is defined by (11).*

Proposition 3 *Let $a_x > f(q), \beta_2 < 2b_x$, $\overline{y} > 0$ and*

$$b_x > \frac{\beta_1\beta_2}{\beta_1 + \beta_2}. \tag{12}$$

Then the reaction function of the first competitor is

$$x = \begin{cases} \varphi_2(y), & 0 \le y < \overline{y}; \\ \varphi_1(y), & \overline{y} \le y \le y_{0,1}; \\ 0, & y > y_{0,1}. \end{cases} \tag{13}$$

Proof. It follows from the assumptions on the parameters that $y_{0,1} > y_{0,2}$, $\overline{y} < \overline{y}_1$, $\overline{y} < y_{0,2}$, $y_{0,2} < \overline{y}_1 < y_{0,1}$, that is the point $\overline{y}_1$ belongs to the interval where only the profit function $\Pi_{x,1}(y)$ is defined. Thus only the point $\overline{y}$ is related to a jump from one function to the other.

The function $\varphi_2(y)$ is defined for $0 < y \le y_3$, $\beta_2 < 2b_x$. It can be shown that $\overline{y} < y_3 < y_{0,2}$.

If $\beta_1 < 2b_x$ then the function $\varphi_1(y)$ is defined for $y_1 \le y < y_{0,1}$. It follows from the assumptions that $y_1 < y_3$, $y_1 < y_{0,1}$ and $y_1 < \overline{y}$; if $\beta_1 > 2b_x$ then $\varphi_1(y)$ is defined for $0 < y < y_{0,1}$.

Thus, the global maximal profit of the first competitor is

$$\Pi_1(y) = \begin{cases} \Pi_{x,2}(y), & 0 \le y < \overline{y}; \\ \Pi_{x,1}(y), & \overline{y} \le y \le y_{0,1}; \\ 0, & y > y_{0,1}; \end{cases}$$

(see fig.2 b), and the reaction function is given by (13).

Proposition 4 *Let $a_x > f(q), \beta_2 < 2b_x$, $\overline{y} < 0$ and*

$$b_x > \frac{\beta_1\beta_2}{\beta_1 + \beta_2}.$$

Then the reaction function of the first competitor is

$$x = \begin{cases} \varphi_1(y), & 0 \le y \le y_{0,1}; \\ 0, & y > y_{0,1}. \end{cases} \tag{14}$$

Proof. It can be easily shown that if $\overline{y} < 0$ then the global maximal profit is

$$\Pi_1(y) = \begin{cases} \Pi_{x,1}(y), & 0 \le y \le y_{0,1}; \\ 0, & y > y_{0,1}; \end{cases}$$

and the reaction function of the first competitor is as (14).

Corollary 2 *If the inequality (12) is not satisfied (which can occur only if $\beta_1 > 2b_x$) but the assumptions $a_x > f(q)$, $\beta_2 < 2b_x$ hold, then $\overline{y} > \overline{y}_1$, i.e. $\overline{y}$ belongs to the interval where the profit function is zero. It can be shown that in this case the reaction function is defined by (14).*

Corollary 3 *If $a_x = f(q)$ then $y_{0,1} = y_{0,2} = q$. If the inequality (9) holds then the reaction function of the first competitor is given by (14), while if the inequality (12) is satisfied then the reaction function is given by (13).*

Using the same arguments for the second competitor one can get similar results for his reaction function putting x instead of y and vice versa, and $\psi_i(x)$ instead of $\varphi_i(y)$, $i = 1, 2$.

3 An Example of the Duopoly Model with Piecewise Linear Demand and Nonlinear Cost

In this section we shall propose a two-parameter family of models defined by the demand function (1) assuming quadratic cost functions, (2) and (3), in order to investigate the dynamics of the piecewise linear models which come out from the process described in the previous section. We shall consider an economic example which generalizes a model proposed by Palander (1936, 1939). The original Palander's example was already considered in Section 5 of the previous Chapter, both in the case of zero cost functions and in the case of a linear cost function. The model we propose here has a demand function given by

$$p = f(x+y) = \begin{cases} 150 - \beta_1(x+y), & 0 \le x+y \le q; \\ 64 - \beta_2(x+y), & q < x+y \le q_1; \\ 0, & x+y > q_1; \end{cases}$$

where

$$q = 86/(\beta_1 - \beta_2), \quad q_1 = 64/\beta_2;$$

so that our linear components of the demand function depend on the parameters β_1 and β_2 which model the slopes of the linear pieces. We shall let these parameters vary in the region denoted by P :

$$P = \{(\beta_1, \beta_2) : 0.043 < \beta_1 < 0.065,\ 0.0022 < \beta_2 < 0.0026\};$$

all giving examples close to the one proposed by Palander.

The cost functions are assumed of quadratic shape, and we shall keep them fixed as follows:

$$\begin{aligned} C_x(x) &= 53x - 0.00128x^2; \\ C_y(x) &= 53y - 0.00125y^2. \end{aligned}$$

In this way we have fixed the following parameters:

$$\begin{aligned} a_1 &= 150,\ a_2 = 64, \\ a_x &= 53,\ b_x = 0.00128, \\ a_y &= 53,\ b_y = 0.00125, \end{aligned}$$

and shall vary the parameters β_1 and β_2 in the region P. We shall see that in spite of the narrow intervals in which these parameters are let to vary, the model reveals a very rich variety of dynamic behavior. It is easy to verify that whichever are the values of β_1 and β_2 in the given ranges, the conditions given in (8) are fulfilled, and that the following also hold:

$$\begin{array}{l} \beta_1 > 2b_x,\ \ \beta_2 < 2b_x, \\ \beta_1 > 2b_y,\ \ \beta_2 < 2b_y, \\ a_x < f(q),\ \ a_y < f(q), \\ b_x < \beta_1\beta_2/(\beta_1+\beta_2), \\ b_y < \beta_1\beta_2/(\beta_1+\beta_2), \\ \overline{y} > 0,\ \overline{x} > 0. \end{array} \tag{15}$$

Thus, from Proposition 1 it follows that the reaction function of the first competitor is

$$x = \varphi(y) = \begin{cases} \varphi_1(y) = s_1 y + d_1, & 0 \le y < \overline{y}; \\ \varphi_2(y) = s_2 y + d_2, & \overline{y} \le y \le y_{0,2} = \frac{97}{\beta_2}; \\ 0, & y > y_{0,2} = \frac{97}{\beta_2}; \end{cases} \tag{16}$$

where

$$
\begin{aligned}
s_1 &= -\frac{\beta_1}{2(\beta_1 - 0.00128)}; \quad d_1 = \frac{97}{2(\beta_1 - 0.00128)};\\
s_2 &= -\frac{\beta_2}{2(\beta_2 - 0.00128)}; \quad d_2 = \frac{11}{2(\beta_2 - 0.00128)};
\end{aligned}
$$

$$
\overline{y} = \frac{97\beta_2 - 11\beta_1}{\left(\sqrt{(\beta_2 - 0.00128)}\sqrt{(\beta_1 - 0.00128)} - 0.00128\right)(\beta_1 - \beta_2)} + q;
$$

while the reaction function of the second competitor is given by

$$
y = \psi(x) = \begin{cases} \psi_1(x) = m_1 y + l_1, & 0 \le x < \overline{x}; \\ \psi_2(x) = m_2 y + l_2, & \overline{x} \le x \le x_{0,2} = \frac{11}{\beta_2}; \\ 0, & x > x_{0,2} = \frac{11}{\beta_2}; \end{cases} \tag{17}
$$

where

$$
\begin{aligned}
m_1 &= -\frac{\beta_1}{2(\beta_1 - 0.00125)}; \quad l_1 = \frac{97}{2(\beta_1 - 0.00125)};\\
m_2 &= -\frac{\beta_2}{2(\beta_2 - 0.00125)}; \quad l_2 = \frac{11}{2(\beta_2 - 0.00125)};
\end{aligned}
$$

$$
\overline{x} = \frac{97\beta_2 - 11\beta_1}{\left(\sqrt{(\beta_2 - 0.00125)}\sqrt{(\beta_1 - 0.00125)} - 0.00125\right)(\beta_1 - \beta_2)} + q.
$$

4 Dynamic Behavior of the Model

As usual, a dynamic model is get by introducing a time lag in the variables x and y, that is

$$
\begin{cases} x_{t+1} = \varphi(y_t); \\ y_{t+1} = \psi(x_t); \end{cases}
$$

so that the time evolution of the model is described by the iterations of *the two-dimensional piecewise linear discontinuous map* $T : \mathbb{R}^2_+ \to \mathbb{R}^2_+$, $\mathbb{R}^2_+ = [0, +\infty) \times [0, +\infty)$, of the form

$$T : \begin{pmatrix} x \\ y \end{pmatrix} \mapsto \begin{pmatrix} \varphi(y) \\ \psi(x) \end{pmatrix}, \tag{18}$$

where the functions $\varphi(y)$ and $\psi(x)$ are given in (16) and (17) respectively.

The study of a two dimensional duopoly model has already been performed in many papers by several authors. Let us recall the paper by Bischi, Mammana and Gardini (2000) which puts in evidence some general results for the class of maps (18), also recalled in Chapter 3. In our case the dynamics of the model can be described by means of the properties of the piecewise linear one-dimensional discontinuous maps $F : \mathbb{R}^+ \to \mathbb{R}^+$ and $G : \mathbb{R}^+ \to \mathbb{R}^+$ given by

$$\begin{aligned} F &: x \mapsto F(x) = (\varphi \circ \psi)(x); \\ G &: y \mapsto G(y) = (\psi \circ \varphi)(y); \end{aligned}$$

and in order to understand the dynamics of the map T, we can consider only the map F from which the properties of the map G can be deduced (for example, if x_i is a periodic point of F then $y_i = \psi(x_i)$ is a periodic point of G).

The map F is given by the following linear functions:

$$x \mapsto F(x) = \begin{cases} (\varphi_2 \circ \psi_1)(x), & 0 \le x < \overline{x},\ \overline{y} \le \psi_1(x) \le y_{0,2}; \\ (\varphi_1 \circ \psi_1)(x), & 0 \le x < \overline{x},\ 0 \le \psi_1(x) < \overline{y}; \\ (\varphi_2 \circ \psi_2)(x), & \overline{x} \le x \le x_{0,2},\ \overline{y} \le \psi_2(x) \le y_{0,2}; \\ (\varphi_1 \circ \psi_2)(x), & \overline{x} \le x \le x_{0,2},\ 0 \le \psi_2(x) < \overline{y}; \\ 0, & 0 \le x < \overline{x},\ \psi_1(x) > y_{0,2}; \\ 0, & \overline{x} \le x \le x_{0,2},\ \psi_2(x) > y_{0,2}; \\ 0, & x > x_{0,2}. \end{cases}$$

It is a discontinuous map with break points $\overline{x}$, $x1 = \psi_1^{-1}(\overline{y})$, $x2 = \psi_2^{-1}(\overline{y})$, $x3 = \psi_1^{-1}(y_{0,2})$, $x4 = \psi_2^{-1}(y_{0,2})$ and $x_{0,2}$ (if the values $\psi_1^{-1}(\overline{y})$, $\psi_2^{-1}(\overline{y})$, $\psi_1^{-1}(y_{0,2})$ and $\psi_2^{-1}(y_{0,2})$ are defined), an example is shown in fig.3.

Depending on the parameters $(\beta_1, \beta_2) \in P$ the map F may be made up by a different numbers of linear pieces.

Let L_1 denote the curve in the $(\beta_1, \beta_2)-$ parameter plane of equation $\psi_1(\overline{x}) = \overline{y}$ (see fig.4). It can be shown that if the parameter point $(\beta_1, \beta_2) \in P$ is above the curve L_1 (i.e. the condition $\psi_1(\overline{x}) < \overline{y}$ is satisfied) then the function $F(x)$ consists of 5 linear pieces (see fig.3):

$$x \mapsto F(x) = \begin{cases} s_2 m_1 x + d_2 - s_2 l_1, & 0 < x \le x1; \\ s_1 m_1 x + d_1 - s_1 l_1, & x1 < x < \overline{x}; \\ s_2 m_2 x + d_2 - s_2 l_2, & \overline{x} \le x \le x2; \\ s_1 m_2 x + d_1 - s_1 l_2, & x2 < x \le x_{0,2}; \\ 0, & x > x_{0,2}. \end{cases}$$

The map F in this case has a fixed point $x_1^* = (d_1 - s_1 l_1)/(1 - s_1 m_1) \in (x1, \overline{x})$ which is attracting in the considered parameter range (being $0 < s_1 m_1 < 1$).

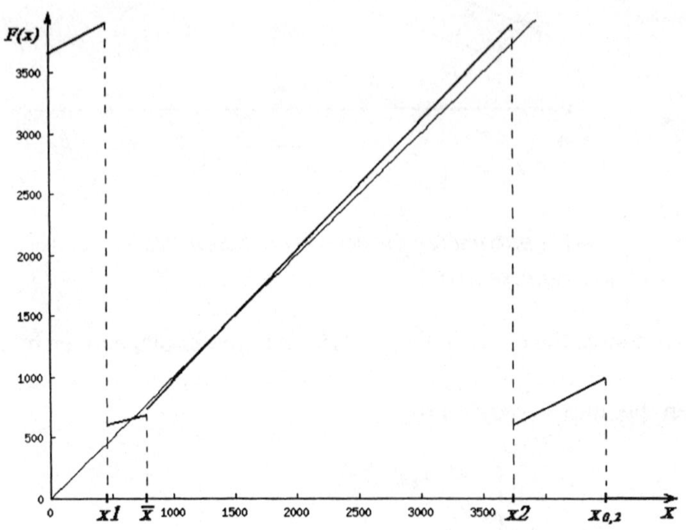

Figure 3: *An example of the map F given by 5 linear functions with break points $x1, \overline{x}, x2$ and $x_{0,2}$* ($\beta_1 = 0.045$, $\beta_2 = 0.00245$).

The map F has one more fixed point $x_2^* = (d_2 - s_2 l_2)/(1 - s_2 m_2) \in (\overline{x}, x2)$ if $F(\overline{x}) < \overline{x}$ and $F(x2) > x2$, or if $F(\overline{x}) > \overline{x}$ and $F(x2) < x2$. Let L_3 denote the curve in the $(\beta_1, \beta_2)-$ parameter plane given by the equation $F(\overline{x}) = \overline{x}$ and L_5 denote the curve of equation $F(x2) = x2$ (see fig.4). The fixed point x_2^* exists and is repelling if the parameter point (β_1, β_2) is below

L_3 and is attracting if (β_1, β_2) is above L_5. The fixed point x_2^* does not exist if (β_1, β_2) is below L_5 and above L_3.

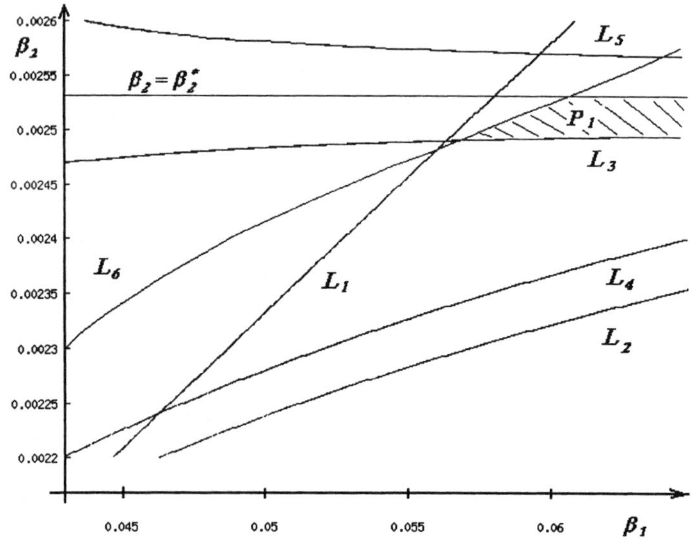

Figure 4: $(\beta_1, \beta_2)-$ *parameter plane where the curves* L_i, $i = 1, ..., 6$ *and the region* P_1 *are represented.*

Let L_2 denote the curve in the (β_1, β_2) parameter plane given by $\psi_2(\overline{x}) = y_{0,2}$. If the parameter point $(\beta_1, \beta_2) \in P$ is below the curve L_2 (i.e. $\psi_2(\overline{x}) > y_{0,2}$) then the map F is given by

$$x \mapsto F(x) = \begin{cases} s_2 m_1 x + d_2 - s_2 l, & 0 < x < \overline{x}; \\ 0, & \overline{x} \leq x < x3; \\ s_2 m_2 x + d_2 - s_2 l_2, & x3 \leq x \leq x2; \\ s_1 m_2 x + d_1 - s_1 l_2, & x2 < x \leq x_{0,2}; \\ 0, & x > x_{0,2}. \end{cases}$$

(see fig.5). We do not consider further this case in the present paper, however we note that it can raise intersting issues from the economical point of view because the presence of a zero branch for $\overline{x} \leq x < x3$ can give rise to attracting cycles with points on the axes. This means that we may have zero quantity produced by one competitor and non-zero by the other in some periods and the opposite situation in other periods, i.e. shifting monopoly.

If the parameter point $(\beta_1, \beta_2) \in P$ is below the curve L_1 but above the curve L_2 (i.e. $\psi_1(\overline{x}) > \overline{y}$ and $\psi_2(\overline{x}) < y_{0,2}$) then the map F is given by

$$x \mapsto F(x) = \begin{cases} f_1(x) = s_2 m_1 x + d_2 - s_2 l_1, & 0 < x < \overline{x}; \\ f_2(x) = s_2 m_2 x + d_2 - s_2 l_2, & \overline{x} \leq x \leq x2; \\ f_3(x) = s_1 m_2 x + d_1 - s_1 l_2, & x2 < x \leq x_{0,2}; \\ 0, & x > x_{0,2}. \end{cases} \tag{19}$$

A trajectory of the map F will never visit the zero branch if the parameter point (β_1, β_2) is above the curve denoted by L_4 (see fig.4) of equation $f_2(x2) = x_{0,2}$ (i.e., if $f_2(x2) < x_{0,2}$ then the interval $(0, x_{0,2})$ is absorbing).

When the inequalities $f_2(x2) > x2$ and $f_2(\overline{x}) > \overline{x}$ are both satisfied (i.e. (β_1, β_2) is below the curve L_5 and above L_3) then the map F has no fixed points (an example is shown in fig.6).

It can be verified that the slopes $s_2 m_1$ and $s_1 m_2$ of the functions $f_1(x)$ and $f_3(x)$ are positive and less then 1. Thus the map F is a contraction on the intervals $(0, \overline{x})$ and $(x2, x_{0,2})$. F is expanding on the interval $(\overline{x}, x2)$ if $s_2 m_2 > 1$, which holds if

$$\beta_2 < \beta_2^* = ((b_x + b_y) + \sqrt{(b_x + b_y)^2 - 3b_x b_y})/3.$$

In the family we are considering we have $\beta_2^* \approx 0.00253$, the line $\beta_2 = \beta_2^*$ is shown in fig.4.

In order to define a particular region, denoted by P_1 in fig.4, let us introduce in the parameter plane another curve, L_6, of equation $f_2(\overline{x}) = f_3(x2)$ (with $f_2(\overline{x}) < f_3(x2)$ below L_6). The three curves L_3, L_6 and $\beta_2 = \beta_2^*$ bound a region P_1 :

$$P_1 = \{(\beta_1, \beta_2) \in P : \overline{x} < f_2(\overline{x}) < f_3(x2),\ \beta_2 < \beta_2^*\}$$

such that for $(\beta_1, \beta_2) \in P_1$ the resulting map F has an invariant absorbing interval $I = I_1 \cup I_2 = [f_3(x2), x2] \cup [x2, f_2(x2)]$. That is, $F(I) = I$ (so that $F(x) \in I\ \forall x \in I$) and the trajectory of any point $x_0 \in [0, +\infty)$ is mapped into I in a finite number of iterations. A qualitative picture of the graph of $F(x)$ in this case is shown in fig.7.

Being F expanding in I_1 any trajectory must have at least one point in the interval I_2, and any point $x_0 \in I_2$ maps into the interval I_1, i.e. $x_1 = f_3(x_0) \in I_1$, then it is necessary to make some iterations by f_2 in order to get the first return in I_2 (i.e, again a point in the interval I_2). Let k be the least integer such that $f_2^k(x_1) \in I_2$, then this number is obtained taking the

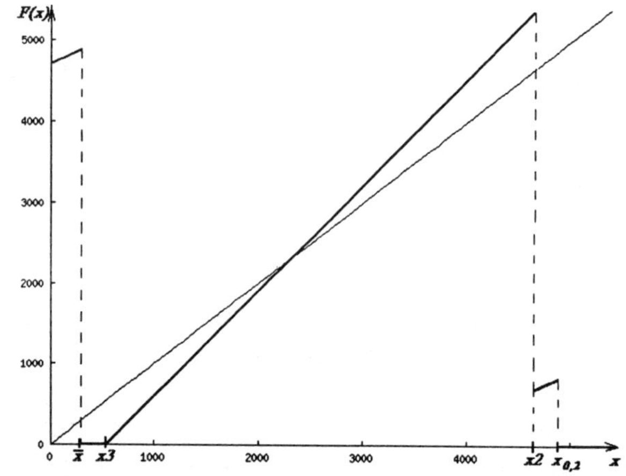

Figure 5: *An example of the function $F(x)$ with zero branches where* $\beta_1 = 0.055,\ \beta_2 = 0.00225$.

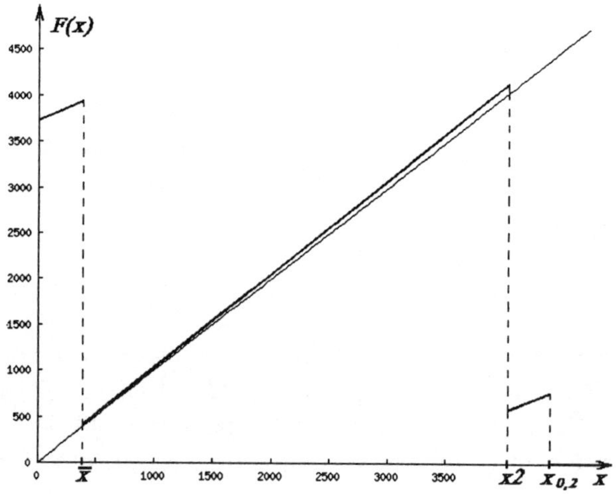

Figure 6: *An example of the function $F(x)$ given by 4 linear pieces with break points* $\overline{x}$, $x2$ *and* $x_{0,2}$ *where* $\beta_1 = 0.06,\ \beta_2 = 0.0025$.

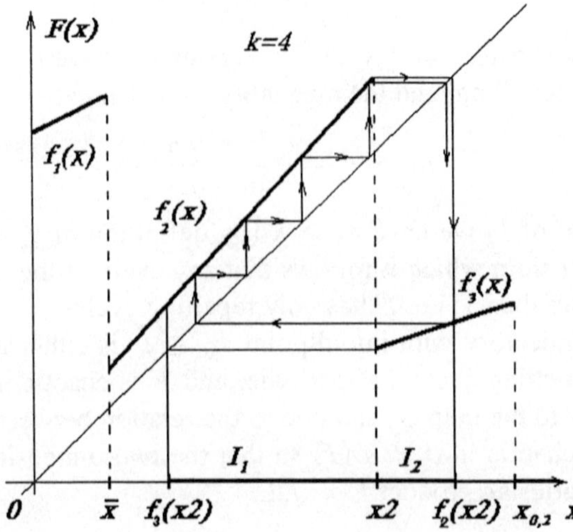

Figure 7: *Qualitative picture of the function* $F(x)$ *at* $(\beta_1, \beta_2) \in P_1$.

trajectory with the initial point $x_0^* = f_2(x2) \in I_2$. That is, k is get as the least integer for which $f_2^k(f_3(x_0^*)) \in I_2$, i.e. $x2 \leq f_2^k(f_3(x_0^*)) \leq f_2(x2)$ (see fig.7). We then prove the following

Proposition 5 *Let* $(\beta_1, \beta_2) \in P_1$. *If*

$$k > n^* = -\frac{\ln s_1 m_2}{\ln s_2 m_2},$$

then the map F *is chaotic in* $I = I_1 \cup I_2 = [f_3(x2), x2] \cup [x2, f_2(x2)]$, *and the map* T *is chaotic in the cartesian product* $I \times J$ *being* $J = \psi(I)$.

Proof. It was shown that the interval I is invariant for $(\beta_1, \beta_2) \in P_1$. In order to prove the proposition it is enough to show that when $k > n^*$ then the map F can have only repelling cycles in I. Let us consider a periodic trajectory, then it must have at least one point in I_2, let $x_0 \in I_2$ and $x_1 = f_3(x_0) \in I_1$, then let n_1 be the integer giving its first return in I_2, i.e. $f_2^{n_1}(x_1) \in I_2$. It follows that the eigenvalue of such a cycle has a factor $\mu_1 = s_1 m_2 (s_2 m_2)^{n_1}$, and the eigenvalue of the cycle is given by the product of similar factors: $\mu_1 \mu_2 ... \mu_r$ for some suitable integer $r \geq 1$ and

$\mu_i = s_1 m_2 (s_2 m_2)^{n_i}$ for $i = 1, ..., r$. It is enough to have $\mu_i > 1\ \forall i$ to state that the cycle is repelling, and the inequality $\mu_i > 1$ is satisfied if

$$n_i > n^* = -\frac{\ln s_1 m_2}{\ln s_2 m_2}.$$

As for any point of I_2 we have $n_i \geq k$ (by definition of k), when $k > n^*$ then $n_i > n^*\ \forall i$ from which it follows that any cycle of the map F has the eigenvalue higher than 1, i.e. F has only repelling cycles.

Thus, any trajectory with initial point $x_0 \in I$ is either asymptotically periodic to a repelling cycle, or aperiodic, and F is chaotic in I. A similar reasoning apply to the map G, and due to the relation between F and G we have that G is chaotic in $J = \psi(I)$ so that the two-dimensional map T is chaotic in the cartesian product $I \times J$. □

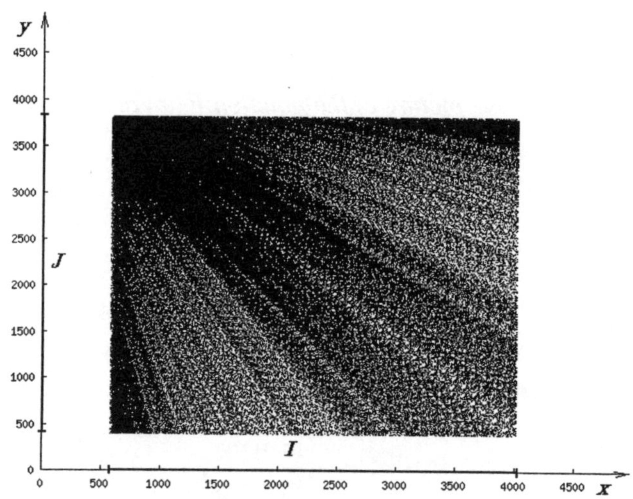

Figure 8: *An example of chaotic attractors of the map T at $\beta_1 = 0.06$, $\beta_2 = 0.0025$.*

As an example let us fix $\beta_1 = 0.06$, $\beta_2 = 0.0025$. For such parameter values $n^* \approx 27.6455$, $k = 78$, so that the assumptions of Proposition 5 are satisfied and the map T has a chaotic attractor, the rectangle $I \times J$ which is shown in fig.8.

The two-dimensional bifurcation diagram of the map F in the (β_1, β_2) - parameter plane is shown in fig.9 where the regions of regular and chaotic behavior are indicated. It is recalled in Chapter 3 that one of the characteristic properties of the class of maps (18) is the multistability, i.e. the coexistence of distinct attractors that may be stable cycles or attracting cyclic chaotic sets. In fig.9 the regions which correspond to coexisting attracting cycles

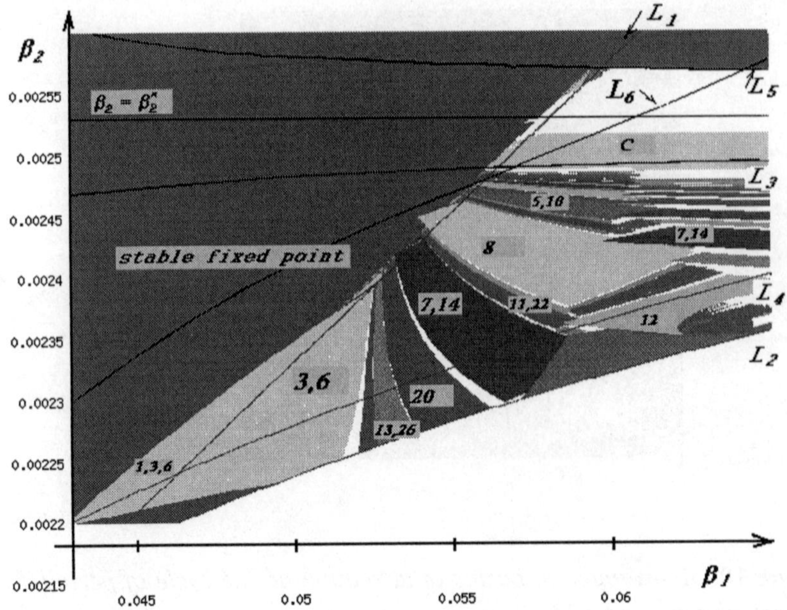

Figure 9: *Two-dimensional bifurcation diagram in the* $(\beta_1\beta_2)$ *parameter plane for the map F where regions corresponding to attracting cycles of the map T are indicated by numbers;region which corresponds to chaotic behavior is indicated by C.*

are indicated by numbers (which denote the periods of the cycles of the two-dimensional map T). From the properties of the maps of type (18) we know that in order to get all the attractors of the map T it is enough to consider first all the attractors of the one-dimensional map F (which, on they turn, are determined by considering the trajectories with initial points equal to the two values associated with all the break points of F), and then, by using the results of Bischi, Mammana, Gardini (2000), we can get all the attractors of T. For instance, the region indicated by $7, 14$ corresponds to the parameter values such that F has an attracting cycle of period 7, so that the map T has *one* attracting cycle of period 7 and *three* coexisting attracting cycles of period 14. In fig.10 we show the basins of attraction in the phase plane (x, y) of the *four* distinct cycles of T for an example in that parameter region, with $\beta_1 = 0.055$ and $\beta_2 = 0.00234$.

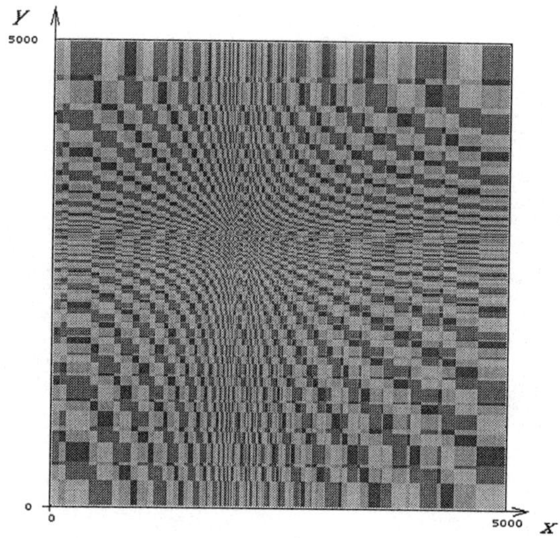

Figure 10: *An example of basins of attraction of one cycle of period* 7 *and three cycles of period* 14.

Also the structure of the basins is typical for this class of maps (18), as we know that the basin boundaries always belong to vertical and horizontal lines.

The region indicated by 20 corresponds to the parameter values such that the map F has an attracting cycle of period 10, so that T has *five* distinct attracting cycles of period 20.

It can be shown that the transition from one region to another is due to a border-collision bifurcation, occurring when some point of an attracting cycle of F coincides with a break point (see, for instance, Nusse and Yorke (1995) where such a bifurcation is descibed for piecewise smooth continuous one-dimensional maps).

If the parameter point (β_1, β_2) belongs to the region denoted by C then the map T has a chaotic attracting set. The region C was obtained by calculating the Lyapunov exponent which is positive for $(\beta_1, \beta_2) \in C$.

An example of a n-piece chaotic attractor is presented in fig.11 where $\beta_1 = 0.065$, $\beta_2 = 0.002488$. Note that in this case $(\beta_1, \beta_2) \in C$ is below the curve L_3 (and thus out of the region called P_1).

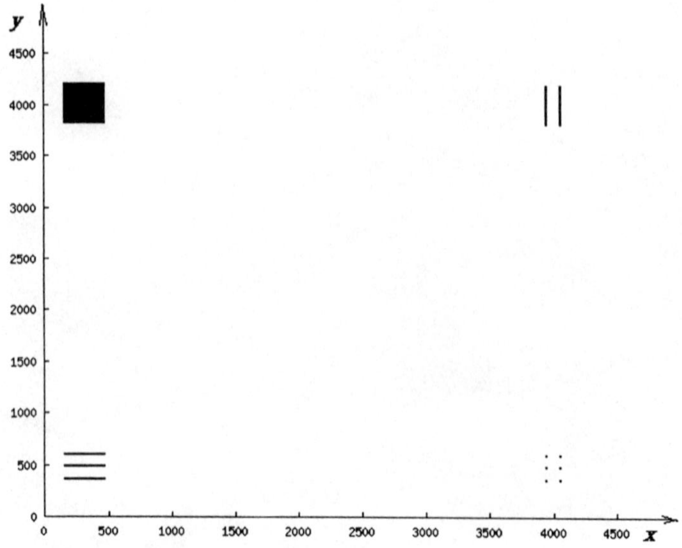

Figure 11: *An example of n-pieces chaotic attractor of the map T at $\beta_1 =$ 0.065, $\beta_2 = 0.002488$.*

References

Bischi, G. I., Mammana, C., and Gardini, L., 2000, "Multystability and cyclic attractors in duopoly games", *Chaos, Solitons & Fractals 11:543-564*

Nusse, H. E. and Yorke, J. A., 1995, "Border-collision bifurcation for piecewise smooth one-dimensional maps", *International Journal Bifurcation and Chaos 5, 1:189-207*

Palander, T., 1936, "Instability in competition between two sellers", in Abstracts of Papers Presented at the Research Conference on Economics and Statistics held by the Cowles Committion at Colorado College, *Colorado College Publications,* General Series No. 208, Studies Series No. 21.

Palander, T., 1939, "Konkurrens och marknadsjämvikt vid duopol och oligopol" ("Competition and market equilibrium in duopoly and oligopoly"), *The Scandinavian Journal of Economics* (then Ekonomisk Tidskrift) 41:124-145, and 222-250

7 A Cournot Duopoly with Bounded Inverse Demand Function

Anna Agliari[1] *and Tönu Puu*[2]

[1]Istituto di Econometria e Matematica AEFA, Università Cattolica di Milano, Italy
[2]Centre for Regional Science, University of Umeå, Sweden

1 Introduction

In 1991 Puu proposed a Cournot duopoly model based on an iso-elastic demand function and constant marginal costs for the competitors. These are probably the simplest economic assumptions under which an oligopoly game easily leads to complex dynamics, as suggested by Rand in 1978. In fact a period-doubling cascade towards chaos was observed by Puu. Further, the extension of the model to cases with three competitors leads to more complex dynamics, such as Neimark-Hopf bifurcations of subcritical types, and multistability, i.e. coexistence of the Cournot equilibrium point with an attracting invariant closed curve. See Agliari *et. al.* 2000 and Puu 2000.

However, the iso-elastic demand function, though well rooted in economic principles has a certain drawback in the oligopoly context. As price and quantity are reciprocal, their product, i.e. total sale values for a monopolist, is a constant. Accordingly, a collusive oligopoly (though normally forbidden by law in most countries), which behaves as a single monopolist, could retain the total income but reduce costs to zero by producing nothing and letting price go to infinity.

The object of the present Chapter is to study a Cournot duopoly model in which this a bit academic but annoying problem is avoided considering a new demand function, such that the price results to be bounded. The two-dimensional model will be described in Sec.3 and, as often occurs in dynamic games, we shall get a two-dimensional noninvertible map whose dynamic behavior will be considered in the consecutive sections. The study is mainly devoted to the determination in the three-dimensional parameter space of the economically meaningful region of the model and of the feasible region in the state space.

As we shall see, reasonable ranges for the parameters may be associated with unfeasible trajectories or meaningful dynamics. Their evolutions as the parameters are changed will be explained by the theory of critical curves and contact bifurcations between basin boundaries and critical curves. As described in the basic references Gumowski & Mira, 1980, Mira *et al.*, 1996, Abraham *et al.*, 1997, and in Chapter 3 of this book, the critical curves give a powerful method to study the global dynamical properties of a noninvertible map. Thus, following this theory, we are able to detect the global bifurcations which cause qualitative changes in the structure of the region of feasible trajectories, leading to the creation of regions with a complex topological structure.

The plan of the work is as follows. After the introduction of the new demand function in Sec.2, in Sec.3 we deduce the duopoly model, described by a two-dimensional map T such that the first component of T^2 depends only on x and the second one only on y ("square separate" property). Sec.3 is organized in two subsections in which we study the main dynamical properties of the model. In particular in section 3.1 a simpler model depending only on two parameters and topologically conjugated to the original one, is introduced and the study of its critical set is also performed. The regions of interest both for the mathematical system (admissible trajectories and regions) and for the economic application (feasible trajectories and regions) are also defined. The local stability analysis of the Cournot equilibrium point is considered in section 3.2, by using the "square separate" property. The dynamical behaviours of the model are described in Sec.s 4 and 5. More precisely, in Sec.4 we study the geometrical structure of the admissible and feasible regions when the Cournot equilibrium point is the unique attracting set of the map. We shall see how the noninvertibility property of the map and its inverse transformations can help us to study the region of the plane in which the model behaves well from an economic point of view, and how global bifurcations (contact bifurcation with critical curves) cause the transitions connected/disconnected in the feasible and admissible regions, as well as to multi-pieces structures. Sec.5 is related to the study of $\omega-$limit sets of the map, and its route to complex dynamics via the usual cascade of period doubling bifurcations. We shall also investigate the multistability, i.e. the coexistence of different attractors, which naturally arises in "square separate" maps, and we shall see that in our model the multistability may also be associated with a quite unusual bifurcation: a *subcritical flip bifurcation* of the Cournot equilibrium point.

2 The demand function

The hyperbola shape for a demand function arises in a natural way if the consumers have preferences in the form of Cobb-Douglas utility functions:

$$U = \prod_{i=1}^{n} q_i^{\alpha_i}$$

where q_i denote quantities of different consumers' goods, and α_i denote some given constants. These by far represent the most commonly used class for both utility functions and production functions. Maximization of U, subject to the budget constraint

$$y = \sum_{i=1}^{n} p_i q_i$$

where p_i are given prices of the goods, and y is a given income to be spent, yields the conditions

$$q_i = \alpha_i y / p_i$$

Hence demand for any commodity is reciprocal to its price. Stated in another words, a given share (the relevant exponent in the utility function) of income is spent on any given commodity.

Then, summing over the m consumers, identified by a suffix j, possibly having different incomes y^j, and different Cobb-Douglas exponents α_i^j , we get aggregate demand:

$$q_i = \sum_{j=1}^{m} q_i^j = \left(\sum_{j=1}^{m} \alpha_i^j y^j \right) / p_i$$

So, this is one of the few cases in economics where the aggregation problem is easily solved, and aggregate demand as well as individual demand both become reciprocal to market price. Rescaling the unit of commodity, we can assume (without loss of generality)

$$\left(\sum_{j=1}^{m} \alpha_i^j y^j \right) = 1$$

so that

$$q_i = 1/p_i \tag{1}$$

Such demand functions lead to nice duopoly and oligopoly systems. See for instance Puu (1991), Agliari *et al.* (2000). We note however a certain problem. If all the competitors but one drop out, there is monopoly, and total revenue becomes constant, more specifically $p_i q_i = 1$, no matter how much or little the firm produces. The monopolist may even charge almost infinite price, selling almost no commodities at all, and yet receive the same revenue. The production costs, on the other hand, increase with the volume produced, so approaching this absurd limit, the monopolist can cut its costs almost to zero and yet retain the entire revenue as net profit. This unrealistic behaviour depends on the structure of the Cobb-Douglas function. Consumers are usually assumed to be able to enjoy any tiny fraction of a commodity as long as it is positive and compensated by a very large amount of some other commodity.

To overcome this difficulty (as our object is to get a more realistic model) we put the demand function in the form

$$(q_i + a_i)(p_i + b_i) = 1 \tag{2}$$

where a_i, b_i are positive constants, so that the hyperbola (1) is translated to a new position, and it intersects the coordinate axes. This is what we in fact do in this Chapter, though concentrating on the first constant (i.e. considering $b_i = 0$). Dealing with just one commodity, we also drop all indices. The other constant is quite easy to deal with, and it has been studied in Agiza (1998).

There is another way to overcome the difficulty noted above, which is by letting a central segment of the hyperbola be replaced by tangential straight line segments at the ends. This case is in fact studied in Chapter 5 (Section 7).

3 The Duopoly Model

Let us consider a Cournot duopoly in which the producers have constant marginal costs and the inverse demand function is given by

$$p = 1/(q + W) \tag{3}$$

obtained by (2) with $a = W$ and $b = 0$. In the case $W = 0$ the demand function reduces to the hyperbola (1) and this case has been alredy studied in Puu (1991). Otherwise the constant W introduces a bound for the price ($1/W$ is the maximum price obtainable).

Denoting by x and y the supplies of the two competitors, by a and b their marginal costs, at equilibrium the profit functions are given by

$$U = \frac{x}{x+y+W} - ax, \qquad V = \frac{y}{x+y+W} - by \tag{4}$$

From (4), it is simple to obtain the reaction functions

$$\begin{cases} x = \sqrt{\frac{y^{(e)}+W}{a}} - y^{(e)} - W \\ y = \sqrt{\frac{x^{(e)}+W}{b}} - x^{(e)} - W \end{cases} \tag{5}$$

where $x^{(e)}$ and $y^{(e)}$ are the expected moves of the competitors.

Assuming expectations *à la Cournot*, that is assuming that each producer expects that the competitor retains his last move, the adjustment process to equilibrium is given by the map

$$T: \begin{cases} x' = \sqrt{\frac{y+W}{a}} - y - W \\ y' = \sqrt{\frac{x+W}{b}} - x - W \end{cases} \tag{6}$$

where the symbol $'$ denotes the unit time advancement operator (i.e. if (x, y) represents the vector of choices at time t, then (x', y') gives the choices at time $(t + 1)$).

In the next sections we shall study the dynamic model defined by T in (6), as the marginal costs and the maximum price are varied: we shall deal with the feasibility of the model in the economic context and with its asymptotic behaviour.

In the case $W = 0$ the reaction function (see Fig.1a) are defined in $\mathbb{R}^+$ and give a positive best response to each quantity value, smaller than the reciprocal of the competitor marginal cost. The case $W > 0$ (see Fig.1b) is more complex because the reaction functions are defined in a set including negative values, i.e. in the set $[-W, +\infty]$. They are positive if the quantity is smaller than the reciprocal of the competitor marginal cost minus the limit price W. We shall consider negative quantities as unmeaningful in the economic context. This suggests us to keep separate comments for the economic model and the mathematical one. We shall consider *admissible points* those which give raise to trajectories in the set of definition of the map, and *feasible points* those having trajectories belonging to $\mathbb{R}^+$. More precisely, denoting by $\mathbb{D}$ the set of definition of the map T in (6), i.e.

$$\mathbb{D} = [-W, +\infty] \times [-W, +\infty]$$

we define the *set S of admissible points* (or admissible trajectories) as

$$S = \{(x, y) \in \mathbb{D} : T^n (x, y) \in \mathbb{D} \quad \forall n > 0\}$$

and the *set F of feasible points* (or feasible trajectories) as

$$F = \left\{(x, y) \in \mathbb{D} : T^n (x, y) \in \mathbb{D} \cap \mathbb{R}^2_+ \quad \forall n \geq 0\right\}.$$

It is easy to see that in the case $W = 0$ the following relation holds

$$F = S \subset \mathbb{D} = \mathbb{R}^2_+$$

whereas in the case $W > 0$ we can only state that

$$F \subseteq S \subset \mathbb{D}$$

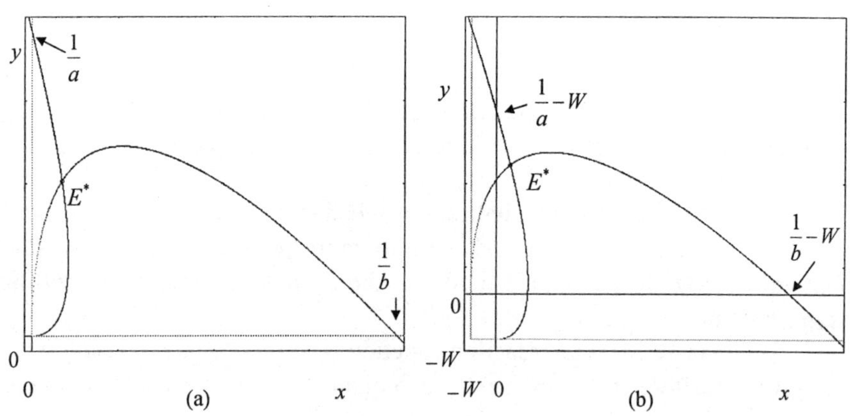

Figure 1: *Reaction curves in the case $W = 0$, (a), and $W > 0$, (b).*

In fact, from the graph of the reaction curves (see Fig.1b), we can deduce that the admissible set S is strictly contained in $\mathbb{D}$, because the functions may have values lower than $-W$. With similar arguments we can say that the feasible region F is strictly contained in S, because the set in which the reaction functions are positive is smaller than S. Then we may expect that the intervals in which the reaction functions are positive are involved, with their cartesian product, in the definition of the feasible region F, and similarly that the intervals in which the functions are greater than $-W$ are involved in the definition of the admissible set S. We shall see in Sec.4 that these regions may be quite more complex. To this end we need to introduce some important properties of the map, useful to simplify the model and its study.

3.1 Properties of the map

Proposition 1 *The map T with parameters (a, b, W) is topologically conjugate to the map $\widetilde{T}$ with parameters $(\tau a, \tau b, W/\tau), \tau > 0$, via the homeomorphism $\Phi(x, y) = (\tau x, \tau y)$.*

Proof. We have

$$\Phi\left(\widetilde{T}(x,y)\right) = \tau\left(\sqrt{\frac{y+\frac{W}{\tau}}{\tau a}} - y - \frac{W}{\tau}, \sqrt{\frac{x+\frac{W}{\tau}}{\tau b}} - x - \frac{W}{\tau}\right) =$$

$$= \left(\sqrt{\frac{\tau y + W}{a}} - \tau y - W, \sqrt{\frac{\tau x + W}{b}} - \tau x - W\right) =$$

$$= T(\Phi(x,y)) \qquad \square$$

This means that the two maps T and $\widetilde{T}$ qualitatively behave in the same way and we can indifferently study one of them, because a simple coordinate transformation permits to extend the results to the other.

In the following we shall study a simpler map, obtained from (6) by using Proposition 1 with $\tau = 1/a$, so that the model reduces to only two parameters $k = b/a$ and $v = W/a$:

$$T: \begin{cases} x' = \sqrt{y+v} - y - v \\ y' = \sqrt{\frac{x+v}{k}} - x - v \end{cases} \tag{7}$$

and we continue to call it T for the sake of simplicity. Moreover, if we assume that producer with marginal cost b is cost leader, we can restrict the study to the parameter space

$$\Omega = \{(k, v) : k \in (0, 1], v \in [0, 1)\}$$

In fact, as we shall see in the following, a necessary condition for the existence of feasible trajectories is that the marginal costs must be smaller of the limit price (which is trivial in the economic context).

The map T in (7) is a noninvertible one, and as described in the Chapter 3 of this book, we shall use this important property in order to obtain global results on the topological structures of the basins of attraction of the existing attracting sets.

We can prove the noninvertibility of T, and describe the *Riemann foliation* of the map, looking for the rank-1 preimages of a given point $(x',y') \in \mathbb{R}^2$ finding the solutions of the algebraic system defining the map, in the unknown (x,y). By defining

$$\begin{aligned} \alpha_\pm &= \frac{1 \pm \sqrt{1-4x'}}{2} \\ \beta_\pm &= \frac{1 \pm \sqrt{1-4y'k}}{2k}, \end{aligned}$$

we have that

- if $x' < 0$ and $y' < 0$, just one solution of the system exists, given by $\left(k\beta_+^2 - v, \alpha_+^2 - v\right)$
- if $x' < 0$ and $0 \leq y' < 1/(4k)$, the system admits two distinct solutions given by $\left(k\beta_+^2 - v, \alpha_+^2 - v\right)$ and $\left(k\beta_-^2 - v, \alpha_+^2 - v\right)$
- if $0 \leq x' < 1/4$ and $y' < 0$, the system admits two distinct solutions given by $\left(k\beta_+^2 - v, \alpha_+^2 - v\right)$ and $\left(k\beta_+^2 - v, \alpha_-^2 - v\right)$
- if $0 \leq x' < 1/4$ and $0 \leq y' < 1/(4k)$, the system admits four distinct solutions given by $\left(k\beta_+^2 - v, \alpha_+^2 - v\right)$, $\left(k\beta_+^2 - v, \alpha_-^2 - v\right)$, $\left(k\beta_-^2 - v, \alpha_+^2 - v\right)$ and $\left(k\beta_-^2 - v, \alpha_-^2 - v\right)$
- if $x' > 1/4$ and $y' > 1/(4k)$, no solution of the system exists.

Then, following the terminology introduced in Mira *et al.*, 1996, we can say that the map T is a $Z_1 - Z_2 - Z_4 - Z_0$ map, which means that in the phase plane there are points with 1, 2 and 4 distinct rank-1 preimages and points with no preimage.

The zones Z_0-Z_4 and Z_0-Z_2 are separated by the critical lines of the map

$$LC = LC^{(a)} \cup LC^{(b)} = \left\{x = \frac{1}{4}\right\} \cup \left\{y = \frac{1}{4k}\right\},$$

i.e. the loci of points having merging rank-1 preimages. Moreover in the map (7), due to the presence of the square root, also the coordinate axes play the role of separation of zones, separating Z_2-Z_4 and Z_1-Z_2. However they are not critical lines, because their points have no merging preimages (as two distinct preimages have the points on the negative half-line, while four distinct preimages those belonging to the positive half-line).

From the behaviour of the preimages, we can understand the action of the map T on its definition set $\mathbb{D}$. To this aim it is useful to introduce the critical set of rank -0 LC_{-1}, locus of the merging preimages of the points of LC ($LC = T(LC_{-1})$). For the map we consider, LC_{-1} is given by the set of points at which the Jacobian determinant vanishes:

$$LC_{-1} = LC_{-1}^{(a)} \cup LC_{-1}^{(b)} = \left\{ y = \frac{1}{4} - v \right\} \cup \left\{ x = \frac{1}{4k} - v \right\}.$$

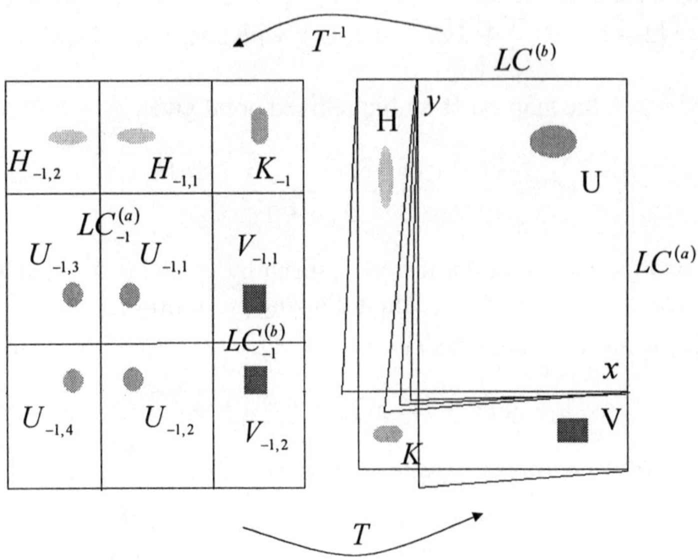

Figure 2: *Riemann Foliation of the map T*

Recalling that the action of a noninvertible map is to fold and pleat the plane, we have that our map T folds twice the plane, along a vertical and a horizontal line, determining the *Riemann foliation* shown in Fig.2.

3.2 Local stability of the fixed point. Another important property of the map T in (7) is that to be "square separate", as it is typical for maps describing Cournot games with naive expectations, as described in Chapter 3 (Section 5) of this book. This property is mainly useful in the study of the attracting sets of the map, because this study can be deduced from that of a one-dimensional map, which is a component of T^2, the second iterate of T. In particular, in our case we shall consider the one-dimensional composite function $x'' = \phi(x)$ of the map in (7) which reads as

$$\phi(x) = \sqrt{\sqrt{\frac{x+v}{k}} - x} - \sqrt{\frac{x+v}{k}} + x \tag{8}$$

It is simple to show that the function in (8) is defined in the interval $\left[-v, (1+\sqrt{1+4vk})/2k\right]$ and it is always positive. Its critical points are a minimum point $c_{-1}^m = 1/(4k) - v$, with critical value $c^m = \sqrt{1/(4k)+v} - 1/(4k) - v$, and two maximum points $c_{-1,a}^M = [2-k-2\sqrt{1+4vk-k}]/(4k)$, $c_{-1,b}^M = [2-k+2\sqrt{1+4vk-k}]/(4k)$, with common critical value $c_M = 1/4$.

For $v > 0$ the map ϕ has a unique fixed point given by

$$x^* = \frac{k - 2v\,(k+1) + k\sqrt{1+4v\,(k+1)}}{2\,(k+1)^2}$$

while two fixed points exist if $v = 0$, given by x^* and 0. By using the first derivative of ϕ, we can deduce the following proposition.

Proposition 2 *The fixed point x^* is locally stable if the inequality*

$$4k(1+2v(k+1)) > (1+k)^2\sqrt{1+4vk\,(k+1)} \tag{9}$$

holds in the parameter space $\Omega = (0,1] \times [0,1)$.

The curve of equation $4k(1+2v(k+1)) = (1+k)^2\sqrt{1+4vk(k+1)}$, separating the regions defined in (9), is a bifurcation curve, crossing which a flip bifurcation occurs, and we shall see that such a flip bifurcation can be of supercritical or subcritical type, depending on the parameter values. The stability region, as well as the region in which the fixed point is positive, are shown in Fig.3. The positivity of the fixed point must be considered, because we are interested in economic meaningful dynamics. It is simple to prove that x^* is positive if $v < k$, and this is the reason why we consider the parameter set $\Omega = (0,1] \times [0,1)$.

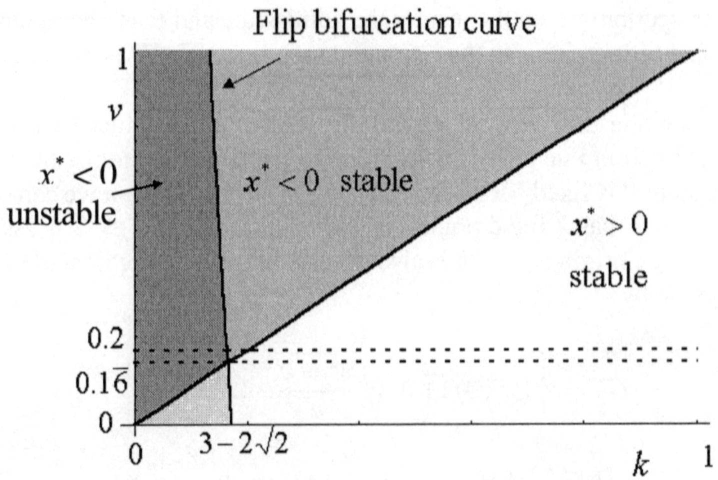

Figure 3: *Stability and positivity regions for the fixed point* x^*.

From Proposition 2 we deduce that if $v > 0$ the map T in (7) has a unique fixed point, given by

$$E^* = \left(x^*, \sqrt{\frac{x^* + v}{k}} - x^* - v\right)$$

which is stable when x^* is stable for the map ϕ. E^* is the Cournot equilibrium point of the game, and we are interested in the region in which it is positive.

When the parameter vector (k, v) crosses the bifurcation curve (which is also shown in Fig.3), and a supercritical flip bifurcation occurs to ϕ, then T undergoes a degenerate bifurcation, as the eigenvalues of the Jacobian matrix of T, evaluated at the fixed point, are 1 and -1. After the bifurcation the map T exhibits an attracting 4-cycle (generated by the attracting 2-cycle of ϕ), a saddle 4-cycle, of mixed type, and the repelling fixed point E^* (see Chapter 3, Sec.5, for major details). Following the cascade of flip bifurcation of ϕ, in Sec.5 we shall study the asymptotic behaviour of T, devoting particular attention to the cycles which belong to the feasible region. Before that, in the following section we shall consider the admissible and feasible regions, and their global bifurcations, in the important case in which E^* is asymptotically stable.

To conclude this section, we briefly recall the particular case $v = 0$, already studied in Puu (1991) with different mathematical tools. In this case also the point 0 is fixed for ϕ, and it is always unstable. Then we can deduce that the map T has 2 fixed points, E^* which is stable if $k > 3 - 2\sqrt{2}$ (see Fig.3), and the origin O which is always unstable. Moreover a saddle 2-cycle exists, given by

$$C_2^* = \left\{ (x^*, 0), \left(0, \sqrt{\frac{x^* + v}{k}} - x^* - v \right) \right\},$$

and the coordinate axes belong to its stable set. When x^* bifurcates for ϕ, the coordinates of the new cycle born via flip bifurcation also interact with C_2^*, generating cycles of mixed type, many of which on the coordinate axes.

4 Admissible and feasible regions

The aim of this section is to study the structure and the topological properties of the admissible and feasible regions defined in Sec.3. In particular we are interested in the feasible region F, given the economic context. The proof of the existence of such a region may be not easy, in general, however we are sure that F is not empty when the Cournot equilibrium point has positive coordinates and is stable. Being interested in the properties of the set F, in the examples of this section we keep a constant value for v, $v = 0.2$, so that the Cournot equilibrium point E^*, when positive, is always stable (see Fig.3).

In the simplest case the feasible set F (and the admissible set S) is a connected region of the plane, as expected. In fact, we can state the following proposition.

Proposition 3 *In the parameter set* Ω,
if $k \geq (\sqrt{1+4v} - 1)/(8v)$ *(or* $v \geq 1/(4k)\,(1/(4k) - 1)$*) then the admissible region* S *is the connected set*

$$S = \left[-v, \frac{1 + \sqrt{1 + 4vk}}{2k} \right] \times \left[-v, \frac{1 + \sqrt{1 + 4v}}{2} \right];$$

if $k \geq 1/(4-4v)$ *(or* $v \leq 1 - 1/(4k)$*) then the feasible region* F *is the connected set*

$$F = \left[0, \frac{1}{k} - v\right] \times [0, 1-v] .$$

In order to prove Proposition 3, we could proceed in a rigorous way, based on the analytical study of the graph of the reaction functions. However we prefer here to follow a more intuitive geometrical approach. An example in which the conditions given in Proposition 3 are fulfilled is shown in Fig.4, where the white points represent the feasible set F while the admissible set S (which clearly includes F) is the union of the light grey and white points, the light grey points being (mathematically) admissible but (economically) unfeasible. Both regions are simply connected, and the set S is the basin of attraction of the Cournot equilibrium point E^*.

Taking into account the Riemann foliation of the map (Fig.2), we observe that a portion of the frontier of the definition set $\mathbb{D}$ belongs to the zones Z_k (k=1,2), that is $\partial\mathbb{D}\cap Z_1$ is the left corner made up of the segments VC and VD shown in Fig.4, while $\partial\mathbb{D}\cap Z_2$ consists of two disjoint segments, CA and DB shown in the same figure. The preimages of these segments give the boundary of the admissible region S. In fact, the segments VC and VD have both one preimage of rank-1, located in the upper right corner of S (the segments $C_{-1,1}V_{-1}$ and $D_{-1,1}V_{-1}$ in Fig.5). The segments CA and DB have both two distinct rank-1 preimages, which merge on $LC_{-1}^{(a)}$ and $LC_{-1}^{(b)}$ respectively, given by the segments $C_{-1,1}A_{-1}$ and $C_{-1,2}A_{-1}$ for CA and the segments $D_{-1,1}B_{-1}$ and $D_{-1,2}B_{-1}$ for DB in Fig.5. All these preimages belong to Z_0, i.e. no further preimages of higher rank exist, and together with the frontier of $\mathbb{D}$, they constitute the boundary of the admissible set S.

Similarly we can obtain the boundary of the feasible set F, by considering the intersections of the region Z_4 with the coordinate axes, which are the segments ξ and ζ shown in Fig.4. In fact these segments ξ and ζ both have four preimages. The rank-1 preimages of ξ are $\xi_{-1,1}$ and $\xi_{-1,2}$, located at opposite side with respect to $LC_{-1}^{(a)}$ and merging on it, as well as $\xi_{-1,3}$ and $\xi_{-1,4}$ (see Fig.5). The rank-1 preimages of ζ are $\zeta_{-1,1}$ and $\zeta_{-1,2}$, located at opposite side of $LC_{-1}^{(b)}$ and merging on it, as well as $\zeta_{-1,3}$ and $\zeta_{-1,4}$. The union of the segments $\xi_{-1,3}$, $\xi_{-1,4}$, $\zeta_{-1,2}$ and $\zeta_{-1,4}$ give a portion of the boundary of the set $\mathbb{D}$ (and their preimages are described above), while the segments $\xi_{-1,1}$, $\xi_{-1,2}$, $\zeta_{-1,1}$ and $\zeta_{-1,3}$ constitute the upper and

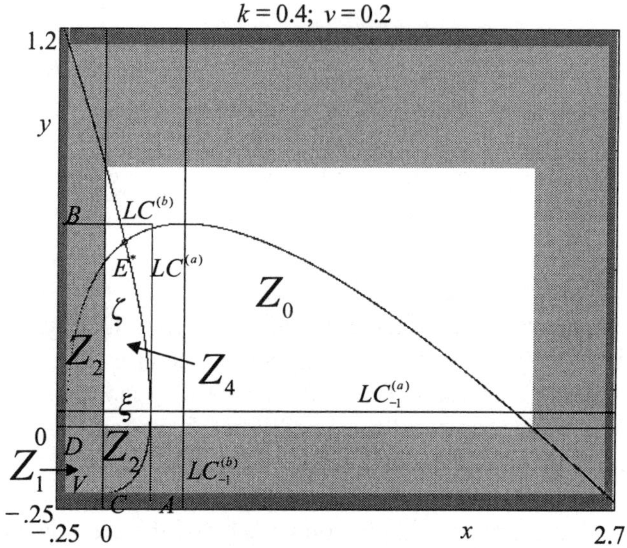

Figure 4: *Connected feasible(white points) and admissible (white and light grey points) regions.*

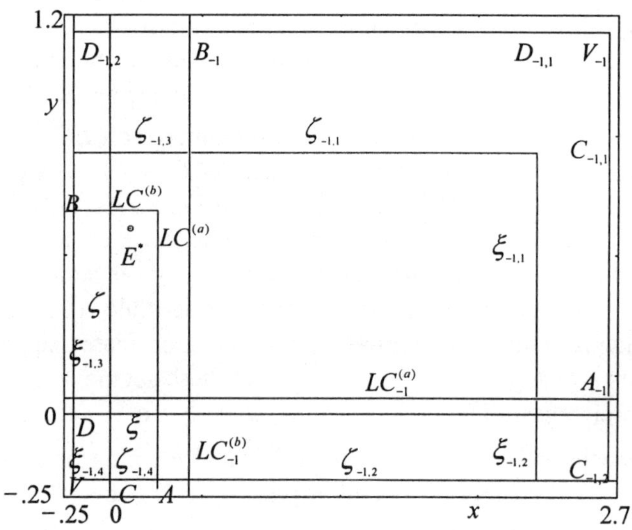

Figure 5: *Admissible region and feasible region obtained by the preimages.*

right portions of the boundary of F (the lower and left boundaries being the coordinate axes). As all the segments of ∂F (except for ξ and ζ) are in Z_0 no further preimages exist.

Thus when the parameters satisfy the conditions given in Proposition 3 the regions F and S are simply connected rectangles, and the geometric construction described above explains the formulas for F and S given in the proposition. The proof of Proposition 3 is complete once we have shown that the limit values given there are bifurcation values which destroy the construction described above. That is, let us consider first the region F. At $k = 1/(4(1-v))$ a global bifurcation occurs after which the region is no longer connected. As we may expect such a global bifurcation is due to a contact between the boundary of the region, ∂F, and the critical curves of the map (i.e. a contact bifurcation, as described in Mira *et al.* (1994), see also Mira *et al.* (1996)). In fact, the portion of $LC^{(b)}$ inside the region F approaches the upper boundary as k is decreased, and at the bifurcation value $k = 1/(4(1-v))$ a portion of $LC^{(b)}$ belongs to ∂F (it merges with the segment $\zeta_{-1,3}$) and for $k < 1/(4(1-v))$ the portion of $LC^{(b)}$ enters the grey region S (an example is shown in Fig.6a). Stated in other words, the upper boundary of F, completely contained in Z_0 for $k > 1/(4(1-v))$, falls on the boundary of the region Z_4 for $k = 1/(4(1-v))$ and for $k < 1/(4(1-v))$ belongs to the region Z_4, and thus has further preimages. That is, the portion of grey points (admissible but unfeasible) now belonging to Z_4, has four distinct rank-1 preimages which, all together, give the single strip of grey points between the two white rectangles shown in Fig.6a, bounded by two vertical lines at $x = x_1$ and $x = x_2$, which belong to Z_0 (so that the boundary is complete) Thus after the bifurcation the region F is no longer a connected set, while it is given by the union of two rectangles $[0, x_1] \times [0, 1-v] \cup [x_2, 1/k - v] \times [0, 1-v]$. This occurs at least as long as the parameter k satisfies

$$\frac{4}{25}(4v+1) \leq k < \frac{1}{4(1-v)}, \tag{10}$$

In such a situation we can deduce that, with fixed maximum price, the strategical set of the player I becomes smaller as the marginal cost decreases. In other words, if player II, cost leader in the market, improves its technology, further reducing the production cost, then the competitor can not produce a value between x_1 and x_2, because otherwise his second move will be out of the market. Obviously, also a larger maximum price, maintaining constant the cost structure, can cause a reduction of the strategical set of the producer with smallest cost. When the value of k decreases further, approaching the other extremum value given in (10), the feasible set F can assume more complex structures (as another global bifurcation occurs). In fact, we can see from Fig.6a that as k decreases the critical curve $LC^{(a)}$ also approaches the boundary of F (as it approaches the line $x = x_1$) and when $k = 16v/25 + 4/25$ another contact bifurcation occurs when $LC^{(a)}$ merges with ∂F and crosses it, because after the bifurcation the line $x = x_1$ belongs to the region Z_4, and thus has further preimages. That is, after this second contact bifurcation further preimages of the grey central strips are created inside the old region F, which thus is further reduced.An example is shown in Fig.6b, where F is made up of six disjoint rectangles (only three are clearly visible in that figure, the other three being very narrow strips).

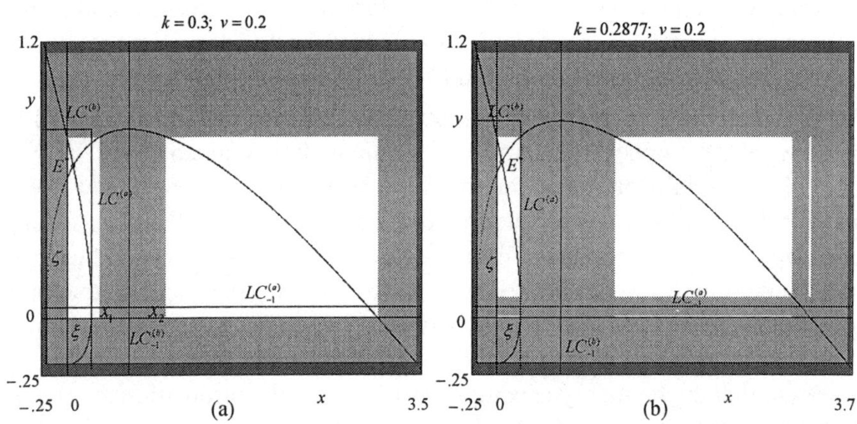

Figure 6: (a) *Disconnected feasible region.* (b) *A more complex structure of the feasible region.*

The contact bifurcations occurring to the admissible region S are clearly of the same kind as those described for F. In Fig.6 we can see that as k is decreased the critical curve $LC^{(b)}$ approaches the upper boundary of S. For $k > \left(\sqrt{1+4v} - 1\right)/(8v)$, S is a simply connected set (a rectangle), but at $k = \left(\sqrt{1+4v} - 1\right)/(8v)$ a contact bifurcation occurs, as $LC^{(b)}$ falls on ∂S (the rank-1 preimages of the segment VD) and then crosses it. Thus for $k < \left(\sqrt{1+4v} - 1\right)/(8v)$ a portion of non admissible region (the dark grey points in Fig.7a) enters the region Z_4, and thus has further preimages, which all together constitute a strip of dark grey points on both sides of $LC_{-1}^{(b)}$, as shown in Fig.7a.

Clearly further contact bifurcations (between LC and ∂S) may also occur, giving rise to more complex structures also for S.

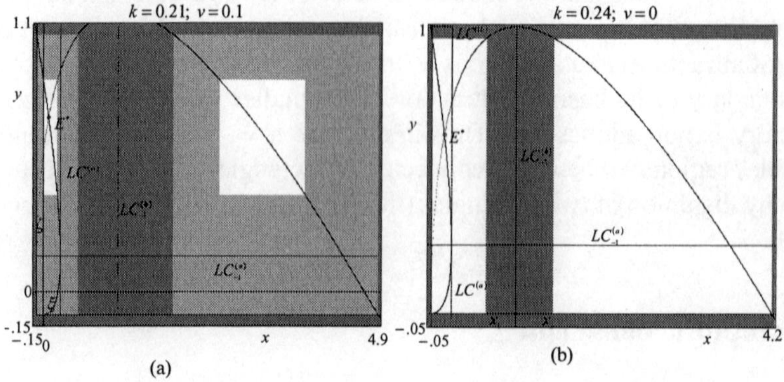

Figure 7: (a) *Disconnected admissible region.* (b) *Disconnected feasible region in the case of* $v = 0$.

A sequence of global bifurcations similar to those described above appears for F and S in correspondence to any value $v > 0$. We recall that in the case $v = 0$ we have $F = S$ so that we don't have to distinguish between them. In this particular case ($v = 0$), when Proposition 3 holds then the feasible set F is a rectangle, whose boundary is the stable manifold of the saddle 2- cycle belonging to the coordinate axes. The vertex of such a rectangle are the four rank-1 preimages of the origin, which is a repelling fixed point. As the value of k decreases below 0.25 (which is the bifurcation value), the feasible set becomes a disconnected set, as in Fig.7b, similar to the previous one (in Fig.7a) but without non admissible points (being the white points all feasible). Moreover, such a structure also persists when the fixed point becomes unstable, for $k < 3 - 2\sqrt{2}$. The cycles and the chaotic attracting sets generated by the usual cascade of flip bifurcations, as we shall see in Sec.5, have a disconnected basin of attraction, given by the union of two rectangles, whose vertex are the preimages of rank-n ($n = 1, 2$), of the repelling fixed point O. At the second bifurcation value for the feasible region, the left extremum of the interval in (10), i.e. $k = 4/25$, we know that a portion of the boundary of the set F coincides with the critical line $LC^{(a)}$. But in this case

($v = 0$) at this bifurcation value for k another important bifurcation occurs. In fact, if we consider the one-dimensional map ϕ in (8), we can observe that the first iterate of its critical value $c^M = 1/4$ is the repelling fixed point 0. This is the homoclinic bifurcation of 0 (as it occurs in the logistic map $f(x) = \mu x(1-x)$ at $\mu = 4$), and the map is chaotic in the whole interval $[0, c^M]$. This means that the chaotic attractor of the two-dimensional map T entirely fills up its immediate basin (the largest connected component of its basin of attraction) and undergoes a *final bifurcation*, having a contact with the boundary of its basin of attraction. For smaller value of k, the generic trajectory is non admissible. Thus in the case $v = 0$ the feasible (and admissible) region can be either connected (a rectangle) or a disconnected set given by the union of two rectangles (more complex structure for F does not exist).

5 Asymptotic behaviour

As we have seen in the previous section, concerning the feasible region F and its global bifurcations, the case $v = 0$ is simpler than generic case, and we shall see in this section that this is no longer true if we look at the asymptotic behaviour. In fact, in the case $v = 0$ the bifurcation diagram of the one dimensional map ϕ in (8), from which we can deduce the asymptotic behaviour of the map T in (7), is "complete". That is, the first flip bifurcation (occurring at $k = 3 - 2\sqrt{2}$) is followed by the usual cascade of period doubling bifurcations as the value of k is decreased. Similarly to what occurs in the well known logistic map, feasible attracting cycles of period 2^k, for any $k > 0$, are detected for ϕ. Each of these cycles corresponds to feasible attracting cycles of the two-dimensional map T. Moreover, we recall that whenever ϕ has some stable cycle of period greater than 2, then T is characterized by multistability. In Fig.8a, obtained for $k = 0.162$, we give an example of such a situation: in that figure we show an enlargement of the feasible region in which two coexisting stable cycles of period eight are present, corresponding to the stable 4-cycle $\{b_1^*, b_2^*, b_3^*, b_4^*\}$ of ϕ. One of these cycles belongs to the reaction curves (the homogeneous one), the other one being a mixed cycle. Also the basins of attraction of the two 8-cycles are represented in Fig.8a by the rectangles of different grey tonalities The structure of the basins is typical of "square separate" maps: each basin is given by the union of disconnected rectangles (and it has a connected closure) whose boundaries are the stable sets of repelling cycles (in our example the unstable fixed point and the unstable 8-cycles). It can be observed, in the figure, that the rectangles accumulate on vertical and horizontal lines through the fixed point E^*, because that lines includes many repelling cycles and theirs stable sets. As k is further decreased, eight pieces and four pieces chaotic attractors are observable, joining in a unique set, corresponding to the different bifurcations of the critical value c^M of ϕ. At $k = 4/25$, the chaotic area fills up its immediate basin (see Fig.8b), whose boundary coincides with critical lines:

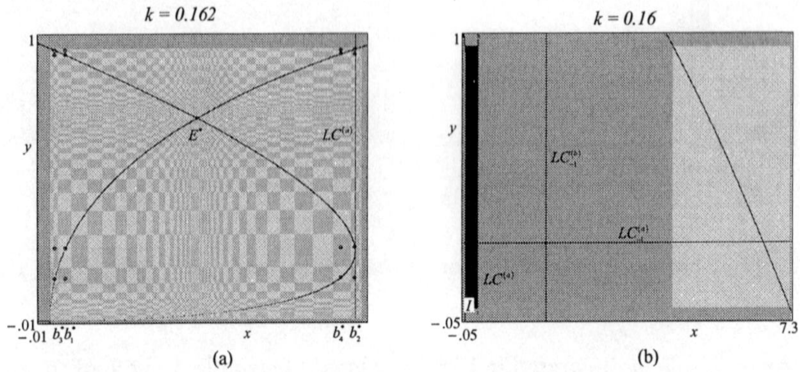

Figure 8: *Case* $v = 0$*:* (a) *Enlargement of the basins of attraction of two coexistent stable 8-cycles.* (b) *Chaotic attracting set at the final bifurcation value.*

as described at the end of the previous section, this is a final bifurcation, after which the generic trajectory is non admissible.

When we consider a positive value of v, we are faced with the problem of admissible but unfeasible trajectories. This in fact reduces the portion of bifurcation diagram we can consider meaningful in the applications, because same chaotic set or same stable cycle may have unfeasible points. In particular, when v is very small, complex dynamics and multistability due to cycles of period $k > 2$ of ϕ can be found in the feasible set, but for $v \geq 0.01$ only cycles of period four and fixed point are feasible and stable. In Fig.9a an example of bifurcation diagram, obtained for $v = 0.003$, is shown. In the enlargement of the lower branch, it is possible to observe that only stable cycles of period eight of ϕ have all the periodic points on the positive half-line, as the other portion of the diagram is below the line $x = 0$ and thus the attracting sets involve negative points. This means that for the Cournot duopoly model we consider, no feasible chaotic trajectory can be detected, then the dynamic behaviour of the map T (when feasible) results in stable cycles of period 16 at most (and clearly we have multistability).

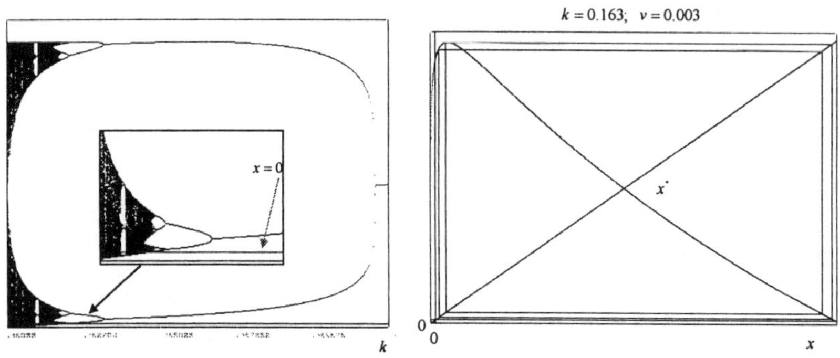

Figure 9: (a) *Bifurcation diagram of the one-dimensional map* ϕ *for* $v = 0.003$. (b) *A stable 4-cycle of* ϕ.

An example is illustrated in Fig.9b, related to a stable 4-cycle of the one-dimensional map ϕ, and the corresponding situation for the two-dimensional map, i.e. the two coexisting 8-cycles, and their basins of attraction, are shown in Fig.10. In particular, in Fig.10a we observe the whole feasible

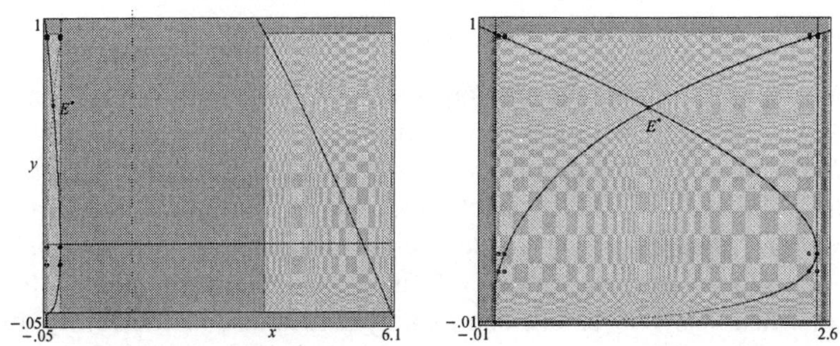

Figure 10: *Case* $v = 0.003$. (a) *The whole admissible set gives the basins of attraction of the two coexisting 8-cycles.* (b) *Enlargement.*

region F, the disconnected set in light grey tonality, due to a sequence of bifurcations as described in Sec.4. The strip in dark grey tonality around F denotes the admissible but unfeasible trajectories, converging to the two cycles. The basins of attraction of the two cycles fill up the whole admissible region. Each cycles has a basin of attraction similar to that described in the case $v = 0$ (compare Fig.10b with Fig.8a). Moreover, looking at the whole basin we may observe that the structure in disconnected rectangles persists in the whole admissible set and that such rectangles accumulate along vertical and horizontal lines through the Cournot equilibrium point and its preimages of increasing rank.

The structure of the basins does not depend on the attracting set, and this can be observed in Fig.11a, in which at a lower value of v, two 8-pieces (feasible) chaotic attractors coexist in the admissible region, and have basins of attraction very similar to the previous ones.

In Fig.11a, the corresponding 4-pieces chaotic attractor of the one- dimensional map ϕ are draft: They are separated in two couples and the two sets in each couple are very near.

The value of k in Fig.11a is close to a bifurcation value and, if we slightly modify this value, the chaotic intervals merge by pairs and ϕ exhibits a 2-pieces chaotic attractor (Fig.11b). This bifurcation has a very important effect on the dynamical behaviour of T, because the multistability disappears, the attracting set is given by a unique 4-pieces chaotic attractor and the "rectangular structure" of the basin is lost, as all the admissible points belong to the (connected) basin of attraction of the chaotic set.

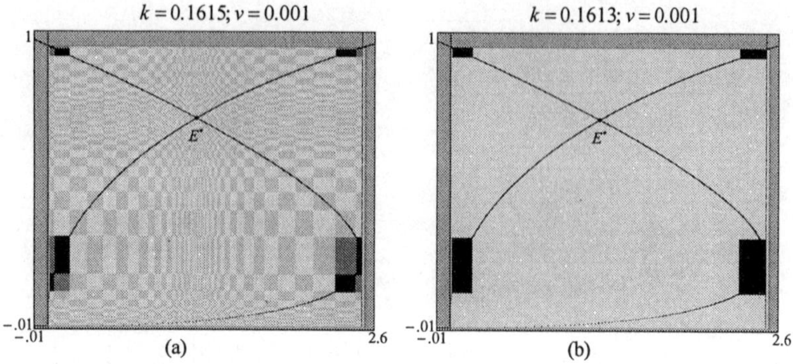

Figure 11: (a) *Two eight-pieces chaotic attractors and their basins of attraction.* (b) *The four-pieces chaotic attractor obtained at a slightly lower value of k.*

As observed above, the introduction of a bound for the price in the isoelastic demand function decreases the complexity of the meaningful dynamics of the adjustment process, at least when the maximum price is a high value. This can be explained observing that when $v > 0$ the admissible (S) and the feasible (F) sets are different, and increasing v, i.e. decreasing the maximum price, some attracting sets have unfeasible points. For example, when $v = 0.1$ also the attracting 2-cycle existing after the flip bifurcation (at $k = k_f$) of the fixed point x^* of ϕ is unfeasible. It is interesting to point out that the flip bifurcation is now of subcritical type, as it is shown in the bifurcation diagram of ϕ in Fig.12. We can see that as k decreases from the regime in which the fixed point is stable, at $k = k_F$ a fold-bifurcation occurs giving rise to the appearance of two cycles of period 2, one stable and one unstable, represented in Fig.12 by a continuous line and a dotted one respectively.

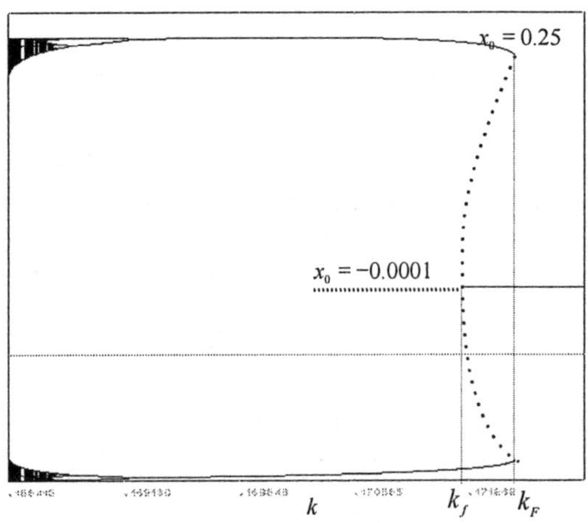

Figure 12: *Bifurcation diagram for the parameter* k, *with* $v = 0.1$.

Thus for lower value of k, and exactly in the interval $k_f < k < k_F$, ϕ has two coexisting attractors, the fixed point x^* and the stable 2-cycle, while the map T (for which at $k = k_F$ a fold-bifurcation occurs) has three coexisting attractors: two stable 4-cycles and the Cournot equilibrium E^*. Their basins of attaction are shown in Fig.13 and the structure is quite clear, recalling the properties of "square separate" maps. In Fig.13a, the disconnected admissible set S is represented in different grey tonalities, the darkest grey points denoting non admissible trajectories. The basin of attraction of the Cournot equilibrium point, $\mathcal{B}(E^*)$, is given by the lighest grey and white points.The other two grey tonalities denote the basins of the two stable 4-cycles. Once more such basins are given by the union of disconnected rectangles. In particular $\mathcal{B}(E^*)$ has six rectangular components, only two of them, the largest ones, contain feasible points (in white). The vertices of the immediate basin of E^*, see Fig.13b, are the points of the unstable 4-cycle born via fold bifurcation together with the stable one. The preimages of increasing rank of these points are the vertices of the other five components of $B(E^*)$. The horizontal and vertical lines connecting these points separate the three basins and are the stable manifolds of the three saddle 4-cycles of mixed type generated by the unstable 2-cycle of ϕ with the stable fixed point x^* and with the stable 2-cycle of ϕ.

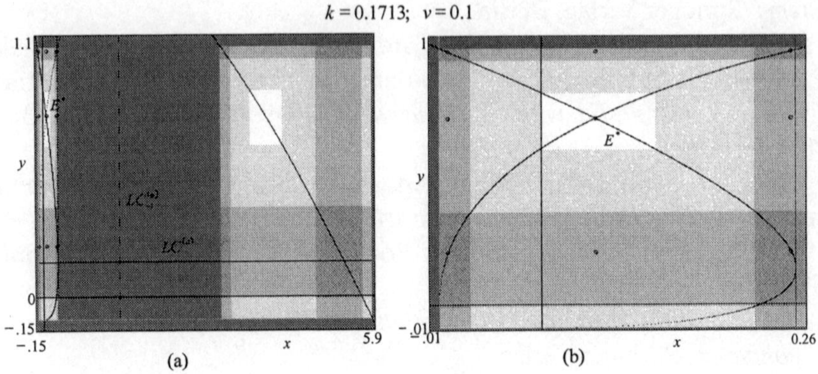

Figure 13: (a) *The admissible region* S *at a value of* k *between* k_f *and* k_F. (b) *Enlargement.*

At $k = k_F$ the flip bifurcation of x^* is subcritical: the fixed point of ϕ merges with the unstable 2-cycle, which disappears for $k < k_f$, leaving an unstable fixed point. Similarly for T: the unstable 4-cycle shrinks around the stable Cournot equilibrium E^*, and at $k = k_F$ merges with it, leaving an unstable equilibrium, a unique attractor (the stable 4-cycle) and a saddle 4-cycle of mixed type.

Acknowledgments. This work has been performed under the auspices of CNR, Italy, and under the activity of the national research project "Dynamic Models in Economics and Finance: Evolution, Uncertainty and Forecasting", MURST, Italy.

References

Abraham, R., Gardini, L. and Mira, C., 1987, *Chaos in discrete dynamical systems (a visual introduction in two dimension)*, Springer-Verlag.

Agiza, H.N., 1998, "Explicit Stability Zones for Cournot Game with 3 and 4 Competitors", *Chaos, Solitons & Fractals*, 9:1955-1966.

Agliari, A., Gardini, L., and Puu, T., 2000, "The Dynamic of a Triopoly Cournot Game", *Chaos, Solitons & Fractals*, 11:2531-2560.

Bischi, G.I., Mammana, C., and Gardini, L., 2000, "Multistability and cyclic attractors in duopoly games", *Chaos, Solitons & Fractals*, 11:543-564.

Gumowski, I. and Mira, C., 1980, *Recurrences and Discrete Dynamical Systems*, Springer Verlag, Berlin.

Mira, C., Fournier-Prunaret, D., Gardini, L., Kawakami, H. and Cathala, J.C., 1994, "Basin bifurcations of two-dimensional noninvertible maps: fractalization of basins", *International Journal of Bifurcation and Chaos,* 4, 343-381

Mira, C., Gardini, L., Barugola, A., and Cathala, J.C., 1996, *Chaotic Dynamics in Two-Dimensional Noninvertible Maps*, World Scientific.

Puu, T., 1991, "Chaos in duopoly Pricing", *Chaos, Solitons & Fractals*, 1:573-581.

Puu, T., 2000, *Attractors, Bifurcations, & Chaos-Nonlinear Phenomena in Economics*, Springer-Verlag.

8 Chaos Synchronization and Intermittency in a Duopoly Game with Spillover Effects

Gian-Italo Bischi[1] *and Fabio Lamantia*[2]

[1]Istituto di Scienze Economiche, Università di Urbino, Italy.
[2]Facoltà di Economia, Università della Calabria, Italy.

1 Introduction

Discrete-time dynamic oligopoly games which exhibit chaotic time patters of the competitors' strategic choices are at the center of a flourishing literature, including the seminal paper by Rand (1978) and several papers after it (see e.g. Dana and Montrucchio, 1986, Puu, 1991, Kopel, 1996), where simple microeconomic situations have been proposed which lead to duopoly games with chaotic dynamics. The main concern, in this stream of literature, is to emphasize the features of the duopoly games which are responsible for the transition from regular to chaotic dynamics and to analyze the bifurcations which cause the loss of stability of the equilibria and the appearance of more complex attractors (see e.g. Bischi et al., 2000).

In this chapter, we stress some peculiar features which characterize different kinds of chaotic behaviors. In particular, given a duopoly game characterized by a chaotic behavior of both the competitors, we try to distinguish different degrees of correlation between the choices of the two players. With the term "correlated chaos" we mean that even if both the players behave chaotically, at each time period their choices, say x_t and y_t, are approximately related by a function $y_t = f(x_t)$ or $x_t = f(y_t)$. This means that the chaotic attractor, in the two-dimensional state space (x, y), is approximately located around a portion of a one-dimensional curve. Instead, "uncorrelated chaos" means that a generic chaotic trajectory fills up a two-dimensional region, so that no relations can be evidenced, at a given time, between the decisions of the two players. With other words, provided that in both cases each player behaves chaotically (so that it is impossible to forecast her/his next period decision) in the presence of *correlated chaos* if one observes the choice of a player then the behavior of the other player at the same time period can be approximately deduced, whereas in the case of *uncorrelated chaos*, even if one observes what a player does at a given time period, nothing can be inferred about the simultaneous choice of the competitor.

An extreme form of correlation is represented by "chaos synchronization", which means that the chaotic trajectories are embedded into a one-dimensional invariant submanifold, say M, and are governed by the one-dimensional dynamical system given by the restriction of the two-dimensional one to M. Indeed, dynamical systems with chaotic trajectories embedded into an invariant submanifold of lower dimensionality than the total phase space have raised an increasing interest in the scientific community (see e.g. Fujisaka and Yamada, 1983, Pecora and Carrol, 1990, Ott and Sommerer, 1994, Ashwin et al, 1996, Buescu, 1997, Hasler and Maistrenko, 1997, just to cite a few).

Another kind of chaotic behavior, which can be considered as intermediate between correlated and uncorrelated chaos, is the so-called *on-off intermittency*, characterized by chaotic time patterns which are synchronized (or quasi-synchronized) for several time periods, but sometimes clusters of large asynchronous fluctuations occur, i.e. sudden bursts away from the submanifold where synchronized dynamics take place. The distribution, over time, of such asynchronous bursts is quite random, but something can be said about their maximum amplitude (see Bischi and Gardini, 2000, Kopel et al., 2000).

In the mathematical and physical literature these phenomena have mainly been studied for coupled chaotic oscillators, where a coupling parameter exists which only influences the dynamics in a direction which is orthogonal with respect to the invariant submanifold where synchronized dynamics take place (a so called *normal parameter*). Indeed, due to the sensitive dependence on initial conditions which characterizes chaotic systems, two identical and independent chaotic oscillators cannot be, in general, synchronized, whereas it is possible to synchronize them, in the long run, provided that some coupling (or interaction) is introduced (Fujisaka and Yamada, 1983, Pecora and Carrol, 1990).

Chaos synchronization and intermittency are generally associated with symmetric dynamical systems, and such a situation is commonly met in dynamic games with identical or quasi-identical competitors, as recently stressed in Bischi et al., 1998, Bischi et al., 1999, Kopel et al., 2000, Bischi and Gardini, 2000. In these models a perfect symmetry (i.e. an absolute identity of the parameters which characterize the players' behaviors) is a very demanding condition, and even the presence of a normal parameter is not so common. However, as we shall see through the numerical explorations presented in this chapter, intermittency phenomena can also be observed with heterogeneous interacting players. Indeed, intermittency phenomena may be seen as a prelude of chaos synchronization as the degree of homogeneity between the players is gradually increased. However, this is not a general rule, as we shall see in the following.

In this chapter we try to investigate how, starting from a condition of heterogeneous players, an increase of the degree of symmetry, gradually leading to conditions of quasi-identical or fully identical players, may induce the appearance of more and more correlated chaotic behaviors. In order to investigate this issue, we consider a Cournot duopoly gâme, recently proposed by Bischi and Lamantia (2001), where the interdependence between the quantity-setting firms is not only related to the selling price, determined by the total production through a given demand function, but also on positive cost externalities due to the effects of know-how spillovers, caused by the ability of a firm to take advantage, for free, of the competitors' Research and Development ($R\&D$) results. In this duopoly game, in the case of identical chaotic players, chaos synchronization can occur, due to the presence of a one-dimensional chaotic attractor embedded into the invariant diagonal. The study of this particular Cournot duopoly game allows us to give examples of different kinds of chaotic behaviors according to the degree of chaos correlation, and we observe the gradual transition between these different kinds as the heterogeneity in the spillover parameters, i.e. the asymmetry in the ability to take advantage of the competitor $R\&D$ results, is varied. However, the spillover parameters are not normal ones, because they influence both the dynamics along the diagonal and the dynamics transverse to it. This implies that it is not possible to use many of the results given in the literature on chaos synchronization. However, following Bischi and Gardini (1998), a global characterization of the phenomena is still possible by the method of critical curves, which allow us to obtain the delimitation of an absorbing area surrounding the one-dimensional attractor on which synchronization occurs (which is often only an attractor in Milnor sense, see Ashwin et al., 1996, Buescu, 1997). In fact, as the time evolution of the duopoly game analyzed in this chapter is represented by the iteration of a noninvertible map, the dynamic phenomena observed, such as chaos synchronization, intermittency and uncorrelated chaos, are confined inside a given absorbing area, whose boundary can be obtained by segments of critical curves, and behaves as a bounded vessel inside which the asymptotic dynamics are trapped (see Mira et al., 1996, Bischi and Gardini, 1998).

The reminder of this chapter is organized as follows: in section 2 we describe the duopoly game, in section 3 we consider the same game in the symmetric case of identical players, in section 4 we describe some numerical explorations through which the concepts outlined above are illustrated.

2 A Cournot duopoly game with spillovers

To illustrate the concepts outlined in the introduction, we consider a Cournot duopoly game, proposed in Bischi and Lamantia (2001), where the interdependence between the quantity-setting firms is not only related to the selling price, determined by the total production through a given demand function, but also on cost-reduction effects related to the presence of the competitor. Such cost reductions are introduced to model the effects of technological and intellectual spillovers among companies, caused by the ability of a firm to take advantage, for free, of the competitors' $R\&D$ results, due to the difficulties to protect know-how or to avoid the movements of skilled workers among competing firms, see e.g. Audretsch and Feldman (1996) Aitken et al. (1997). The results of $R\&D$ are generally assumed to lead to costs reductions (see e.g. D'Aspremont and Jaquemin, 1988). So, spillover effects can be seen as a positive cost externality, which we model by assuming the following cost function of firm i:

$$C_i(q_1,q_2) = \frac{c(q_i) + R\&D_i}{1 + \gamma_{ii} g_i\left(R\&D_i\right) + \gamma_{ij} h_j\left(R\&D_j\right)}; \; i,j = 1,2 \; ; \quad j \neq i \tag{1}$$

where $c\left(q_i\right)$ represents the production costs of firm i, an increasing function of its own production q_i, $R\&D_i$ represents the $R\&D$ expenses of firm i, g_i and h_j are increasing functions and the positive parameters γ_{ii} and γ_{ij} give a measure of the cost reduction related to its own and competitor's $R\&D$ respectively. A particular choice is proposed in Bischi and Lamantia (2001), where a very simple cost function is obtained by assuming linear production costs, $c_i(q_i) = k_i q_i$, $R\&D$ expenses proportional to the production, $R\&D_i = s_i q_i$, h_i linear and $g_i \equiv 0$. The last assumption captures the fact that only external spillovers are considered cost-reducing, because the benefits from its own $R\&D$ are assumed to be balanced by induced costs, such as higher salaries required by more skilled workers or expenses of a firm to avoid spillovers. However, we assume that $R\&D$ are necessary in a high-tech market, where without $R\&D$ the produced goods become obsolete. With these assumptions, and a linear demand function

$D(p) = a - b(q_1 + q_2)$, the profit of firm i becomes

$$\pi_i(q_1, q_2) = q_i [a - b(q_1 + q_2)] - \frac{c_i q_i}{1 + \gamma_{ij} q_j} \; ; \; i, j = 1, 2, i \neq j \qquad (2)$$

In a fully rational Cournot duopoly game, each player decides its own production in order to maximize the expected profit, on the basis of the following two assumptions:

(i) each firm knows beforehand its rival's production decision;

(ii) each firm has a complete knowledge of the profit function.

From the first order conditions $\partial \pi_i / \partial q_i = 0$, we get

$$q_i = r_i(q_j) = \frac{1}{2b} \left(a - bq_j - \frac{c_i}{1 + \gamma_{ij} q_j} \right) \; i, j = 1, 2 \; ; \; \; j \neq i. \qquad (3)$$

A simple check of the second derivatives testifies that these solutions represent profit maxima, provided that the quantities are non negative. Hence the portions inside the positive orthant of the functions $q_1 = r_1(q_2)$ and $q_2 = r_2(q_1)$, are the two reaction curves. Every intersection between the two reaction curves, being an optimal choice for both firms, is a Cournot-Nash Equilibrium, characterized by the fact that no firm has an incentive to unilaterally deviate from its chosen strategy given the choice of its rival. It is immediate to realize that the introduction of spillover effects in the cost functions has the effect of changing the reaction curves from lines to strictly concave curves, which are unimodal for sufficiently high values of spillover parameters. In Bischi and Lamantia (2001) it is proved that at most one Nash Equilibrium exists, and in order to investigate its stability the Nash Equilibrium is considered as the outcome of a dynamic adjustment process occurring when less that fully rational players play the game repeatedly (see e.g. Fudenberg and Levine, 1998, or Binmore, 1992, ch.9 for such evolutionary interpretation of the stability of a Nash equilibrium). This means that the players generally do not reach a Nash equilibrium immediately, but play the game repeatedly in order to approach it. Several kinds of boundedly rational adjustment processes may be considered, all sharing the same Nash equilibrium but with different methods to update productions when the system is out of it. One kind of dynamic adjustment, proposed by Cournot (1838), is based on the assumption that the two firms have a global knowledge of the profit function, so that they are able to compute their best reply to the expected production choice of the competitor, but the two firms are not so rational to be able to know in advance the competitor's choices, and at each time step they adopt a very simple (or *naive*) expectation, by guessing that the production of the other firm will remain the same as in the previous period. So the repeated game is defined by the recurrence

$$q_i(t+1) = r_i\left(q_j(t)\right) \;\; i,j = 1,2 \;\; i \neq j \quad . \tag{4}$$

As proved in Bischi and Lamantia (2001) this adjustment mechanism, with reaction functions (3), leads to global stability of the Nash equilibrium, i.e. an asymptotic convergence to it for any initial condition in the strategy space. A second kind of dynamic adjustment is proposed in Bischi and Lamantia (2001) where the firms are assumed to be even less rational, in the sense that they don't have a complete knowledge of the profit function, and consequently they use a simpler (and less expensive) "rule of thumb" (see e.g. Baumol and Quandt, 1964) in their decision-making processes, known in the literature as gradient dynamics (or myopic adjustment, see e.g. Sacco, 1991, Varian, 1992, flam, 1993). This gives rise, for certain sets of parameters, to periodic and chaotic dynamics around the Nash equilibrium, and this will be the object of our studies.

According to this kind of dynamic adjustment, the two players are assumed to update their production strategies at discrete time periods on the basis of a local estimate of the marginal profit $\partial\pi_i/\partial q_i$: At each time period t a firm decides to increase (decrease) its production for period $t+1$ if it perceives positive (negative) marginal profit on the basis of information held at time t, according to the following dynamic adjustment mechanism (see e.g. Bischi and Naimzada, 1999)

$$q_i(t+1) = q_i(t) + \alpha_i(q_i(t))\frac{\partial \pi_i}{\partial q_i}\left(q_1(t), q_2(t)\right) \quad ; \qquad i = 1,2 \tag{5}$$

where $\alpha_i(q_i)$ is a positive function which gives the extent of production variation of ith firm following a given profit signal. Notice that the two producers are not requested to have a complete knowledge of the demand and cost functions, since they only need to infer how the market will respond to small production changes by an estimate of the marginal profit, which may be obtained by brief experiments of small (or local) production variations performed at the beginning of period t (Varian, 1992). Of course, this local estimate of expected marginal profits is much easier to be obtained than a global knowledge of the demand function (involving values of q_i that may be very different from the current ones). In the following we assume linear functions $\alpha_i\left(q_i\right) = v_i q_i$, $i = 1,2$, since this assumption captures the fact that *relative* production variations are proportional to marginal profits, i.e.

$$\frac{q_i(t+1) - q_i(t)}{q_i(t)} = v_i \left(\frac{\partial \pi_i}{\partial q_i} \right).$$

With these assumptions, and the profit functions given in (2), we obtain a discrete dynamical system of the form $(q_1(t+1), q_2(t+1)) = T(q_1(t), q_2(t))$, where the map $T : \mathbb{R}^2 \to \mathbb{R}^2$ is given by

$$T : \begin{cases} q_1' = q_1 + v_1 q_1 \left[a - 2bq_1 - bq_2 - \frac{c_1}{1+\gamma_{12} q_2} \right] \\ q_2' = q_2 + v_2 q_2 \left[a - 2bq_2 - bq_1 - \frac{c_2}{1+\gamma_{21} q_1} \right] \end{cases} \tag{6}$$

3 Gradient dynamics with identical players

We now consider the symmetric case of identical players, i.e. firms which have the same speeds of reaction, the same production costs and the same ability to take advantage from $R\&D$ spillover:

$$v_1 = v_2 = v; \;\; c_1 = c_2 = c; \;\; \gamma_{12} = \gamma_{21} = \gamma \tag{7}$$

Under this assumption, the map (6) remains the same if the variables q_1 and q_2 are interchanged, i.e. after a reflection through the diagonal (line of identical productions)

$$\Delta = \left\{ (q_1, q_2) \in \mathbb{R}^2 | q_1 = q_2 \right\} \; . \tag{8}$$

This symmetry property implies that the diagonal is mapped into itself, i.e., $T(\Delta) \subseteq \Delta$, which corresponds with the obvious statement that, in a deterministic framework, identical competitors, starting from identical initial conditions, behave identically for each time, and the trajectories embedded into Δ, characterized by $q_1(t) = q_2(t)$ for every t, are governed by the one-dimensional map (restriction of T to the invariant submanifold Δ) $f = T|_\Delta : \Delta \to \Delta$, given by:

$$q' = f(q) = q \left(1 + av - 3vbq - \frac{vc}{1 + \gamma q} \right) \tag{9}$$

In Bischi et al. (1999) this one-dimensional model has been considered as the model of a *representative agent* whose dynamics summarize the common behavior of the two synchronized competitors. The map (9) is a unimodal map whose iteration generates chaotic trajectories for sufficiently high values of the common speed of adjustment v. So, synchronized chaos occurs. Moreover, a trajectory starting out of Δ, i.e. with $q_1(0) \neq q_2(0)$, is said to synchronize if $|q_1(t) - q_2(t)| \to 0$ as $t \to +\infty$.

A question which naturally arises is whether trajectories starting from different initial conditions will synchronize in the long run, so that the asymptotic behavior is governed by the simpler one-dimensional model. This question can be reformulated as follows. Let A_s be an attractor of the one-dimensional map: Is it also an attractor for the two-dimensional map? Of course, an attractor A_s of (9) is stable with respect to perturbations along Δ, so an answer to the question raised above can be given through a study of the stability of A_s with respect to perturbations transverse to S (*transverse stability*).

A second question concerns the behavior of the dynamical system when quasi-identical players are considered, i.e. with small differences among the parameters. Of course, in this case the invariance property of the diagonal Δ is lost, and some different dynamic scenarios may replace synchronization. For example, the attractor A_s embedded inside Δ may be replaced by another attractor close to it, where correlated chaos occurs, or a greater attractor may suddenly appear, surrounding the portion of Δ where A_s was located, inside which endless intermittency occurs. To distinguish between these two dynamic scenarios we need at least two steps:

a) a study of the *transverse Lyapunov exponents* for the symmetric system, by which the "average" local behavior of the trajectories in a neighborhood of the invariant set A_s can be understood (see below);

b) a study of the global dynamic behavior of the two-dimensional map, in order to see, in the case A_s is a Milnor (but not asymptotic) attractor if a greater (two-dimensional) Lyapunov attractor exists around A_s, where transient dynamics and intermittency phenomena are trapped (see Bischi et al., 1998, or Bischi and Gardini, 2000, for details).

In order to compute the transverse Lyapunov exponents for the model (6) with identical players we consider the Jacobian matrix along the invariant diagonal

$$DT(q,q) = \begin{bmatrix} 1 + av - 5vbq - \frac{vc}{1+\gamma q} & vq\left(\frac{c\gamma}{(1+\gamma q)^2} - b\right) \\ vq\left(\frac{c\gamma}{(1+\gamma q)^2} - b\right) & 1 + av - 5vbq - \frac{vc}{1+\gamma q} \end{bmatrix}$$

whose eigenvalues are $\lambda_{||} = 1 + av - 6vbq - vc/(1+\gamma q)^2$, with eigenvector along Δ, and $\lambda_{\perp} = 1 + av - 4vbq - vc(1-2\gamma q)/(1+\gamma q)^2$, with eigenvector orthogonal to Δ. So, the transverse Lyapunov exponents, computed along a generic trajectory embedded into Δ, are:

$$\Lambda_{\perp} = \lim_{N\to\infty} \frac{1}{N} \sum_{t=0}^{N} \ln |\lambda_{\perp}(q(t))| . \tag{10}$$

where $\{q(t) = f^t(q(0)), t \geq 0\}$ is a trajectory embedded in A_s. Indeed, as in a chaotic attractor A_s infinitely many periodic points are nested (all unstable along Δ), for a chaotic set $A_s \subset \Delta$, infinitely many transverse Lyapunov exponents can be obtained, because if $q(0)$ is a k−periodic point then $\Lambda_{\perp} = \ln |\lambda_{\perp}^k|$ and the k−cycle is transversely stable if $\Lambda_{\perp} < 0$, whereas if $q(0)$ belongs to a generic aperiodic trajectory embedded inside the chaotic set A_s then $\Lambda_{\perp}$ is the *natural transverse Lyapunov exponent* $\Lambda_{\perp}^{nat}$. Hence a spectrum of transverse Lyapunov exponents can be defined, $\Lambda_{\perp}^{\min} \leq ... \leq \Lambda_{\perp}^{nat} \leq ... \leq \Lambda_{\perp}^{\max}$ (see e.g. Ashwin et al., 1996, Buescu, 1997) where $\Lambda_{\perp}^{nat}$ expresses a sort of "weighted balance" between the transversely repelling and transversely attracting cycles. If $\Lambda_{\perp}^{\max} < 0$, i.e. all the cycles embedded in A_s are transversely stable, then A_s is asymptotically stable, in the usual Lyapunov sense, for the two-dimensional map T. However, it may occur that some cycles embedded in the chaotic set A_s are transversely unstable, i.e. $\Lambda_{\perp}^{\max} > 0$, while $\Lambda_{\perp}^{nat} < 0$. In this case, A_s is no longer Lyapunov stable, but it continues to be a *Milnor attractor* i.e. it attracts a positive (Lebesgue) measure set of points of the two-dimensional phase space. In the latter case intermittency phenomena can occur, according to the global dynamic properties of the map. In fact, the trajectories that are locally repelled along (or near) the local unstable manifolds of the transversely repelling cycles may be reinjected towards A_s by the global (nonlinear) action of the map, so that the dynamics of such trajectories are characterized by some bursts far from Δ before synchronizing on it. This is a rough explanation of the origin of intermittency phenomena (see Ott and Sommerer, 1994, or Bischi and Gardini, 1998, 2000, for more detailed explanations).

The distinction between asymptotic stability and stability in the weaker Milnor sense may also help to understand the effect of a symmetry breaking given by the introduction of small heterogeneities between the two players, i.e. some parameters' mismatch starting from the homogeneity condition (7). Indeed, as conjectured in Bischi et al.(1999) starting from a condition where chaos synchronization occurs, i.e. $\Lambda_{\perp}^{nat} < 0$, a parameter mismatch which breaks the symmetry (and hence destroys the invariance of Δ) can lead to different situations according to the sign of a $\Lambda_{\perp}^{\max}$, i.e. according to the fact that A_s is an asymptotic attractor or only an attractor in Milnor sense. In particular, if A_s is an asymptotic attractor ($\Lambda_{\perp}^{\max} < 0$) then we expect that after a parameter's mismatch the one-dimensional "synchronized attractor" A_s is replaced by a similar attractor, on which quasi-synchronized dynamics occur, i.e. correlated chaos. Instead, if A_s is attractor only in Milnor sense ($\Lambda_{\perp}^{\max} > 0$, i.e. at least a transversely repelling cycle is embedded inside A_s) then a bigger two-dimensional chaotic area may suddenly appear, which is an attractor in the usual Lyapunov sense, inside which endless on-off intermittency is observed. When the time evolution of the duopoly game is obtained by the iteration of a noninvertible map, as generally occurs in problems of chaos synchronization, the boundary of such chaotic area can be often obtained by the method of *critical curves*, as described in Bischi and Gardini, 1998, 2000, see also Chapter 3 of this book.

Indeed, the map T defined in (6) is a noninvertible map, because given a point $(q_1', q_2') \in \mathbb{R}^2$ its preimages are computed by solving (6) with respect to q_1 and q_2, which gives a sixth degree algebraic system which may have up to six real solutions. For a given set of parameters, the critical curves of the map (6) can be easily obtained numerically. In fact, being the map (6) continuously differentiable, the set LC_{-1} can be obtained numerically as the locus of points (q_1, q_2) for which the Jacobian determinant det DT vanishes, and the critical curves LC, which separate regions Z_k whose points have different numbers of preimages, are obtained as $LC = T(LC_{-1})$.

4 Players' heterogeneity, correlated chaos, intermittency and synchronization

In this section we consider the dynamic duopoly game described in section 2 in order to illustrate, through numerical explorations guided by the theoretical background of chaos synchronization and critical curves, some of the topics outlined above, such as uncorrelated, correlated and synchronized chaos, intermittency phenomena and the related concepts of homogeneity and heterogeneity of the two players.

In the numerical explorations given below, the parameters $a = 10$, $b = 0.5$, $v_1 = v_2 = v = 0.32$, $c_1 = c_2 = c = 2$ are fixed, and identical for the two players, and different dynamic situations are obtained by tuning the two spillover parameters γ_{12} and γ_{21}. first of all, we consider the case of homogeneous players (7): in fig. 1a we show the bifurcation diagram, for γ ranging from 0.05 to 0.55, for the restriction (9) of the map T to the invariant diagonal; in fig. 1b is the plot of a numerical computation of the transverse Lyapunov exponent (10) $\Lambda_\perp$ as a function of γ in the same range. Each point of the graph is obtained by iterating the map (starting from an initial condition on the diagonal) 10,000 times to eliminate transient behavior, and then averaging over another 50,000 iterations. Of course, we cannot say that the graph in fig. 1b represents $\Lambda_\perp^{nat}$ because when the parameter γ is inside a periodic window of the bifurcation diagram, the corresponding trajectory is captured by the stable cycle, so the computation of (10) gives the transverse Lyapunov exponent of that cycle. However, the global shape of the graph in fig. 1b can give us a qualitative idea of the values of $\Lambda_\perp^{nat}$ as a function of the parameter γ, because the value $\Lambda_\perp^{nat}$ is well approximated by the value of the transverse Lyapunov exponent computed along a cycle, provided that the period of the cycle is sufficiently high. It is plain that the periodic windows of the cycles of period 5 and 3 and other low-period stable cycles, which are clearly visible in the bifurcation diagram, correspond to peaks of the graph in fig. 1b which cannot be peaks of $\Lambda_\perp^{nat}$. Such a difficulty is due to the fact that γ is not a normal parameter, so that as γ varies also the dynamics along Δ change, as clearly shown in the bifurcation diagram of fig. 1a. However, we can guess that around $\gamma = 0.14$ we have $\Lambda_\perp^{nat} < 0$ and probably also $\Lambda_\perp^{\max} < 0$, so that the chaotic attractor $A_s \subset \Delta$, on which synchronized dynamics occur, is asymptotically stable. Moreover, also around $\gamma = 0.4$ a small neighborhood exists where $\Lambda_\perp^{nat} < 0$, but in this case $\Lambda_\perp^{\max} > 0$, i.e. transversely unstable cycles exist embedded inside chaotic attractor A_s, which is, consequently, only an attractor in the weaker Milnor sense.

Our first numerical simulation is shown in fig. 2, where we consider a situation with a marked heterogeneity (or asymmetry) in the spillover parameters, i.e. only player 2 has the ability to take advantage from the *R&D* results of the competitor, being $\gamma_{12} = 0$ and $\gamma_{21} = 0.8$. For this set of parameters the time evolution is chaotic, as clearly appears by looking at the plot, in the phase space (q_1, q_2), of a generic trajectory starting from an

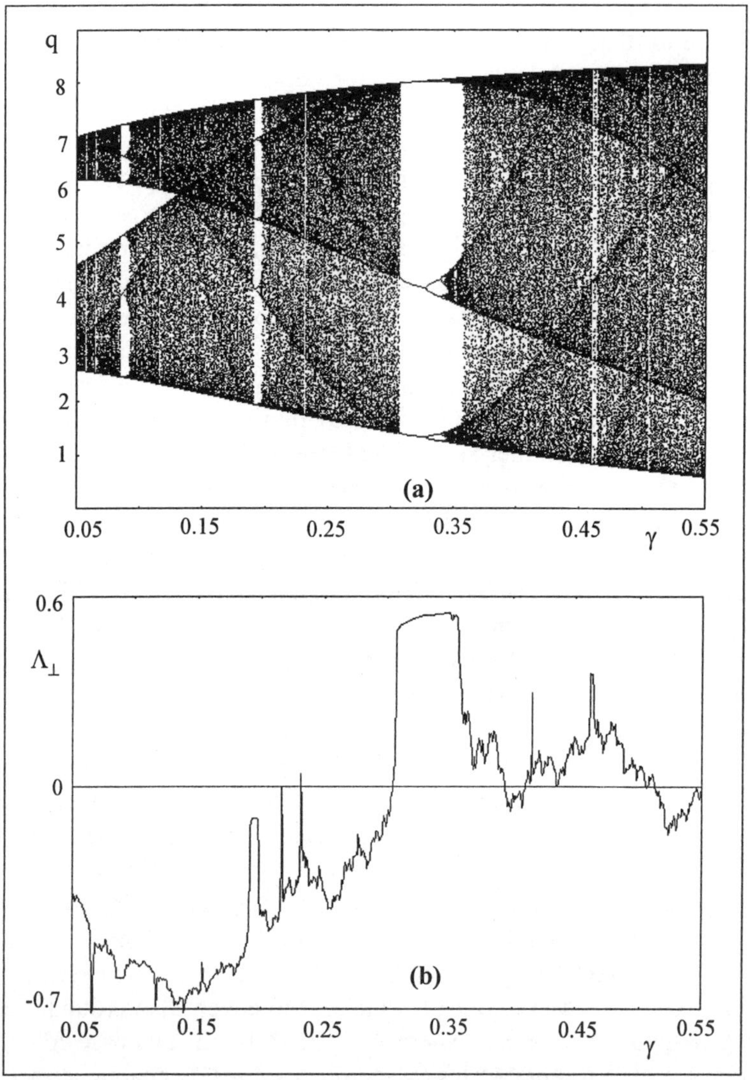

Figure 1: *Symmetric case of identical players, with parameters* $a = 10$, $b = 0.5$, $v_1 = v_2 = v = 0.32$, $c_1 = c_2 = c = 2$. *(a) Bifurcation diagram for the restriction of the map* T *to the invariant diagonal; (b) numerical computation of the transverse Lyapunov exponent* $\Lambda_\perp$ *for* γ *ranging from* 0.05 *to* 0.55.

initial condition in the white region (the grey region represents the set of initial conditions which generate unbounded trajectories).

The iterated points of such a trajectory erratically fill up a quite big chaotic area, and the corresponding time series $q_1(t)$ and $q_2(t)$ are represented versus time in fig. 2c for $0 \leq t \leq 200$.

This is a typical example of uncorrelated chaos, because no correlation can be evidenced between the two simultaneous choices $q_1(t)$ and $q_2(t)$. This can be better appreciated if we represent the difference $(q_1(t) - q_2(t))$ versus time, as in fig. 2d. It can be seen that, often, $(q_1(t) - q_2(t)) < 0$, i.e. $q_2(t) > q_1(t)$, as a consequence of the greater ability of firm 2 in taking advantage from spillover effects. However, some periods such that $q_1(t) > q_2(t)$ exists, and no rules seem to correlate production variations of producer 1 to production variations of producer 2 and vice-versa.

Let us remark that, as the time evolution of the repeated duopoly game is represented by the iteration of a noninvertible map, the boundary of the chaotic attractor can be obtained by segments of critical curves, as shown in fig.2b (see Mira et al., 1996, Bischi and Gardini, 1998, Puu, 2000). This trapping region (also called absorbing area) gives an upper bound for the oscillations of $q_1(t)$ and $q_2(t)$. A practical procedure to obtain such boundaries makes use of the concept of critical curves and can be outlined as follows: starting from a portion of LC_{-1}, say ω, approximately taken in the region occupied by the area of interest, its images by T of increasing rank are computed until a closed region is obtained. The length of the initial segment must be taken, in general, by a trial and error method, although several suggestions are given in Mira et al., 1996. In order to obtain the boundary of the chaotic area $\mathcal{A}$ shown in fig. 2a, eight images of the generating arc $\omega = \mathcal{A} \cap LC_{-1}$ are sufficient, hence in fig. 2b we have $\partial\mathcal{A} \subset \bigcup_{k=1}^{8} T^k(\omega)$.

In fig. 3 we consider a more homogeneous situation with respect to the firms' ability to take advantage from the spillover effects. Indeed, in the numerical simulations performed in fig. 3 we used the same set of parameters as in fig.2, but more homogenous spillover parameters, namely $\gamma_{12} = 0.35$, $\gamma_{21} = 0.45$. As expected, the chaotic attractor shown in fig.3a is more symmetric with respect to the diagonal, i.e. time periods with $q_2(t) > q_1(t)$ are balanced by more or less equally probable periods characterized by $q_2(t) < q_1(t)$. Moreover, the fact that the chaotic attractor is larger than the one shown in fig.2, suggests that greater differences between the two production choices should be expected at a given time period. However, this statement is quite misleading, because the density of the iterated points inside the chaotic area is mainly concentrated along the diagonal $q_1 = q_2$, i.e.

a generic trajectory inside the chaotic area visits much more often the region around the diagonal with respect to the portions of the chaotic area which are far from it.

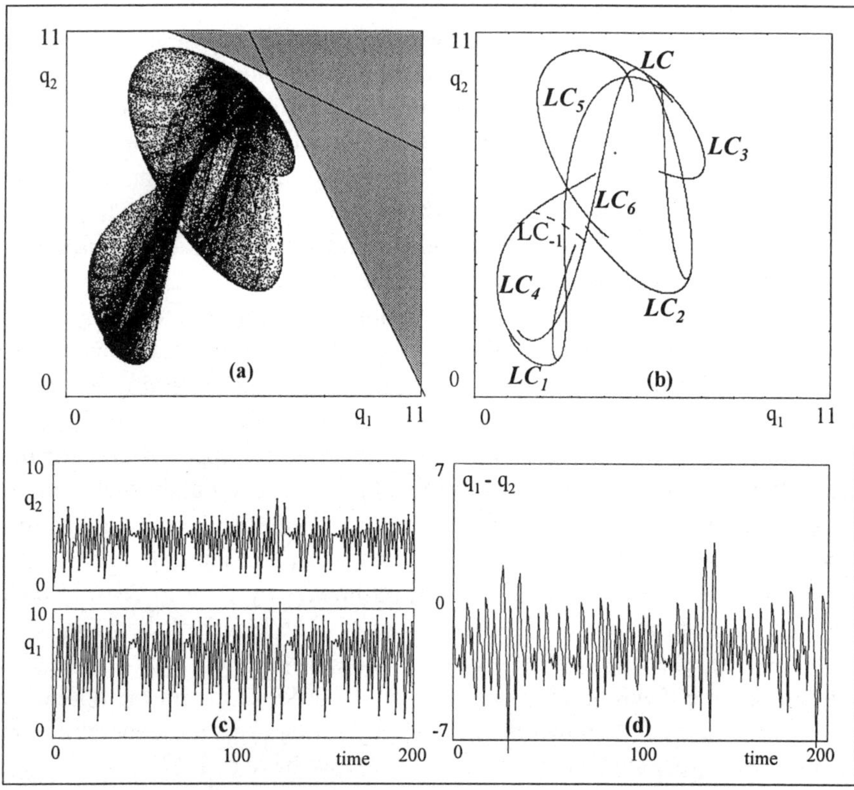

Figure 2: *Numerical simulation obtained with the same set of parameters as in fig. 1 and* $\gamma_{12} = 0$ *and* $\gamma_{21} = 0.8$. *(a) chaotic attractor represented in the phase space* (q_1, q_2); *(b) boundary of the chaotic attractor obtained by segments of critical curves*; *(c)* $q_1(t)$ *and* $q_2(t)$ *are represented versus time for* $0 \le t \le 200$; *(d) the difference* $(q_1(t) - q_2(t))$ *is represented versus time for the same time range as in (c).*

This property reveals the occurrence of *intermittency* phenomena, as clearly appears from the representation of $(q_1(t) - q_2(t))$ versus time, shown in fig. 3b, where it is evident that several time periods exist at which the difference $(q_1(t) - q_2(t))$ is close to zero, i.e. the production choices of the two firms are similar (both low or both high). However, very frequent "bursts" of asynchronous productions occur along the time evolution. With other words, a typical time evolution is characterized by sequences of time periods, of the order of e.g. 20 or 30 time periods, at which the two firms produce almost the same output, interspersed with clusters of large asynchronous fluctuations.

If homogeneity is further increased, the time intervals of synchronized behavior become longer, i.e. the bursts of asynchronous productions are more rare, and when the two firms become perfectly homogenous, so that the diagonal Δ becomes an invariant submanifold, then chaos synchronization can be observed if $\Lambda_\perp^{nat} < 0$. More precisely, if $\Lambda_\perp^{nat} < 0$ and also $\Lambda_\perp^{\max} < 0$ a fast chaos synchronization occurs, as a consequence of the fact that A_s is an asymptotic attractor, whereas if $\Lambda_\perp^{nat} < 0$ and $\Lambda_\perp^{\max} > 0$ then synchronization is still obtained in the long run, but after a transient characterized by several bursts. Instead, if $\Lambda_\perp^{nat} > 0$ then such bursts never stop (i.e. asymptotic synchronization does not occur).

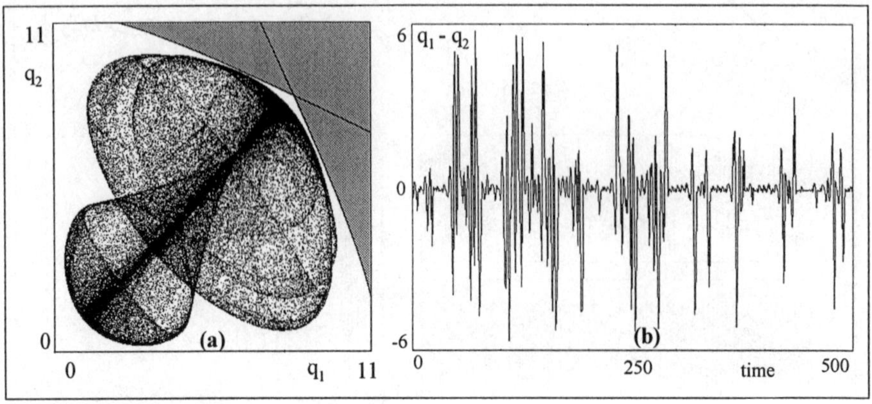

Figure 3: *Numerical simulation obtained with the same set of parameters as in fig.2, but more homogenous spillover parameters,* $\gamma_{12} = 0.35$, $\gamma_{21} = 0.45$ *(a) A trajectory is represented in the phase space* (q_1, q_2) ; *(b) the difference* $(q_1(t) - q_2(t))$ *is represented versus time for* $0 \leq t \leq 500$.

In the case shown in fig. 4, obtained with $\gamma_{12} = \gamma_{21} = 0.4$, we have $\Lambda_\perp^{nat} < 0$ and $\Lambda_\perp^{\max} > 0$. In this case, a generic trajectory, starting in the basin of bounded trajectories, synchronizes along the chaotic one-dimensional attractor $A_s \subset \Delta$ after a transient characterized by some asynchronous bursts. The presence of this transient is seen in the phase space (fig. 4a) where several points out of Δ can be seen, and it is more clearly evidenced in the representation, versus time, of $q_1(t)$ and $q_2(t)$ (fig. 4c) and of the difference $(q_1(t) - q_2(t))$ (fig. 4d).

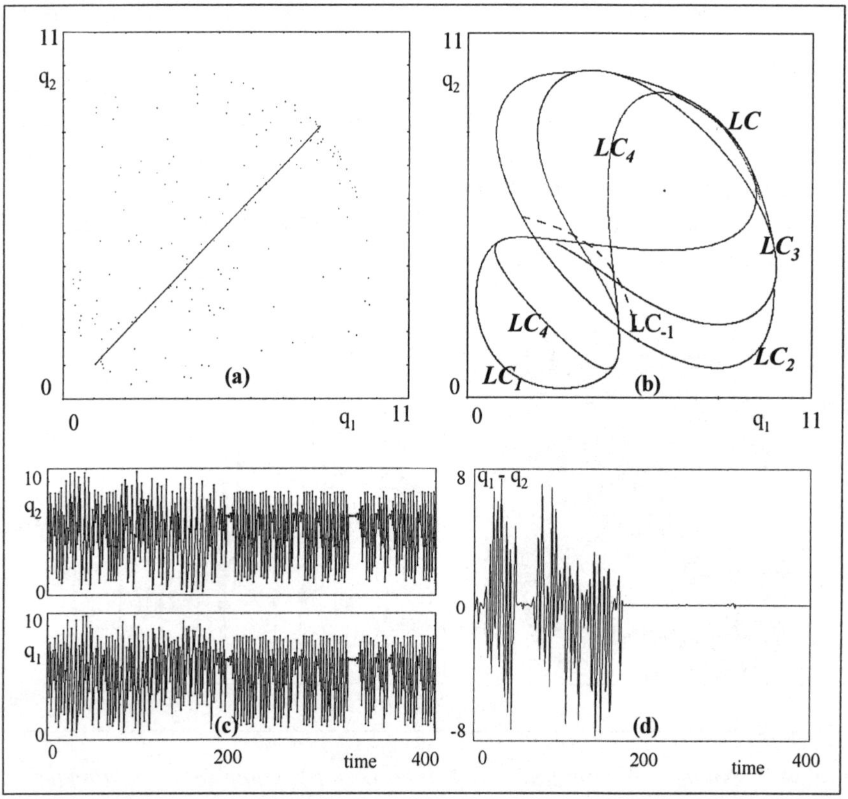

Figure 4: *Numerical simulation obtained with the same set of parameters as in fig. 1 and* $\gamma_{12} = \gamma_{21} = 0.4$ *(a) A typical synchronizing trajectory is represented in the phase space* (q_1, q_2) ; *(b) boundary of the chaotic attractor obtained by segments of critical curves; (c)* $q_1(t)$ *and* $q_2(t)$ *are represented versus time for* $0 \leq t \leq 400$; *(d) the difference* $(q_1(t) - q_2(t))$ *is represented versus time for the same time range as in (c).*

These time sequences clearly show that after the early 200 iterations the two players' production choices become perfectly synchronized, even if each of them exhibit a chaotic behavior. Of course, such an asymptotic synchronization occurs because the natural transverse Lyapunov exponent is negative for $\gamma = 0.4$ (see fig. 1b), i.e. the one-dimensional chaotic attractor embedded into the invariant diagonal is an attractor. However, being $\Lambda_{\perp}^{\max} > 0$ it is only an attractor in Milnor sense, i.e. transversely repelling cycles are embedded in the chaotic attractor (even if the attracting ones prevail) and this gives an explanation of the intermittent transient. We observe, however, that the locally repelled trajectories cannot reach the basin of infinity due to the presence of an absorbing area $\mathcal{A}$, bounded by segments of critical curves (shown in fig. 4b). Loosely speaking $\partial\mathcal{A}$ behaves as a bounded vessel for the intermittency phenomena, and the local unstable sets of these transversely repelling cycles embedded inside A_s are folded back (reinjected) by the folding action of the critical curves that form $\partial\mathcal{A}$. Moreover, as remarked above, the presence of such an absorbing area gives us the possibility to define an upper bound for the asynchronous bursts. Instead, the length of the transients and the kind of intermittency which characterizes the trajectories before synchronization, cannot be, in general, forecasted. To explain this statement, in fig.5 we show the versus time representations of the early 400 points of trajectories obtained with the same set of parameters as in fig. 4, but starting from different initial conditions. The trajectory shown in figures 4a,c,d was obtained starting from the initial condition $(q_1(0), q_2(0)) = (5, 6)$, whereas starting from other initial conditions different transient behaviors are observed, characterized by longer or shorter asynchronous bursts, even if synchronization always occurs in the long run. For example, the trajectories used to obtain the $(q_1(t) - q_2(t))$ sequences shown in figures 5a and 5b are obtained starting from $(q_1(0), q_2(0)) = (5.7, 6.2)$ and $(q_1(0), q_2(0)) = (5.5, 6.5)$ respectively. It is interesting to note that sometimes synchronization seems to be reached, but after several time periods other bursts still occur. For example, in fig. 5b, after about 90 time periods the evolution of the system seems to have reached almost complete synchronization. Instead, after 80 time periods of almost perfect synchronization, the trajectory then moves again far away from the diagonal, and the two competitors now act again in a different fashion. Such an intermittent behavior is typical of the convergence to a non-topological Milnor attractor. The pattern of the time series resembles that of a system which is subject to exogenous random shocks, even if the dynamical system that generates such a pattern is completely deterministic. We now consider what happens if we introduce a *small heterogeneity*, due to

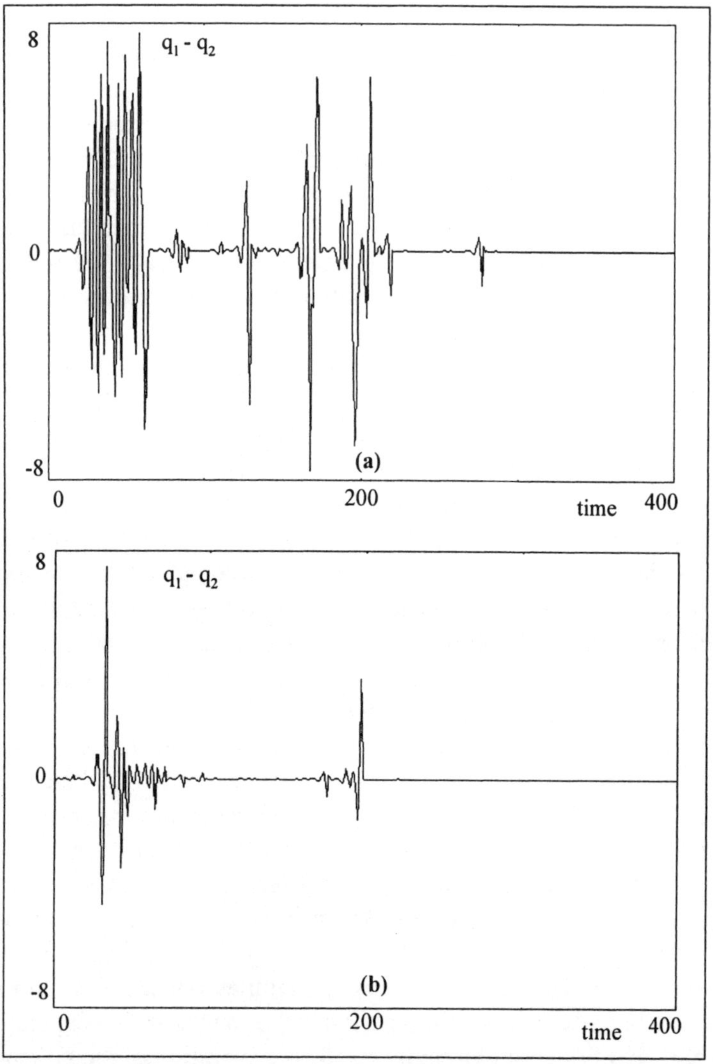

Figure 5: *With the same set of parameters as in fig.4, two sequences* $(q_1(t) - q_2(t))$*, are shown, during the transient with* $0 \leq t \leq 400$*, obtained starting from different initial conditions: (a)* $(q_1(0), q_2(0)) = (5.7, 6.2)$; *(b)* $(q_1(0), q_2(0)) = (5.5, 6.5)$

a small parameter mismatch. For example, if we change one of the spillover parameters of just 1% with respect to the homogenous situation of fig.4, i.e.

$$\gamma_{21} = \gamma_{12} + 0.005 \tag{11}$$

where $\gamma_{12} = 0.4$, we obtain the result that bursts never stop, and an endless on-off intermittency is obtained, with bursts whose amplitude is determined by the absorbing area located around the diagonal (fig. 6).

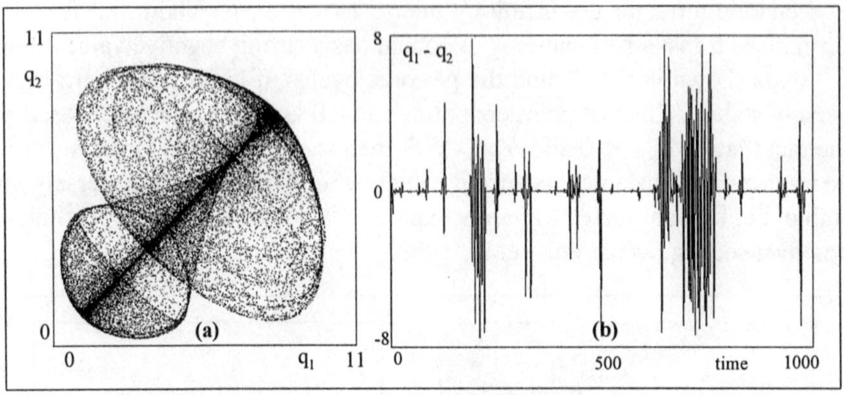

Figure 6: *Symmetry breaking obtained starting from the set of parameters used in fig.4, and with quasi-identical spillover parameters:* $\gamma_{12} = 0.4$, $\gamma_{21} = 0.405$. *(a) Chaotic attractor in the phase space; (b) for the same trajectory shwn in (a), the difference* $(q_1(t) - q_2(t))$ *is represented versus time for* $0 \leq t \leq 1000$.

Such a parameters' mismatch causes the destruction of the invariance of Δ, due to the fact that the map is no longer symmetric (this kind of perturbation has been called *symmetry breaking* in Bischi et al. 1999). The fact that the diagonal is no longer an invariant set causes the disappearance of the one-dimensional Milnor attractor A_s along the diagonal, and such a small perturbation may lead to quite different dynamics, since after the symmetry breaking synchronization can no longer occur, and the bursts never stop, and the generic trajectory fills up the absorbing area. However, if the attractor A_s existing along Δ before the parameters' mismatch is a *topological attractor*, that is $\Lambda_{\perp}^{\max} < 0$, then the introduction of small heterogeneities does not have such a disruptive effect. In this case the symmetric model still serves as a good approximation of the behavior of the two firms, and correlated chaos is obtained. Such a situation is shown in fig. 7. In fig.7a identical firms are considered, with the same values of a, b, c, v as in the previous numerical simulations, but with a lower value of γ, $\gamma = 0.15$. As stated before, while commenting fig.1, this set of parameters belongs to a region of the parameter space where the one-dimensional chaotic attractor embedded into the diagonal is an asymptotic attractor, i.e. $\Lambda_{\perp}^{\max} < 0$. We remark that this is not easy to be proved in general, since the cycles included in a chaotic attractor are infinitely many. However, we claim the fact that the natural transverse Lyapunov exponent has a strong negative value, about -0.6, as shown in fig. 1, and the periodic cycles of lower period are transversely stable. This last point constitutes a well known conjecture, based on the fact that if $\Lambda_{nat} < 0$ and $\Lambda_{\max} > 0$, then some low period cycles should be transversely unstable, because if a cycle of high period is transversely unstable, i.e. its transverse Lyapunov exponent is positive, also Λ_{nat} should be positive, see e.g. Maistrenko et al. 1998.

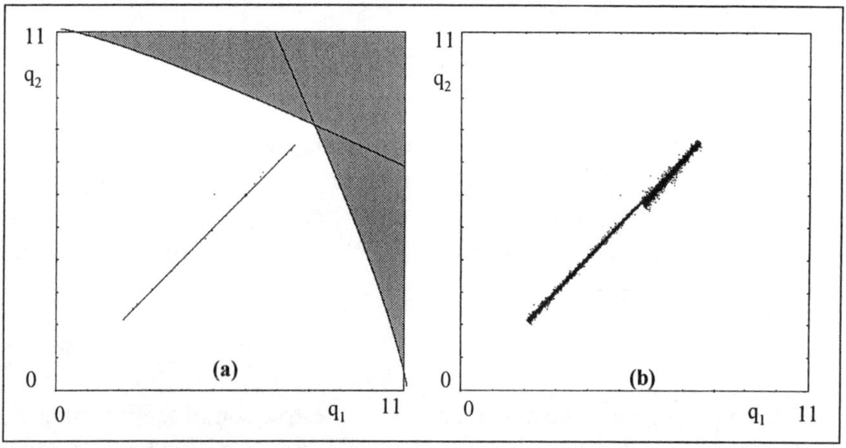

Figure 7: (a) *With the same set of parameters as in fig. 1, and* $\gamma_{12} = \gamma_{21} = 0.15$, *a typical synchronizing trajectory is represented in the phase space* (q_1, q_2); *(b) chaotic attractor after a symmetry breaking obtained with* $\gamma_{12} = 0.15$, $\gamma_{21} = 0.17$.

The fact that the chaotic set on which synchronized dynamics occur is an asymptotic attractor implies two things: first, the synchronization of trajectories starting out of it (in its basin) is very fast; second, if we introduce a symmetry breaking due to a small parameter mismatch (such as $\gamma_{12} = 0.15$ and $\gamma_{12} = 0.17$) the resulting trajectories are "almost synchronized" (see fig.7b), i.e. small heterogeneities imply small production differences during the time evolution. This is a typical example of correlated chaos. A comparison of figures 6 and 7 leads us to the conjecture that only if the attractor of the symmetric model is a topological attractor, then the introduction of a small heterogeneity would still lead to almost synchronized trajectories. Otherwise, in the case of a Milnor (not asymptotic) attractor endless intermittency phenomena characterize the chaotic evolutions of the two players.

Acknowledgments. This work has been performed under the auspices of CNR, Italy, and under the activity of the research project "Nonlinear Dynamics and Stochastic Models in Economics and finance", MURST, Italy.

References

Aitken, B., Hanson, G. H. and Harrison, A. E., 1997, "Spillovers, foreign investment, and export behavior", *Journal of International Economics*, 43: 103-132.

Ashwin, P., Buescu, J. and Stewart, I., 1996 "From attractor to chaotic saddle: a tale of transverse instability", *Nonlinearity*, 9: 703-737.

Audretsch, D.B. and Feldman, M.P., 1996, "R & D Spillovers and the Geography of Innovation and Production", *American Economic Review*, 86: 630 - 640.

Baumol, W.J. and Quandt, R.E., 1964, "Rules of Thumb and optimally imperfect decisions", *The American Economic Review*, 54: 23-46.

Binmore, K., 1992, *Fun and Games*, D.C. Heath & C..

Bischi, G.-I. and Gardini, L., 1998 "Role of invariant and minimal absorbing areas in chaos synchronization", *Physical Review E*, 58: 5710-5719.

Bischi, G.-I., Gallegati, M. and Naimzada, A., 1999, "Symmetry-Breaking bifurcations and representative firm in dynamic duopoly games", *Annals of Operations Research*, 89: 253-272.

Bischi, G.-I., Stefanini, L. and Gardini, L., 1998, "Synchronization, intermittency and critical curves in duopoly games", *Mathematics and Computers in Simulations,* 44: 559-585.

Bischi, G.-I., Mammana, C. and Gardini, L., 2000, "Multistability and cyclic attractors in duopoly games", *Chaos, Solitons and Fractals*, 11: 543-564.

Bischi, G.-I. and Naimzada, A., 1999, "Global Analysis of a Duopoly game with Bounded Rationality", *Advances in Dynamic Games and Applications*, 5:361-385, Birkhauser.

Bischi, G.-I. and Gardini, L., 2000, "Global Properties of Symmetric Competition Models with Riddling and Blowout Phenomena" *Discrete Dynamics in Nature and Society* 5: 149-160

Bischi, G.-I. and Lamantia, F., 2001, "Nonlinear Duopoly Games with Positive Cost Externalities due to Spillover Effects" *Chaos, Solitons & Fractals* (in press)

Buescu, J., 1997, *Exotic Attractors*, Birkhäuser.

Cournot, A., 1838, *Récherches sur les principes matématiques de la théorie de la richesse*, Hachette, Paris.

Dana, R.A. and Montrucchio, L., 1986, "Dynamic complexity in duopoly games", *Journal of Economic Theory*, 40, 40-56.

D'Aspremont, C. and Jaquemin, A., 1988, "Cooperative and Noncooperative R&D in Duopoly with Spillovers", *American Economic Review*, 78, 1133-1137.

Flam, S.D., 1993, "Oligopolistic Competition: from Stability to Chaos", in: F. Gori, L. Geronazzo and M.Galeotti (Eds.) *Nonlinear Dynamics in Economics and Social Sciences*, Lecture Notes in Economics and Mathematical Systems, 399, 232-237, Springer-Verlag.

Fudenberg, D. and Levine, D., 1998, *The Theory of Learning in Games*, MIT Press.

Fujisaka, H. and Yamada, T., 1983, "Stability theory of synchronized motion in coupled-oscillator systems", *Progress of Theoretical Physics*, 69 (1) pp. 32-47.

Kopel, M., 1996, "Simple and Complex Adjustment Dynamics in Cournot Duopoly Models", *Chaos, Solitons & Fractals*, vol.7 n.12, 2031-2048.

Kopel, M., Bischi, G.-I. and Gardini, L., 2000, "On new phenomena in dynamic promotional competition models with homogeneous and quasi-homogeneous firms" in *Interaction and Market Structure. Essays on Heterogeneity in Economics* D. Delli Gatti, M. Gallegati and A.P. Kirman (Eds.), Springer-Verlag, pp. 57-87.

Hasler, M. and Maistrenko, Yu., 1997, "An introduction to the synchronization of chaotic systems: coupled skew tent maps", *IEEE Trans. Circuits Syst.*, vol. 44 (10) pp. 856-866.

Maistrenko, Yu., Maistrenko, V.L., Popovich, A. and Mosekilde, E., 1998, "Transverse instability and riddled basins in a system of two coupled logistic maps", *Physical Review E*, 57 (3), 2713-2724.

Mira, C., Gardini, L., Barugola, A., and Cathala, J.C., 1996, *Chaotic Dynamics in Two-Dimensional Noninvertible Maps*, World Scientific.

Ott, E. and Sommerer, J.C., 1994, "Blowout bifurcations: the occurrence of riddled basins and on-off intermittency", *Phys. Lett. A*, 188, 39-47.

Pecora, L.M. and Carrol, T.L., 1990, "Synchronization in chaotic systems", *Physical Review Letters*, 64 (8) pp. 821-824.

Puu, T., 1991, "Chaos in Duopoly Pricing", *Chaos, Solitons & Fractals*, vol.1 n.6, 573-581.

Puu, T., 2000, *Attractors, Bifurcations, and Chaos*, Springer-Verlag.

Rand, D. "Exotic phenomena in games and duopoly models", *J. Math. Econ.*, 5, 173-184, (1978).

Sacco, P.L., 1991, "Adaptive Response and Cournotian Behavior", Economic Notes, 20, 474-496, (1991).

Varian, H.R., 1992, *Microeconomic Analysis*, 3rd edition, W.W. Norton & Company.

9 Intrinsic Stability, Multi Stability and Instability

Shahriar Yousefi[1]

[1] Department of Statistics and Demography, University of Southern Denmark

1 Introduction

Standard economic theories on market structure can be divided in two categories depending on their underlying assumptions about the degree of strategic (and structural) dependence among the interacting agents (firms, producers etc.) who are operating on the supply side of the economy. The first category includes theories about extreme scenarios of markets with perfect competition and markets with no competition. The second category includes theories about markets with a few agents competing for the lion's share of the market. Theories on oligopolistic competition refer to such scenarios. The main feature of oligopoly markets and oligopolistic competition is the immense significance of strategic dependence among agents who are interacting on the supply side of the economy. Obviously, due to the presence of the interaction among agents, the decision of any agent is affected by the decisions of the other agents. Therefore, any agent in its decision making process should make some assumption about the reaction of its rivals to its actions. Incorporating these issues in a (mathematical) modelling framework, implies that in an oligopolistic market, the best response of a given agent should be defined in terms of his expectations of rivals' actions which is again defined in terms of their expectations of his actions. Among the important factors in this process is the number of interacting agents and the dominating cost structure that ultimately affects the competitive dynamics of an oligopolistic market structure.

As regards the number of agents, various studies focus on the duopolistic competition which is the simplest form of oligopoly, since there are only two agents involved in the competitive process. As regards the cost structure, assumptions generally depend on the nature of analysis and the issues that the modellers intend to highlight.

The literature on oligopolistic market structures (and competition) is quite sizeable and its origin can be traced to the pioneering work of Cournot in 1830s followed by critical reviews of Bertrand in early 1880s and Edgeworth in late 1890s. In a nutshell, while Cournot's model was based on competition in quantity between interacting agents operating on the supply side of the economy, Bertrand's emphasis was on price competition and Edgeworth argued that since agents make both price and quantity decisions, much attention should be given to capacity restrictions in modelling the competitive characteristics of oligopolistic markets.

Over the course of time many theories about oligopolistic competition were introduced resulting in a very sizeable literature ranging, from theories on static one shot competition to significant contributions in game theory, industrial economics, behavioural models, to dynamic models of strategic decisions. Some twenty years ago Friedman referred to this diversity of ideas and eloquently stated that "*...the boundaries of oligopoly theory are drawn somewhat arbitrarily, reflecting the judgement of those who have worked in the area about the ways in which useful and interesting results may be obtained. Everyone must judge for himself whether the choices made are justified, and, hopefully, some will be encouraged to break new and important ground*" (Friedman 1977, p. 5).

Since the focus of the present effort is on complex dynamics, let us take the liberty of concentrating on this part of literature for a while. This brief introduction is followed by elaboration on a few contributions that represent a degree of complexity not present in standard dynamic models of oligopoly. A detailed analysis of one of these models proposed by Puu (1997) is the main ingredient of the present effort.

2 Complex dynamics and oligopolistic competition

Apparently, Rand (1978) represents the first case of application of chaos theory in modelling economic phenomena in general and oligopolistic competition in particular. Rand's results concern a particular interpretation of Cournot's duopoly model from the 1830s. Although his analysis was associated with some ad hoc assumptions with no particular economic substance, nevertheless, it revealed that the dynamic Cournot model could lead to the possible emergence of chaos. Rand's results initiated a few other studies with particular focus on the framework of competitive games and the potential associations with nonlinear dynamics.

For instance Dana and Montrucchio (1986) used the notion of perfect equilibrium with stationary Markovian strategies (MPE) to model the dynamics of response in a Cournot adjustment process. Establishing the conditions by which an MPE exists seems to be a fairly difficult task. Therefore this immense obstacle is circumvented in the analysis by arguing that for a sufficiently small discount rate any pair of twice differentiable functions could be an MPE of some game. Consequently, in cases where an MPE exists the corresponding sequence of agents' actions is the trajectory of a map with similarities to Cournot dynamics. For discount rates converging zero, the set of MPEs converge to the best reply function of the one shot game corresponding to the simple case investigated by Rand. Furthermore, they argue that the degree of complexity of the dynamics of Cournot competition depends on the composition of reaction functions.

These results initiated a critical study by van Witteloostuijn and van Lier (1990) in which they questioned the lack of economic rationale in Rand (1978) and Dana and Montrucchio (1986). For instance they pointed out a simple problem that occurs in both models due to the hill shaped structure of the underlying reaction curves, namely the zero production of one firm in response to zero production of the other firm. In other words, a monopolist produces zero units in reaction to zero production of the competitor! Witteloostuijn and van Lier (1990) raise this issue and by discussing various scenarios argue that the general results should also hold for the case with positive output for the monopolist.

A different contribution to the literature on complex dynamics in oligopolistic competition was provided by Keenan and O'Brien (1993). They incorporated the spatial aspects of competition among a fixed number of agents positioned around a circular city. The analysis is quite in line with Bertrand's ideas of price competition but formally based on a cellular automata framework where agents on every site are associated with an identical reaction function defined in terms of the prices of his two immediate neighbours. By describing different scenarios, they illustrate a certain collusive mode (seemingly arising from a process of spontaneous self-organization) interacting in a chaotic manner with a competitive one. A particular aspect of Keenan and O'Brien's contribution is their introduction of nonlinearities in preferences and technologies as a formidable source for generating complex dynamics.

Among a few other interesting results that are worth to be mentioned here are Kopel (1996), Bischi et al. (1996) and Rassenti et al. (2000). Since space limits force heuristic brevity, let us confine the discussion to a few words about these contributions and proceed further with the main task of the present effort. As regards Kopel (1996), the main interesting feature is the emphasis on the cost structure and its influence on the dynamics of Cournot competition viewed in the framework of a system of coupled maps. As regards Bischi et al. (1996), the main insight provided by this study is associated with its mathematical theme which is predominantly an investigation of the transversal stability, bifurcation properties and the basin of attraction of a system of coupled oscillators. A brief discussion on the existence of critical curves is also provided (for a study with particular focus on the method of critical lines and a clear emphasis on duopolistic competition see Puu 2000 ch. 7.). As regards Rassenti et al. (2000), it is one of a few studies that actually incorporate simulation experiments in the analysis and provide a sort of "experimental reasoning" for explaining out of equilibrium rationally bounded behaviour among a group of agents interacting in a Cournot oligopoly framework.

3 Puu's interpretation of Cournot oligopoly

From the middle of the 1990s a particular dynamic interpretation of Cournot's competitive framework suggested by Tonu Puu (see Puu 1991, 1997, 2000) exerted a certain influence on some studies of complex dynamics in oligopolistic competition (for a few recent contributions see Ahmed and Agiza 1998, Ahmed et al. 1999, 2000, Agiza 1998, Agiza et al. 1999, Bischi et al. 1999, Agliari et al. 2000, Huang 2001).

Puu's main contribution was to introduce a set of reasonable economic assumptions concerning the cost and demand structure (i.e. constant marginal costs and an isoelastic demand function) into the framework of Cournot's sequential optimal response strategies and used the framework of this suggested model to address the emergence of deterministic complex dynamics in various scenarios such as duopolistic and oligopolistic competition.

Let us confine our efforts to the simplest case of two competitors (duopoly) and proceed by a brief description of Puu's duopoly model. To begin with, let us assume that demand is always equal to supply (consisting of the supply of two competitors, given by x and y) and constant marginal costs are given by a and b. In addition to these, assume an isoelastic demand function such that price is reciprocal to demand given by $p = 1/(x+y)$.

The profit functions for the two competing agents are then given by $\pi_1(x,y) = px - ax$ and $\pi_2(x,y) = py - by$ respectively, and the profit maximisation process leads to the reaction functions given by

$$\pi'_{1,2} = 0 \Rightarrow x = \sqrt{y/a} - y \text{ and } y = \sqrt{x/b} - x.$$

The so called Cournot point is the intersection of these reaction functions. Introducing dynamics, leads to a set of independent 2D maps given by:

$$\begin{cases} x_{t+1} = \sqrt{y_t/a} - y_t \\ y_{t+1} = \sqrt{x_t/b} - x_t \end{cases} \quad (1)$$

System (1) indicates that firms in each period, only rely on their knowledge about their own marginal cost and the competitor's behaviour in the previous period. Assuming that the competitors adjust their previous decisions by different adjustment speeds given by λ and μ (both parameters belong to the unit interval), the previous 2D system is extended to a system of nonlinear coupled maps given by

$$\begin{cases} x_{t+1} = x_t + \lambda\left(\sqrt{y_t/a} - y_t - x_t\right) \\ y_{t+1} = y_t + \mu\left(\sqrt{x_t/b} - x_t - y_t\right) \end{cases} \quad (2)$$

Puu (1997) used system (2) to establish the stability condition of the Cournot point and visualized the emergence of different types of complex dynamics such as fixed points, cycles and chaos. Furthermore, he extended the framework of the analysis to three competing oligopolists and proceeded with addressing similar dynamic features of the model.

4 Asymptotic stability

Numerical simulations indicate that, for significantly large portions of the parameter space, the system given by (2), leads to negative quantities (in other words trajectories leave the positive part of the real plane). Obviously, negative quantities do not make sense (from an economic viewpoint) in the present context and therefore system (2) should always be stated as:

$$\begin{cases} x_{t+1} = \max\left\{0, x_t + \lambda\left(\sqrt{y_t / a} - y_t - x_t\right)\right\} \\ y_{t+1} = \max\left\{0, y_t + \mu\left(\sqrt{x_t / b} - x_t - y_t\right)\right\} \end{cases} \tag{3}$$

It is intuitively clear that the asymptotic behaviour of system (3) is significantly different from the asymptotic behaviour of system (2).

From an economic viewpoint it is interesting to determine how the parameters of the model (3) influence the amplitude and the periodicity of different cycles in the competition process. A substantial analytical investigation of this issue is hindered by significant analytical difficulties. From an analytical viewpoint, system (3) is a nonsmooth map with a formidable nonlineraity dominating its dynamics. Due to the non-smoothness, presence of nonlinearities and inherent discontinuities in the derivatives of the system (3), limited information can be extracted by reliance on standard analytical methods such as the Hopf bifurcation theorem, the Poincaré-Bendixon theorem or the Hartman-Grobman theorem. Consequently, issues such as the structural stability, the long-term behaviour/stability and the bifurcation properties of the system cannot be studied through standard analytical methods.

Among the other analytical difficulties that emerge from the study of system (3), the problem of verification/quantification of the existence of sensitivity to initial conditions in the system by the means of the estimation of the largest Lyapunov exponent is worth noting.

Besides a few exceptions, these issues have not been previously addressed in the literature (in my reading). These few exceptions concern a paper by Muller (1995) followed by three recent studies by Stefanski and Kapitaniak (2000), Kunze (2000) and Galvanetto (2000) concerning the estimation of largest Lyapunov exponent of nonsmooth systems. But even these types of study cannot be easily applied to address the problem at hand. This is mainly due to various technical difficulties in implementing these routines and a series of difficulties associated with their basic assumptions.

Consequently, the only remaining practical remedy is to rely on numerical methods and to specify parameter combinations (λ, μ) that lead to cycles of different higher orders.

5 Numerical Results

Numerical investigations of system (3), reveal a few interesting features of the model. The obtained results seem to be fairly robust on four issues. First, as regards the cost structure, numerical investigations indicate that the emergence of complex dynamics is associated with a highly asymmetric cost structure (marginal costs). Furthermore, for fixed marginal costs, the emergence of complex dynamics is associated with relatively high adjustment speeds. Next, as regards the various bifurcation scenarios, the obtained results indicate that the dominant bifurcation types are Hopf bifurcations and period adding bifurcations accompanied by occasional occurrences of saddle node bifurcations and period doubling bifurcations. Finally, as regards the asymptotic behaviour of the system, seemingly a large set of different scenarios, such as fixed point, periodic motions of different types, limit cycles, strange attractors and coexistence of different attractors (multi stability) dominates the dynamics of the system.

The general dynamics of the system have a tendency to settle down in stable regimes (predominantly fixed points corresponding to stable Cournot points). With respect to multistability and the presence of coexisting attractors, it is worth noting that the numerical simulations illustrate a rich mixture of different conceivable scenarios. Seemingly there is a significant number of parameter combinations (λ, μ) that eventually lead to coexisting attractors. These parameter combinations are predominantly located in the boundaries of different stability regions in parameter space.

In order to proceed further, consider system (3) and assume $a = 10$ and $b = 1$. Fig. 1 illustrates the 2D bifurcation plane for system (3) in (λ, μ) space. It is worth noting that Hopf bifurcations usually occur in the boundaries between P_1 and the black region.

Saddle node bifurcations occur in the white region when different nodes of periodic cycles become unstable and give birth to more complicated attractors. In order to visualize the occurrence of period adding bifurcations a further specification of significant periodic regions in parameter space is needed. The largest of these regions are illustrated in Fig. 2.

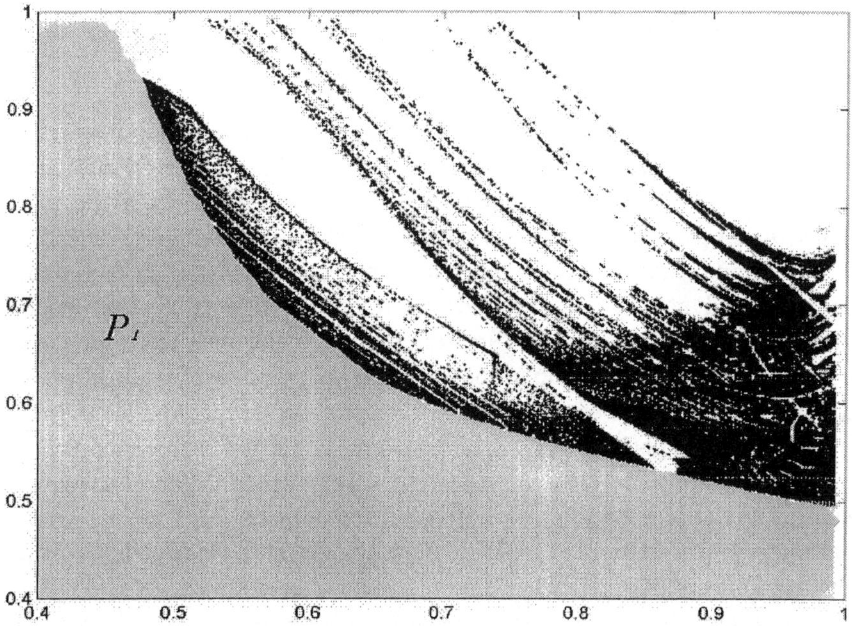

Fig. 1. Parameter combinations (λ,μ) *that lead to fixed point attractors are denoted as* P_1 , *cycles of high order* (>500) *are given in black and other periodic cycles are given in white (for* a =10 *and* b =1).

Apparently, there are no stable period-2 and period-3 cycles associated with the system. Furthermore, in the upper right corner of Fig. 2 the sequence of period adding bifurcations from period-4 cycles to period-8 cycles and further is quite visible. The same procedure takes place for various other periodic cycles as well. For instance period addings from period-5 cycles to period-10 and other take place in a small region in the middle of the region denoted by P_5. In general the visualization of these phenomena for other cases demands a higher degree of graphical resolution and finer steps in simulations.

It is worth noting that Fig. 1 and Fig. 2 illustrate to what extent the dynamics of the system (3) tend to settle down in a stable mode. In this regard note

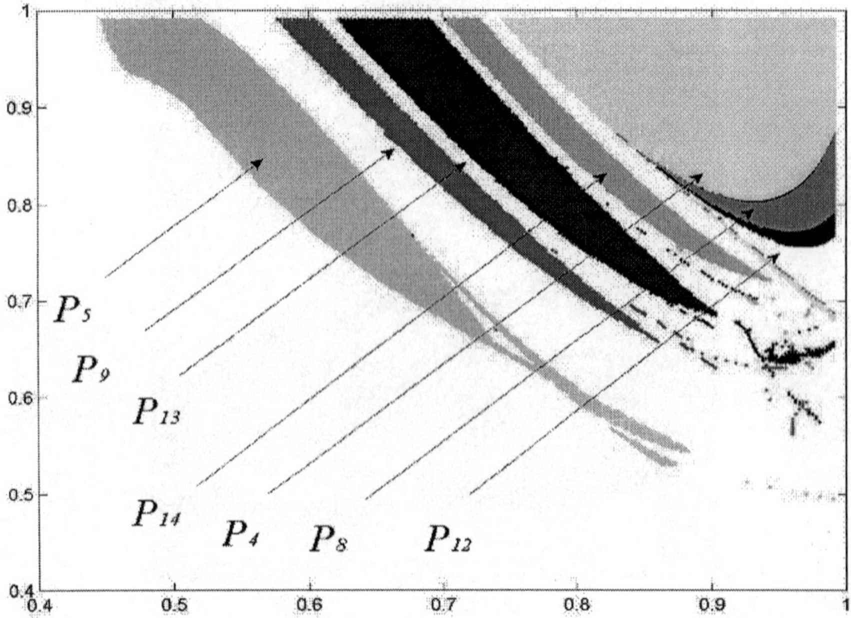

Fig. 2. Large regions in parameter (λ, μ) space that lead to dominant periodic cycles for $a = 10$ and $b = 1$.

that besides P_1, the parameter space consists of a series of "peninsulas" of stability that extend into a "sea" of instability. In this process, apparently all of these stability regions extend in the same direction and none of the large regions cross other regions in another direction. Consequently, besides few occasional cases, one should expect that the emergence of multistability is associated with the parameter combinations that belong to the boundaries between different regions.

Despite the fact that a more qualified assessment of this hypothesis demands further elaboration on the detailed structure of the boundary regions, the present result is sufficient to point out a rich mixture of different conceivable scenarios concerning the presence of different coexisting attractors. Fig. 3- Fig. 6 illustrate a few of these scenarios.

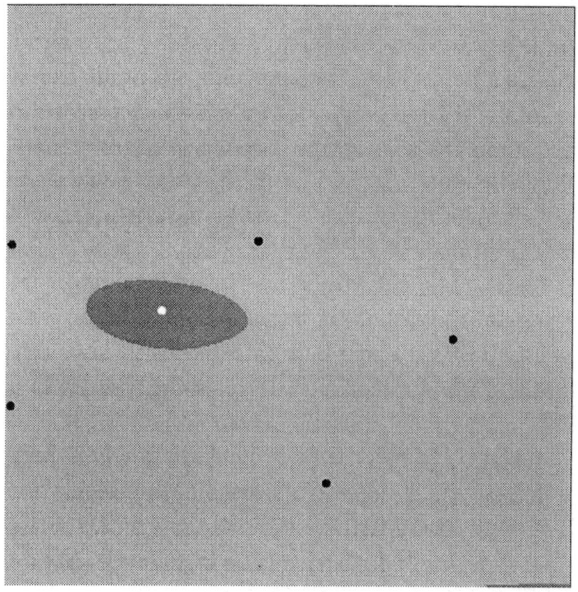

Fig. 3: Coexistence of fixed point and stable period-5 cycle.

Fig. 4: Coexistence of fixed point and 2 stable period-5 cycles.

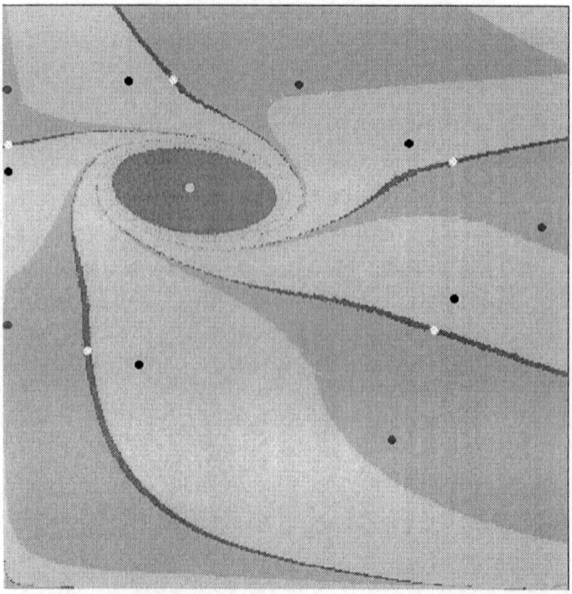

Fig. 5: Coexistence of fixed point and 3 stable period-5 cycles.

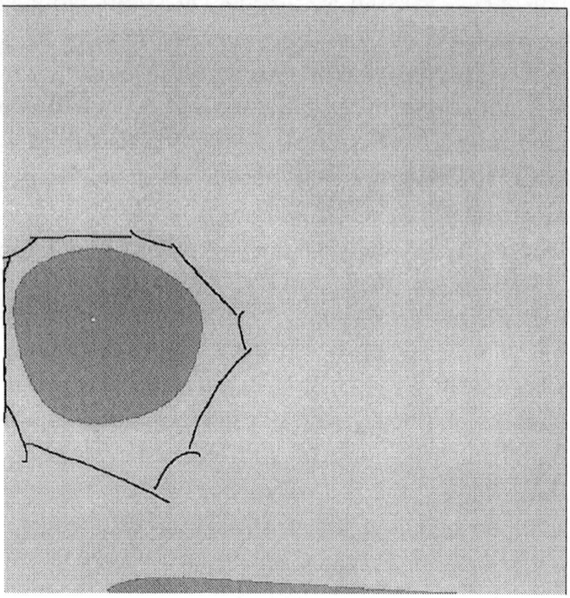

Fig. 6: Coexistence of fixed point and chaotic attractor.

6 Concluding remarks

The present study elaborated on some aspects of the asymptotic behavior of Puu's chaotic duopoly. A few analytical difficulties in addressing this issue were underlined while the emphasis was deliberately on investigating the limitations and restrictive features of the model. Despite the presence of discontinuities and the nonsmothness of the system, the obtained results illustrate a fairly regular pattern coexisting with a wealth of different complex scenarios.

References

Ahmed E, Agiza H.N, 1998, Dynamics of a Cournot Game with n-Competitors, *Chaos, Solitons and Fractals* 9 (9): 1513-17.

Ahmed E, Agiza H.N, Hassan S.Z., 1999, On Modelling Advertisement in Cournot Duopoly, *Chaos, Solitons and Fractals* 10 (7):1179-84.

Ahmed E, Agiza H.N, Hassan S.Z., 2000, On Modifications of Puu's Dynamical Duopoly, *Chaos, Solitons and Fractals* 1 (7), pp. 1025-28.

Agiza HN, 1998, "Explicit Stability Zones for Cournot Game with 3 and 4 Competitors", *Chaos, Solitons and Fractals* 9 (12), pp. 1955-66.

Agiza H.N., Bischi G.I., Kopel M., 1999, Multistability in a dynamic Cournot game with three oligopolists., *Mathematics and Computers in Simulation* 51 (1-2): 63-90.

Agliari A, Gardini L, Puu T., 2000, The Dynamics of a Triopoly Cournot Game, *Chaos, Solitons and Fractals* 11(15): 2531-60.

Bischi, G.I., Stefanini, L., Gardini, L., 1998, Synchronization, Intermittency and Critical Curves in a Duopoly Game, *Mathematics and Computers in Simulation* 44: 559-85.

Bischi GI, Gallegati M, Naimzada A, 1999, Symmetry Breaking Bifurcations and Representative Firm in Dynamic Duopoly Games, *Annals of Operations Research* 89: 253-72.

Dana, R.A., Montrucchio, L., 1986, Dynamic Complexity in Duopoly Games, *Journal of Economic Theory* 40: 40-56.

Friedman, J.W., 1977, *Oligopoly and Theory of Games*, North Holland.

Galvanetto, U., 2000, Numerical computation of Lyapunov exponents in discontinuous maps implicitly defined, *Computer Physics Communication* 131 (1-2):1-9.

Huang W.H., 2001, Theory of Adaptive Adjustment, *Discrete Dynamics in Nature and Society* 5 (4): 247-63.

Keenan, D.C., O'Brien, M.J., 1993, Competition, Collision and Chaos, *Journal of Economic Dynamics and Control* 17: 327-53.

Kopel, M., 1996, Simple and Complex Adjustment Dynamics in Cournot Duopoly Models, *Chaos, Soliton and Fractals* 7(12): 2031-48.

Kunze, M., 2000, On Lyapunov Exponents for Non Smooth Dynamical Systems With an Application to a Pendulum With Dry Friction, *Journal of Dynamics Differential Equations* 12: 31-116 .

Muller, P.C., 1995, Calculation of Lyapunov Exponents for Dynamic Systems with Discontinuities, *Chaos, Solitons and Fractals* 5(9):1671-81.

Puu, T., 1991, Chaos in Duopoly Pricing, *Chaos, Solitons and Fractals* 1: 573-81.

Puu, T., 1997, *Nonlinear Economic Dynamics, Springer*.

Puu, T., 2000, *Attractors, Bifurcations and Chaos, Nonlinear Phenomena in Economics, Springer*.

Rand, D. 1978. Exotic Phenomena in Games and Duopoly Models, *Journal of Mathematical Economics* 5(2): 173-84.

Rassenti, S., Reynolds, S.S., Smith, V.L., Szidarovszky, F., 2000, Adaption and Convergence of Behavior in Repeated Experimental Cournot Game. *Journal of Economic Behavior and Organization* 41: 117-46.

Stefanski, A., Kapitaniak, T., 2000, Using Chaos Synchronization to Estimate the Largest Lyapunov Exponent of Nonsmooth Systems, *Discrete Dynamics in Nature and Society* 4(3): 207-15.

van Witteloostuijn, A. van Lier, A., 1990, Chaotic Patterns in Cournot Competition, *Metroeconomica* 41(2): 161-85.

10 The Interaction of Uncertainty and Information Lags in the Cournot Oligopoly Model

Carl Chiarella[1], *Ferenc Szidarovszky*[2] *and Peiyuan Zhu*[1]

1 School of Finance and Economics, University of Technology, Sydney
2 Systems and Industrial Engineering Department, University of Arizona

1 Introduction

The Cournot oligopoly model is one of the classical models of the economic dynamics literature. The pioneering work of Cournot (1838) initiated a large sequence of studies on static and dynamic models. Okuguchi (1976) discusses single-product models and also gives a comprehensive literature review, particularly of stability conditions. Okuguchi and Szidarovszky (1999) extend the analysis to allow for oligopolies consisting of multi-product firms. Many existing oligopoly models assume perfect knowledge by firms of the market demand functions as well as of the firm's own cost function. While perfect knowledge of the cost functions seems to be a realistic assumption, that of the demand functions may be less so. There have been only some limited attempts at modelling inaccurate knowledge of the demand function in oligopolies. Cyert and DeGroot (1971, 1973) have examined duopoly models. Kirman (1975) considered differentiated products and linear demand functions which firms misspecify and attempt to estimate. The resulting process may converge to the full information equilibrium or to some other equilibria. He also analysed how the resulting equilibrium values are affected (Kirman, 1983). Gates et al. (1982) have considered linear demand functions and differentiated products and assumed some economically guided learning process. Shapiro (1989) has discussed the use of trigger price strategies. Szidarovszky and Okuguchi (1990) have analysed the asympotic stability of oligopolies with perceived marginal costs. Léonard and Nishimura (1999) considered a discrete dynamic duopoly with linear cost functions and illustrated how the fundamental dynamic properties of the model are drastically altered as a result of the lack of full information on the demand functions.

The inaccurate knowledge of the demand function could be modelled by introducing a statistical error function. An alternative approach, and one we adopt here, is to consider a parametric family of demand function and assume that all firms select their perceived demand functions from this family, whose exact parameter values they are unable to assess. Léonard and Nishimura (1999) have used a one-parameter family and assumed that all firms know the general shape of the demand function, but they do not know its scale exactly. This special model with linear cost functions has the mathematical advantage that there is a close relation between the true and believed best responses. Jin (2001) has used a similar idea to analyse the stability of monopolistic competition under bounded rationality. In his analysis firms hold some (usually incorrect) estimate of the slope of the demand function.

In this paper we will embed the Léonard and Nishimura technique for modelling uncertainty about the market demand function within a continuous time-scale model of an n-firm oligopoly. We will show that under a certain concavity assumption the resulting dynamic system is always asymptotically stable, however the equilibrium might differ from the full information equilibrium. We then go on to show that instability may occur when continuously distributed time lags are introduced. Such lags in dynamic economic models have been discussed in general by Invernizzi and Medio (1991), and in the special case of dynamic oligopolies by Chiarella and Khomin (1996), and Chiarella and Szidarovszky (2001). The assumption of the existence of time lags in acting upon information about rivals' output is more realistic than the assumption of instantaneous reaction to information. This paper extends earlier work of Chiarella and Szidarovszky (2001) by focusing on firm differences, particularly with respect to the parameters capturing uncertainty, speeds of output adjustment and time delays.

The plan of the paper is as follows. Section 2 sets out the structure of the n-firm oligopoly and the way in which uncertainty about demand is modelled. Section 3 undertakes a local stability analysis in the situation involving no information lags. Section 4 outlines how information lags are modelled and how the dynamics of firms' output can be modelled as a set of integro-differential equations. The characteristic equation that determines local stability properties is derived. Section 5 reviews some results on local stability obtained by Chiarella and Szidarovszky (2001) in the case of identical firms. Section 6 shows how the set of integro-differential equations for output dynamics can be reduced to a set of ordinary differential equations that become the basis for the numerical simulations of later sections. Section 7 considers a specific reaction function that allows some analysis of the bifurcation behaviour in the case of identical firms. This example is used in the subsequent numerical simulations. Section 8 proposes one systematic way of investigating firm differences, focusing on differences in average information time lags, in speeds of output adjustment and in the parameter for measuring mis-specification of the demand function. The triopoly case is investigated numerically and some results are reported for oligopolies having a larger number of firms. Section 9 concludes and makes suggestions for future research.

1 The Model

2.1 The Oligopolistic Structure. Consider an *n*-firm Cournot oligopoly without product differentiation. Assume the demand function is $q=D(p)$ at each point in time. The cost function of each firm is assumed to be linear, $c_j q_j$ with $c_j > 0$ for $j=1,2,\dots,n$. The profit of firm j in the full information case is given as

$$\Pi_j = q_j f\left(q_j + Q_j\right) - c_j q_j \tag{1}$$

where $f = D^{-1}$ is the inverse demand function and $Q_j = \sum_{l \neq j} q_l$ is the output of the rest of the industry.

The following assumptions are made on the inverse demand function:

(A) f is twice continuously differentiable on R_+ with $f' < 0$.

(B) For all $q>0$ and $Q>0$; $f'(Q) + q f''(Q) < 0$.

(C) $\lim_{Q \to \infty} f(Q) = 0$.

For any $Q_j > 0$, the best response of firm j is obtained by maximizing Π_j. Since the second derivative of Π_j with respect to q_j is $2f'(Q) + q_j f''(Q)$, assumptions (A) and (B) imply that Π_j is strictly concave in q_j and (C) implies that there is a unique non-negative best response of firm j which is characterised by the relations

$$f\left(q_j + Q_j\right) + q_j f'\left(q_j + Q_j\right) - c_j = 0 \tag{2}$$

if $q_j > 0$, and

$$f\left(Q_j\right) - c_j \le 0 \tag{3}$$

for q_j=0. Let $R_j\left(Q_j\right)$ denote the best response of firm *j*. If $R_j\left(Q_j\right) > 0$, then equation (2) is satisfied for $q_j = R_j\left(Q_j\right)$, and simple differentiation shows that

$$R_j'\left(Q_j\right) = -\frac{f' + q_j f''}{2f' + q_j f''} \quad . \tag{4}$$

As a consequence of assumptions (A) and (B), we can assert that $-1 < R_j'\left(Q_j\right) < 0$, which shows that R_j is strictly decreasing in Q_j.

A Nash-equilibrium is a vector $\underline{q}^* = \left(q_1^*, \ldots, q_n^*\right)$ such that for all *j=1,2,...,n,*

$$q_j^* = R_j(Q_j^*)$$

with

$$Q_j^* = \sum_{l \neq j} q_l^* .$$

That is, each component q_j^* of the equilibrium is the best response of firm *j* to the equilibrium output Q_j^* of the rest of the industry.

2.2 Modelling Uncertainty. Assume now that firm j (j=1,2,...,n) believes that the demand function is $q=\varepsilon_j D(p)$ where ε_j is a positive parameter smaller or larger than 1 that captures the extent to which firm j under- or over- estimates demand. Equivalently, firm j believes that the price function is $p=q/\varepsilon_j$, where q is the output of the industry. At each time period each firm observes the market price and, because of the error in the perceived demand function incorrectly infers the output of the rest of the industry. Firm j could pose itself the following question:

What value for $Q_j^e(t)$ should have been used in order to make the demand calculated according to our demand function be consistent with the observed market price p(t)?

The answer would be the quantity $Q_j^*(t)$ satisfying

$$f\left(\frac{1}{\varepsilon_j}\left(q_j(t)+Q_j^*(t)\right)\right)=p(t). \tag{5}$$

In other words

$$Q_j^*(t)=\varepsilon_j D\left(p(t)\right)-q_j(t).$$

Firm *j* would calculate $Q_j^*(t)$ according to (5) by making use of the observed price *p(t)*. But in the market the price is arrived at according to the true demand function i.e.

$$p(t)=f\left(q_j(t)+Q_j(t)\right). \tag{6}$$

Equations (5) and (6) and the fact that *f* is strictly monotonic imply that the value of $Q_j^*(t)$ calculated by firm *j* in fact satisfies

$$Q_j^*(t)=\left(\varepsilon_j-1\right)q_j(t)+\varepsilon_j Q_j(t). \tag{7}$$

The simplest assumption that firm j can make about output of the rest of the industry is to assume that in the next instant of time

$$Q_j^e(t) = Q_j^*(t). \tag{8}$$

Therefore the expected profit of firm j at time period t is given as

$$\bar{\Pi}_j = q_j f\left(\frac{1}{\varepsilon_j}\left(q_j + Q_j^e(t)\right)\right) - c_j q_j, \tag{9}$$

and the best response of this firm in this case is characterised by the relations

$$f\left(\frac{1}{\varepsilon_j}\left(q_j + Q_j^e(t)\right)\right) + q_j f'\left(\frac{1}{\varepsilon_j}\left(q_j + Q_j^e(t)\right)\right)\frac{1}{\varepsilon_j} - c_j, \tag{10}$$

for $q_j > 0$, and

$$f\left(\frac{1}{\varepsilon_j} Q_j^e(t)\right) - c_j \leq 0 \tag{11}$$

for $q_j = 0$. Comparing equations (2) and (3) with (10) and (11) shows that in this case the best response of firm j is given as

$$\varepsilon_j R_j\left(\frac{1}{\varepsilon_j} Q_j^e(t)\right). \tag{12}$$

Notice that for $\varepsilon_j = 1$, this function reduces to the best response of firm j with full information.

3 Stability Analysis

Assuming a continuous time scale and that each firm adjusts its output in the direction of its best response (12) with $Q_j^e(t)$ given by equations (8) and (9), a nonlinear dynamic system for the evolution of outputs is obtained, viz.

$$\dot{q}_j = k_j \left(\varepsilon_j R_j \left(\frac{\varepsilon_j - 1}{\varepsilon_j} q_j + Q_j \right) - q_j \right), \qquad (j = 1, 2, \ldots, n) \quad (13)$$

where, the speed of adjustment of firm j , $k_j > 0$ is a given constant. Notice that for $\varepsilon_j \neq 1$ the equilibrium of this system is usually different from the equilibrium of the oligopoly with full information.

Consider now a positive equilibrium of system (13). The main objective of this section is to prove that this equilibrium is always locally asymptotically stable. The proof is taken from Chiarella and Szidarovszky (2001) and is based on showing that all eigenvalues of the Jacobian at the equilibrium have negative real parts.

Proposition I . *The equilibrium of the differential equation system (13) is always locally asymptotically stable.*

Proof. Simple differentiation shows that the Jacobian at the equilibrium has the form:

$$\underline{J} = \begin{pmatrix} k_1\left((\varepsilon_1 - 1)R_1' - 1\right) & k_1\varepsilon_1 R_1' & \cdots & k_1\varepsilon_1 R_1' \\ k_2\varepsilon_2 R_2' & k_2\left((\varepsilon_2 - 1)R_2' - 1\right) & \cdots & k_2\varepsilon_2 R_2' \\ \vdots & \vdots & \ddots & \vdots \\ k_n\varepsilon_n R_n' & k_n\varepsilon_n R_n' & \cdots & k_n\left((\varepsilon_n - 1)R_n' - 1\right) \end{pmatrix}$$

$$= \underline{D} + \underline{a}\,\underline{1}^T,$$

where $\underline{1}^T = (1, 1, \ldots, 1)$,

$$\underline{D} = diag\left(-k_1\left(R_1' + 1\right), \ldots, -k_n\left(R_n' + 1\right)\right), \text{ and}$$

$$\underline{a} = \left(k_1\varepsilon_1 R_1', k_2\varepsilon_2 R_2', \cdots, k_n\varepsilon_n R_n'\right)^T.$$

The characteristic polynomial of $\underline{J}$ has the form[1]

$$\varphi(\lambda)=det\left(\underline{D}+\underline{a}\,\underline{1}^T-\lambda\underline{I}\right)=det\left(\underline{D}-\lambda\underline{I}\right)det\left(\underline{I}+\left(\underline{D}-\lambda\underline{I}\right)^{-1}\underline{a}\,\underline{1}^T\right)$$
$$=\prod_{j=1}^{n}\left(-k_j\left(R_j'+1\right)-\lambda\right)\cdot\left[1+\sum_{j=1}^{n}\frac{k_j\varepsilon_jR_j'}{-k_j\left(R_j'+1\right)-\lambda}\right]. \tag{14}$$

Set $\gamma_j=-k_j\left(R_j'+1\right)$ and order these values such that $\gamma_1<\gamma_2<...<\gamma_r$. For $k=1,2,...,r$, let I_k denote the set of firms such that $-k_j\left(R_j'+1\right)=\gamma_k$. If set I_k has m_k elements, then

$$\varphi(\lambda)=\prod_{k=1}^{r}(\gamma_k-\lambda)^{m_k}\left[1+\sum_{k=1}^{r}\frac{\delta_k}{\gamma_k-\lambda}\right], \tag{15}$$

where $\delta_k=\sum_{j\in I_k}k_j\varepsilon_jR_j'$. From relation (4) we see that both γ_k and δ_k are negative for all k.

If $m_k=1$, then γ_k is not an eigenvalue. If $m_k\geq 2$, then γ_k is an eigenvalue with multiplicity m_k-1. The remaining eigenvalues are the solutions of the equation

$$g(\lambda)=\sum_{k=1}^{r}\frac{\delta_k}{\gamma_k-\lambda}=-1,$$

which is equivalent to a polynomial equation of degree r. Clearly,

$$\lim_{\lambda\to\pm\infty}g(\lambda)=0,\qquad \lim_{\lambda\to\gamma_k\pm0}g(\lambda)=\pm\infty,$$

and

$$g'(\lambda)=\sum_{k=1}^{r}\frac{\delta_k}{\left(\gamma_k-\lambda\right)^2}<0,$$

implying that g is locally strictly decreasing. The graph of $g(\lambda)$ is shown in Figure 1. There is a root before γ_1, and one root between

[1] We have used the well-known fact that for all n-element real vectors $\underline{u}$ and $\underline{v}$, $det\left(\underline{I}+\underline{u}\,\underline{v}^T\right)=1+\underline{v}^T\underline{u}$.

each pair γ_{k-1}, γ_k of poles (k=2,3,...,r). Therefore there are r real negative roots, so all eigenvalues of $\underline{J}$ are real and negative. Hence the equilibrium is locally asymptotically stable.

The equilibrium of an oligopoly with full information is also locally asymptotically stable by selecting $\varepsilon_1 = \varepsilon_2 = ... = \varepsilon_n = 1$. This observation shows that the asymptotic behaviour of the equilibrium does not change by introducing inaccurate knowledge of the demand functions. As we will show in the following sections, instability may be brought about by the introduction of information lags.

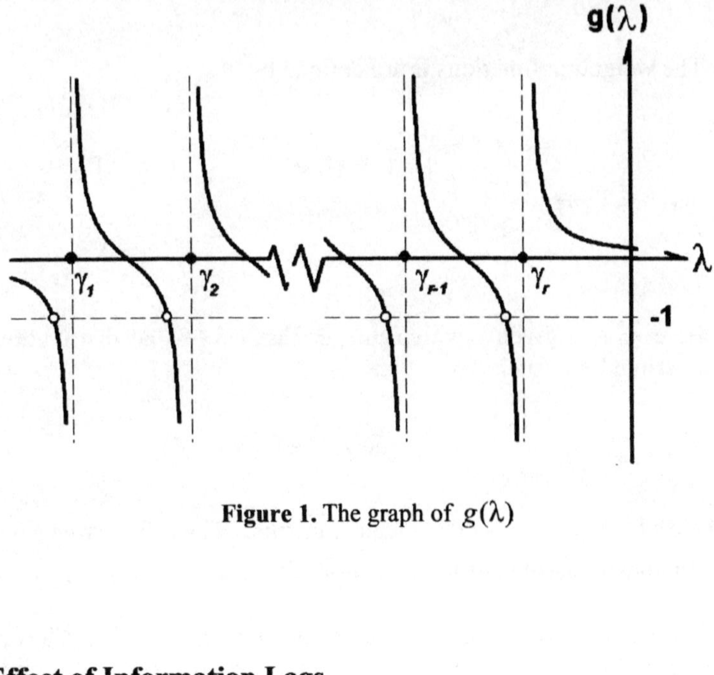

Figure 1. The graph of $g(\lambda)$

4 Effect of Information Lags

4.1 The Information Lag Structure. In this section we assume that firms may not react only to the market price just immediately observed, but rather may use a history of past prices in trying to infer the output of the rest of the industry in the next instant of time. We model this effect by using continuously distributed time lags. This type of lag was introduced into oligopoly models and analysed as discrete lags by Russel, Rickard and Howroyd (1986) and then put into a more general continuously distributed time lag framework by Chiarella and Khomin (1996) and Chiarella and Szidarovszky (2000). Instead of assuming that firm j uses equation (8) to infer the output of the rest of the industry, we assume that firm j uses a continuously weighted average of past $Q_j^*(t)$ (which depend on past prices and past values of its own output) according to

$$Q_j^e(t) = \int_0^t w(t-s, T_j, m_j) Q_j^*(s)\, ds .$$

The weighting functions w are defined by[2]

$$w(t-s,T,m) = \begin{cases} \dfrac{1}{m!}\left(\dfrac{m}{T}\right)^{m+1} (t-s)^m e^{-\frac{m(t-s)}{T}} & \text{if } m \geq 1 \\ \dfrac{1}{T} e^{-\frac{(t-s)}{T}} & \text{if } m = 0. \end{cases} \tag{16}$$

Here m is a given positive integer that causes the distribution to peak around T in such a way that

$$\lim_{m \to \infty} w(t,T,m) = \delta(t-T), \tag{17}$$

where $\delta(\bullet)$ is the Dirac delta function. Thus for large m the continuously distributed lag acts more like a discrete time lag at time T.

The positive parameter T measures the average time lag. Thus as T becomes smaller, more weight is placed on the most recent observations and the limiting property is

$$\lim_{T \to 0} w(t,T,m) = \delta(t). \tag{18}$$

---

[2] The weighting functions (16) were also used by Chiarella and Khomin (1996) and Chiarella and Szidarovszky (2001).

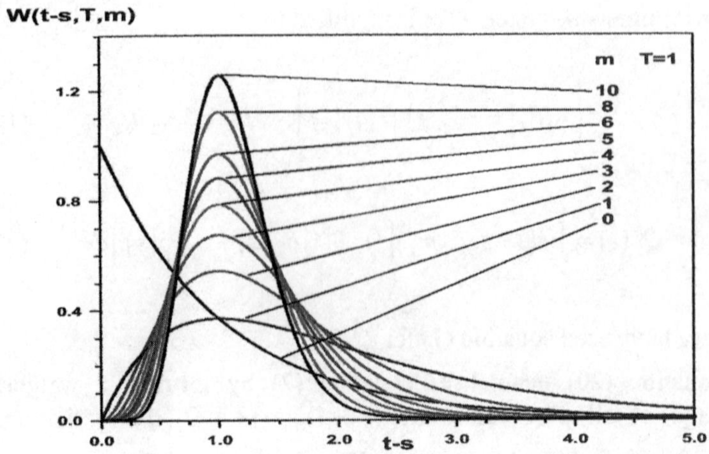

Figure 2a. The weighting function for varying m

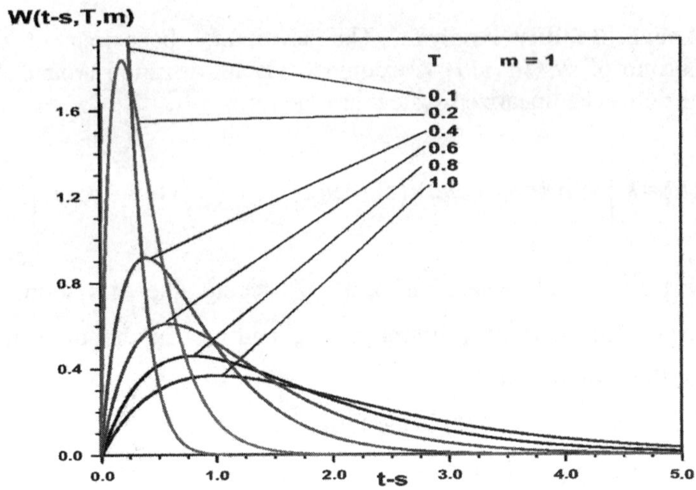

Figure 2b. The weighting function for varying T

Figure 2a (2b) illustrates the limiting behaviour $m \to \infty (T \to 0)$. In this framework system (13) is modified to

$$\dot{q}_j = k_j \left(\varepsilon_j R_j \left(\frac{1}{\varepsilon_j} Q_j^e(t) \right) - q_j(t) \right), \quad j = 1, 2, \ldots, n, \tag{19}$$

where

$$Q_j^e(t) = \int_0^t w(t-s, T_j, m_j) \left[(\varepsilon_j - 1) q_j(s) + \varepsilon_j Q_j(s) \right] ds \tag{20}$$

and we have used equation (7) for $Q_j^*(t)$.

Equation (20) generalises equation (7) by forming a weighted average of all past values of $(\varepsilon_j - 1) q_j(\bullet) + \varepsilon_j Q_j(\bullet)$; in the limit $T_j \to 0$ equation (7) is recovered. Thus in the limiting case $T_j \to 0$, for all j, the system (19) and (20) becomes the system (13) because of property (18).

4.2 Local Stability Analysis. The asymptotic behaviour of the equilibrium of system (19) is examined via linearization around the equilibrium. The linearized system has the form

$$\dot{q}_{j\delta}(t) = k_j \left[\gamma_j \int_0^t w(t-s, T_j, m_j) \left((\varepsilon_j - 1) q_{j\delta}(s) + \varepsilon_j Q_{j\delta}(s) \right) ds - q_{j\delta}(t) \right] \tag{21}$$

for j=1,2,...n and where $q_{j\delta}$ and $Q_{j\delta}$ denote the deviation of q_j and Q_j from their equilibrium levels, and γ_j is the derivative of R_j at the equilibrium.

We seek solutions of (21) in the form

$$q_{j\delta}(t) = v_j e^{\lambda t} \qquad (j = 1, 2, \ldots, n). \tag{22}$$

Substituting this form into equation (21) and letting $t \to \infty$, we have (for $j = 1, 2, \ldots, n$.)

$$\left[\lambda - k_j\gamma_j\left(\varepsilon_j - 1\right)\int_0^\infty w\left(s,T_j,m_j\right)e^{-\lambda s}ds + k_j\right]v_j$$
$$-\left[k_j\gamma_j\varepsilon_j\int_0^\infty w\left(s,T_j,m_j\right)e^{-\lambda s}ds\right]\sum_{l\neq j} v_l = 0, \tag{23}$$

Introduce the notation

$$A_j\left(\lambda\right) = \lambda - k_j\gamma_j\left(\varepsilon_j - 1\right)\left(\frac{\lambda T_j}{r_j} + 1\right)^{-(m_j+1)} + k_j$$

with

$$r_j = \begin{cases} m_j & \text{if } m_j > 0, \\ 1 & \text{if } m_j = 0. \end{cases}$$

Furthermore let

$$B_j(\lambda) = -k_j\gamma_j\varepsilon_j\left(\frac{\lambda T_j}{r_j} + 1\right)^{-(m_j+1)}.$$

Then equation (23) can be simplified to

$$A_j(\lambda)v_j + B_j(\lambda)\sum_{l\neq j} v_l = 0, \qquad j = 1,2,\ldots,n, \tag{24}$$

which is equivalent to the determinantal equation

$$\det\begin{pmatrix} A_1(\lambda) & B_1(\lambda) & \cdots & B_1(\lambda) \\ B_2(\lambda) & A_2(\lambda) & \cdots & B_2(\lambda) \\ \vdots & \vdots & \ddots & \vdots \\ B_n(\lambda) & B_n(\lambda) & \cdots & A_n(\lambda) \end{pmatrix} = 0. \tag{25}$$

Notice that the structure of this matrix is the same as that of the Jacobian $\underline{J}$ introduced in section 3, so we can use the same method to obtain the equivalent closed form expression:

$$P_n\left(\lambda\right) \equiv \prod_{j=1}^{n}\left(A_j\left(\lambda\right) - B_j\left(\lambda\right)\right)\left[1 + \sum_{j=1}^{n}\frac{B_j\left(\lambda\right)}{A_j\left(\lambda\right) - B_j\left(\lambda\right)}\right] = 0. \tag{26}$$

The solution of this equation in the general case can be performed in a similar manner to that shown in Chiarella and Szidarovszky

(2001). However unlike the no-time lag case of section 4, no general definite results can be obtained. Therefore instead of a general analysis, some special cases will be examined.

5 The Symmetric Case

In this section we seek to gain some insights into the stability properties of the model by focusing on the case of identical firms. Thus we assume that $c_i = c$, $\varepsilon_i = \varepsilon$, $T_i = T$, $m_i = m$, $l_i = l$, and $k_i = k$ for $i=1,2,\ldots,n$. Then $R_1 = R_2 = \ldots = R_n = R$ and therefore $\gamma_1 = \gamma_2 = \ldots = \gamma_n = \gamma$ at a symmetric equilibrium. In this case

$$A_1(\lambda) = \ldots = A_n(\lambda) = \lambda - k\gamma(\varepsilon - 1)\left(\frac{\lambda T}{r} + 1\right)^{-(m+1)} + k,$$

and

$$B_1(\lambda) = \ldots = B_n(\lambda) = -k\gamma\varepsilon\left(\frac{\lambda T}{r} + 1\right)^{-(m+1)},$$

so that equation (24) reduces to

$$(\lambda + k)\left(\frac{\lambda T}{r} + 1\right)^{m+1} - k\gamma(n\varepsilon - 1) = 0. \tag{27}$$

Consider first the case when m=0.

Then r=1 and we have a quadratic equation for λ:

$$\lambda^2 T + \lambda(1 + kT) + k\left(1 - \gamma(n\varepsilon - 1)\right) = 0. \tag{28}$$

Since k, T, n, ε are positive and $\gamma \in (-1,0)$, both $1+kT$ and $1 - \gamma(n\varepsilon - 1) = (1+\gamma) - \gamma n\varepsilon$ are positive showing the local asymptotic stability of the equilibrium.

Consider next the case when m=1.

Then equation (27) is the cubic equation

$$(\lambda + k)(\lambda^2 T^2 + 2\lambda T + 1) - k\gamma(n\varepsilon - 1) = 0. \tag{29}$$

Limit cycles may be born if there is a pure complex eigenvalue $\lambda = i\alpha$. Substituting $\lambda = i\alpha$ into the cubic equation and equating the real and imaginary parts we see that

$$\alpha^2 = \frac{k\left[1-(n\varepsilon-1)\gamma\right]}{T(2+kT)} = \frac{2kT+1}{T^2},$$

so pure complex roots exist if and only if

$$\gamma^* = -\frac{2}{n\varepsilon-1}\left[\frac{1}{kT}+kT+2\right]. \tag{30}$$

Figure 3 shows the value of γ as a function of kT, when $n\varepsilon - 1 > 0$. This is a realistic assumption, since typically one would expect $\varepsilon \approx 1$.

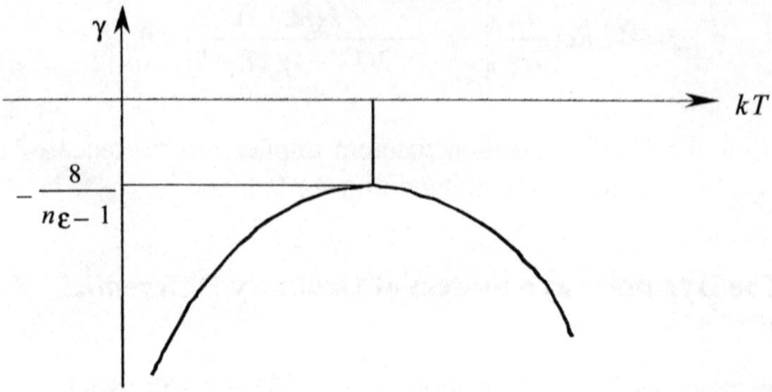

Figure 3. The graph of γ as the function of kT

Notice that the value of γ has to be between -1 and 0 (see equation (4)), so realistic γ values can be obtained only if $-\frac{8}{n\varepsilon-1} > -1$ which holds if and only if

$$n > \frac{9}{\varepsilon}. \tag{31}$$

The result (31) is interesting in that it shows that, at least in the case of identical firms, the minimum size oligopoly for which limit cycle behaviour is possible increases (decreases) with the degree to which demand is under (over)-estimated.

The existence of a pure complex root alone does not guarantee the existence of limit cycles. In applying the Hopf bifurcation theorem (see for example, Guckenheimer and Holmes, 1983) we select γ as the bifurcation parameter and consider the eigenvalue λ as a function of γ. Differentiating equation (29) with respect to γ gives the derivative

$$\frac{d\lambda}{d\gamma}=\frac{k(n\varepsilon-1)}{3T^2\lambda^2+2T\lambda(2+kT)+(2Tk+1)}.$$

The real part of this derivative at the pure complex root $\lambda = i\alpha$ is

$$Re\left(\frac{d\lambda}{d\gamma}\right)\Bigg|_{\gamma=\gamma^*}=-\frac{k(n\varepsilon-1)}{2(kT+1)(kT+5)}<0.$$

Thus the Hopf bifurcation theorem implies the existence of limit cycles for γ in the neighbourhood of $\gamma=\gamma^*$.

6 The Dynamics as a System of Ordinary Differential Equations

The integro-differential equation system (19) and (20) may be written as the system of ordinary differential equations. To this end we note first that $Q_j^e(t)$ in equation (20) may be written

$$Q_j^e(t)=(\varepsilon_j-1)q_j^{(m_j,m_j)}(t)+\varepsilon_j\sum_{i\neq j}q_i^{(m_j,m_j)}(t)$$

where for each firm j we need to define the set of subsidiary quantities (for $\nu_j = j,i$)

$$q_{\nu_j}^{(l_j,m_j)}(t)=\int_0^t\frac{1}{l_j!}\left(\frac{r_j}{T_j}\right)^{l_j+1}(t-s)^{l_j}e^{-\frac{r_j}{T_j}(t-s)}q_{\nu_j}(s)ds \tag{32}$$

where

$$r_j=\begin{cases} m_j, & m_j\geq 1\\ 1, & m_j=0\end{cases}$$

and $l_j = m_j, m_j-1,\cdots,0$. When $\nu_j = j(i)$ the jth firm is taking weighted averages of its own (rivals') output.

Using the results in Appendix 1 we have that for each firm j the integro-differential equation system (19), (20) is replaced by the set of ordinary differential equations

$$\dot{q}_j(t)=k_j\left(\varepsilon_j R\left(\frac{\varepsilon_j-1}{\varepsilon_j}q_j^{(m_j,m_j)}(t)+\sum_{i\neq j}q_i^{(m_j,m_j)}(t)\right)-q_j(t)\right), \quad (33)$$

$$\dot{q}_j^{(l_j,m_j)}(t)=\frac{r_j}{T_j}\left(q_j^{(l_j-1,m_j)}(t)-q_j^{(l_j,m_j)}(t)\right), \quad \left(\text{for } l_j=m_j,m_j-1,\cdots,1\right), \quad (34)$$

$$\dot{q}_j^{(0,m_j)}(t)=\frac{r_j}{T_j}\left(q_j(t)-q_j^{(0,m_j)}(t)\right), \quad (35)$$

$$\dot{q}_i^{(l_j,m_j)}=\frac{r_j}{T_j}\left(q_i^{(l_j-1,m_j)}(t)-q_i^{(l_j,m_j)}(t)\right), \quad \left(\text{for } l_j=m_j,m_j-1,\cdots,1\right), \quad (36)$$

$$\dot{q}_i^{(0,m_j)}=\frac{r_j}{T_j}\left(q_i(t)-q_i^{(0,m_j)}(t)\right). \quad (37)$$

For the last two equations (36), (37) the subscript i takes values in the set $\{1,2,\cdots,n;\ i\neq j\}$.

Thus for firm j we have a set of $n\ m_j+n+1$ ordinary differential equations[3], and so for the system of n firms we have in all a set of $n(n+1)+n\sum_{j=1}^{n}m_j$ ordinary differential equations.

Finally we observe from equation (32) that $q_{\nu_j}^{(l_j,m_j)}(0)=0$ for $\nu_j=j$ and i and $l_j=m_j,m_j-1,\cdots,0$. In the numerical simulations

---

[3] This number is arrived at as follows:- 1 equation for the adjustment of $q_j(t)$; (m_j+1) equations for the subsidiary quantities $q_j^{(l_j,m_j)}(t)$ where firm j averages over its own past outputs; $(n-1)(m_j+1)$ equations for the subsidiary quantities $q_i^{(l_j,m_j)}(t)$ where firm j averages over the outputs of the *(n-1)* rival firms.

discussed in later sections we take the initial values $q_j(0)$ in the neighbourhood of the steady state of the economy.

7 A Specific Example

We generate a specific numerical example by considering directly the reaction function for firm j given by

$$R_j(Q_j)=\begin{cases}0 & \text{if } Q_j \geq A/c_j \\ \mu_j\left(\sqrt{\dfrac{AQ_j}{c_j}}-Q_j\right) & \text{otherwise}\end{cases}$$

which is illustrated in Figure 4. Here $A(>0)$ and $c_j(>0)$ are constants.

This choice of reaction function is in fact based on a hyperbolic inverse demand function $f(Q)=A/Q$, with a linear function cost function c_jQ, but the parameter μ_j is incorporated to allow the height of the reaction function to be varied without changing the intercept on the Q axis. The μ_j could be thought of as arising from the introduction of some positive externality into the cost function, as discussed for example by Kopel (1996). Note that when $\mu_j=1$ we obtain the reaction function used by Chiarella and Khomin (1996).

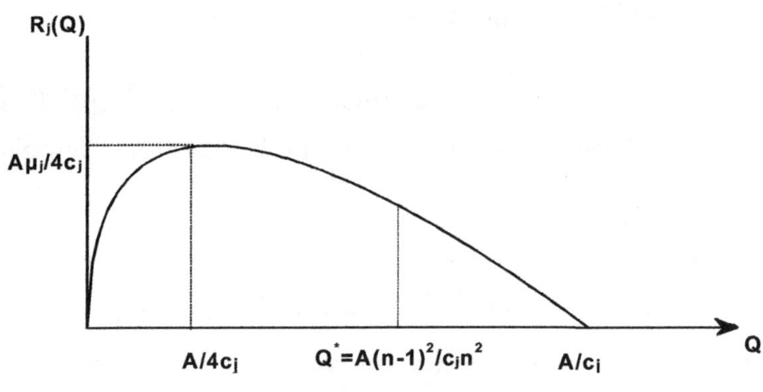

Figure 4. Best Response of Firm j

At a symmetric equilibrium $Q_j = (n-1)q$ and $Q = nq$. The symmetric equilibrium with incorrect knowledge of the demand function is the solution of the equation

$$q = \varepsilon R\left(\frac{\varepsilon-1}{\varepsilon}q + (n-1)q\right) = \varepsilon R\left(\frac{(n\varepsilon-1)q}{\varepsilon}\right),$$

which follows from equation (13). In our case this equation has the particular form

$$q = \varepsilon\mu\left(\sqrt{\frac{A(n\varepsilon-1)q}{c\varepsilon}} - \frac{(n\varepsilon-1)q}{\varepsilon}\right).$$

The unique solution is given by

$$q(\varepsilon) = \frac{A(\varepsilon n-1)\mu^2\varepsilon}{c(\mu\varepsilon n-\mu+1)^2}.$$

The bifurcation parameter is

$$\gamma^*(\varepsilon) = R'(Q) = \sqrt{\frac{A}{c}}\frac{\mu}{2\sqrt{Q}} - \mu$$

where we have used equation (7) and (8) to find $Q_j^e(t)$ at the equilibrium. So

$$\gamma^*(\varepsilon) = \frac{\mu+1-\mu n\varepsilon}{2(n\varepsilon-1)} \tag{38}$$

For the case in which m=1 we obtain the value of ε at which the Hopf bifurcation occurs by equating $\gamma^*(\varepsilon)$ in equation (38) with critical value given by equation (30). This critical value turns out to be

$$\varepsilon = \frac{4(1/kT + kT)+9}{\mu n} + \frac{1}{n}. \tag{39}$$

We have simulated the system (33)-(37) in the symmetric case for n=3, T=1, k=3, m=1, A=1, c=1 and μ=2. Figure 5a shows a bifurcation diagram of q for a range of values of ε, with limit cycle motion indeed being born at around ε=4.06 as predicted by equation (39). The bifurcation diagram also shows that for this example the limit cycle persists for a wide range of ε. For comparison we have calculated the corresponding bifurcation diagram for the case m=50.

In the case m=1, limit cycle motion occurs at an unrealistically high value of uncertainty about the demand function. The corresponding bifurcation value close to 1 in the case m=50 is more realistic as we would expect firms to not be over-estimating or under-estimating demand by too much. We recall that the higher value of m indicates that the weighting fuction is more tightly distributed around the average information time lag, in this case T=1.

In order to illustrate the effect of the cost externality factor μ we display in figure 5b the bifurcation diagrams for the cases $\mu=1$, $\mu=2$ and $\mu=3$. This sequence of diagrams indicates how the increase in μ lowers the critical value of ε at which bifurcation occurs. Furthermore higher values μ of seem to be associated with greater complexity.

The calculations of this section give us some insight into the bifurcation behaviour of the *n* firm oligopoly. In subsequent sections we shall see how these insights carry over to the situation of non-identical firm.

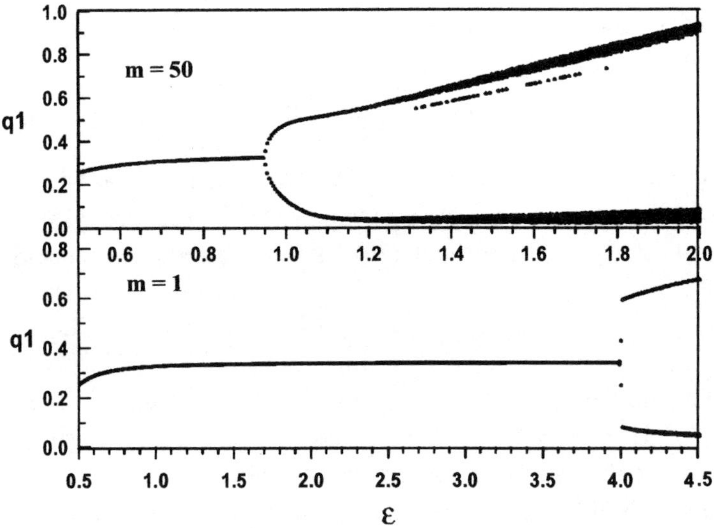

Figure 5a. Bifurcation diagrams as functon of ε for symmetric firm case - here $\mu = 2$ and m=1, m=50

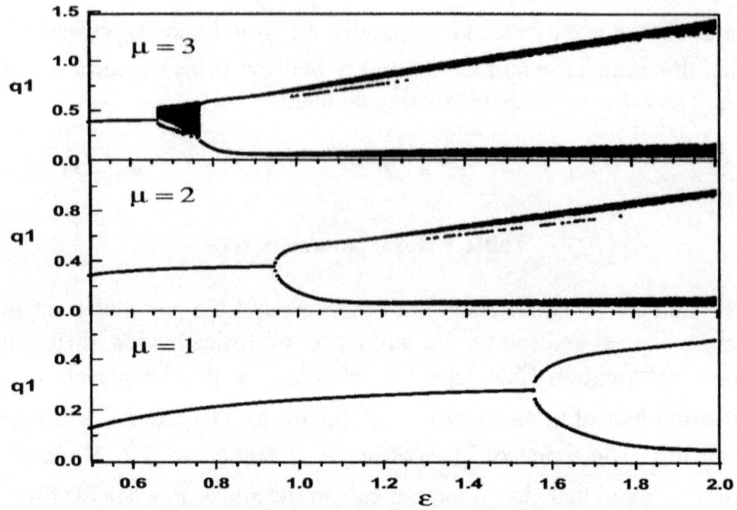

Figure 5b. Bifurcation diagrams as functon of ε for svmmetric firm case – different u and m=50

8 Dynamic Behaviour as a Function of Firm Differences

Firm differences can be modelled in a wide variety of ways. Here we have chosen to focus on the differences brought about by differences in the average time lags T_j, the speeds of output adjustment k_j and the degree of demand function uncertainty ε_j. We allow for differences in T_j according to

$$T_1 = T, \qquad T_j = \beta_T^{(j-1)} T \qquad (j = 2, \cdots, n)$$

By varying the β_T parameter we can allow for a wide degree of average time lag differences amongst the firms.

Similarly we allow for differences in k_j according to

$$k_1 = k, \qquad k_j = \beta_k^{(j-1)} k \qquad (j = 2, \cdots, n)$$

and letting β_k vary.

The differences in ε are calculated by specifying a central value ε_0 and a range $(\varepsilon_0 - \varepsilon^*, \varepsilon_0 + \varepsilon^*)$ within which the values of the ε of the n firms are evenly spread. Typically we would take ε_0 close to 1 so that this scheme would have roughly half the firms overestimating demand and the rest underestimating demand.

$k = 3$	$T = 1$	$\varepsilon_0 = 1.2$	$A/c = 1$	$m_j = 50, all$ j

Table 1 Basic parameter set

The numerical experiments are based around the parameter set in table 1. Figure 6 displays the influence of firms having different average information time lags via variation in β_T. Displayed is a bifurcation plot of q_1 as a function of β_T for n=3 (figure 6a) and n=9 (figure 6b). The value of μ was set to 2 and ε to 1.2. However keeping in mind that $\beta_T = 1$ corresponds to the situation when all firms have identical information lags, we see that increasing β_T does lead to greater complexity. When contrasting figures 6a and 6b we see that increasing the number of firms brings about greater complexity. Figure 7 displays the phase plot for the n=3 firm case at the value β_T =2. As suggested by figure 6a, the motion is rather complex, but whether it is high order cycles or chaotic would require further numerical investigation.

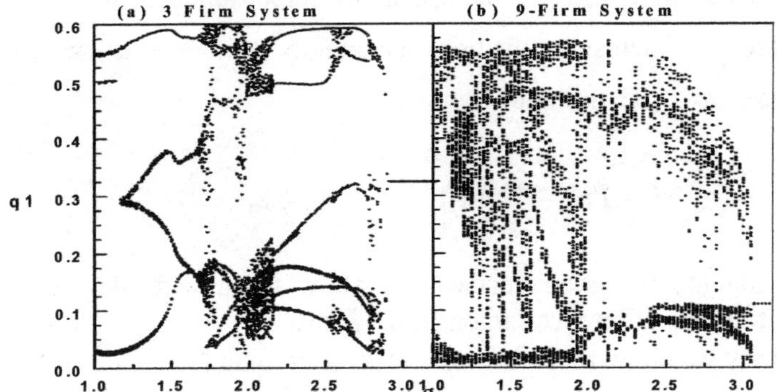

Figure 6. Bifurcation diagrams as functon of β_T, capturing differences in information time lags

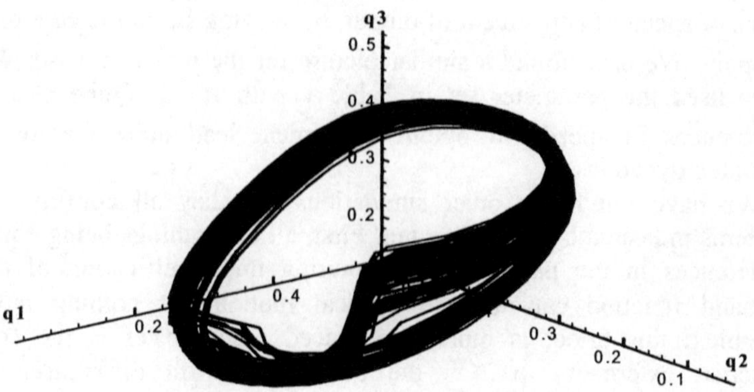

Figure 7. Phase plot for the triopoly when $\beta_T = 2$

Figure 8 considers a triopoly and displays a bifurcation diagram for q_1 as a function of ε^* at different values of μ. We have found very little change in this figure as the number of firms is increased. We see that the dynamics move from stability to cycles of various orders as μ increases. In order to focus on the behaviour of trajectories we choose the value ε^*=0.2, though figure 8 suggests that at a given μ the qualitative dynamic behaviour is relatively independent of ε^*. Figure 9a displays the phase plot for the n=3 firm case, and figure 9b for the n=9 firm case. In these calculations, we set μ=3, ε_0=1 and ε^*=0.2. In each case the axes chosen represent the firm with ε closest to the central value ε_0 (q_2 in figure 9a; q_5 in figure 9b) and the firms with ε furthest from the central value ($q_{1,}q_3$ in figure 9a; $q_{1,}q_9$ in figure 9b). The phase plots show that the firms that have the largest mis-specification of the demand fuction exhibit the largest output fluctuations, and this seems independent of the number of firms.

Figure 10 displays the effect of increasing the differences amongst firms of speed of adjustment of output, by varying β_k in the case of a triopoly. We have found a similar picture for the n=9 firm case. We have used the parameter set in Table 1 with $\mu = 2$. Quite clearly differences in speeds of output adjustment lead quite readily to complex dynamics.

We have run many other simulations but they all confirm the patterns indicated by the foregoing. First, all other things being equal differences in the parameter ε capturing mis-specification of the demand function can lead to cyclical motion but nothing more complex; this effect is more pronounced the greater is the cost function externality μ. On the other hand firm differences in information lags and speeds of output adjustment do seem to play a role in bringing about complex dynamics. Furthermore in all situations increasing the cost function externality μ increases the likelihood of complexity. Lack of time and space prevented us from investigating firm differences in μ.

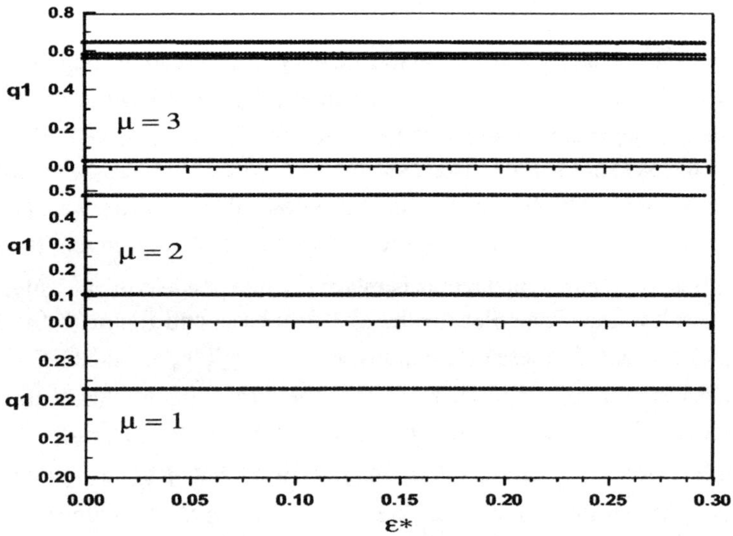

Figure 8. Bifurcation diagrams as a function of ε^* - different values of μ

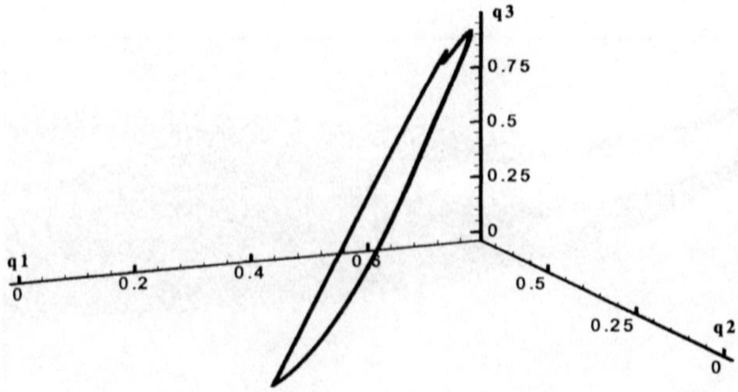

Figure 9a. Phase plot for the 3 firm case with different ε

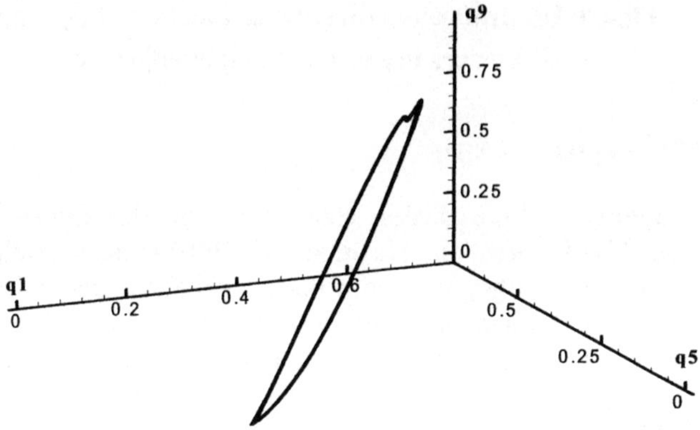

Figure 9b. Phase plot for the 9 firm case with different ε

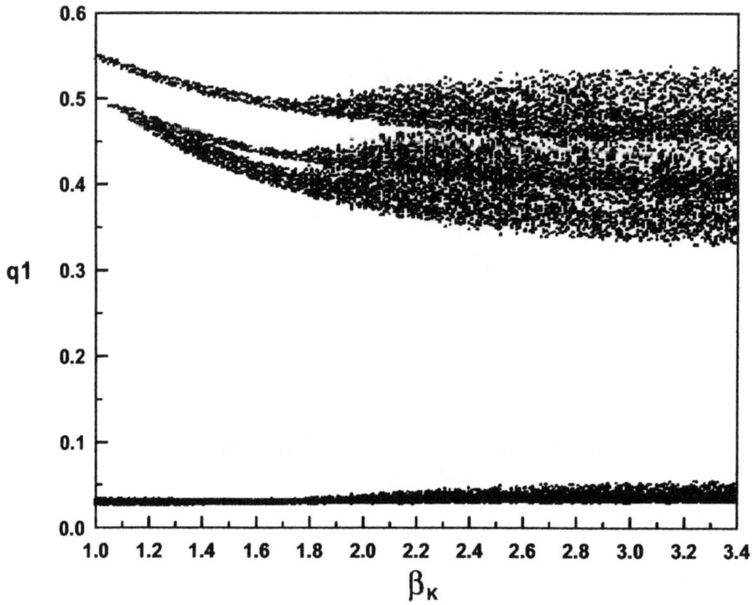

Figure 10. Bifurcation diagram as function of β_k, capturing differences in speeds of output adjustment

9 Conclusions

The continuous time scales counterpart of the dynamic model introduced by Léonard and Nishimura (1999) has been studied in this paper for *n*-firm oligopolies with linear cost functions and imperfect knowledge of the demand functions. In the full information case as well as in the case of imperfectly assessed demand functions the equilibrium is always locally asymptotically stable. That is, if the initial outputs are close to the equilibrium, then the outputs converge to the equilibrium values as $t \to \infty$. If there is a continuously distributed time lag in implementing information on the output of the rivals (as inferred from past market prices), then the interplay between these time lags, speeds of output adjustment and the imperfect knowledge of the demand functions can cause local instability of the steady state equilibrium to occur. There then arises the possibility of some sort of stable cyclical attractor with self-sustaining fluctuations of output as the equilibrium outcome of the oligopoly. This possibility will depend on the eigenvalue structure of the Jacobian matrix of the system dynamics for which we obtained an expression in the general case (equation (26)), however it seems impossible to obtain any analytical results in general. We adopted the strategy of considering the special case of the symmetric oligopoly to obtain some analytical results. We have considered via use of the Hopf bifurcation theorem how the change from local stability to local instability can give rise to the birth of limit cycle attractors for output. For one particular lag structure we obtained some insight into the relation between the size of the oligopoly (n), the mean time lag (T) and the degree of demand function uncertainty (ε) that can lead to the birth of limit cycle dynamics for output.

Our analytical results were local and in order to obtain some feel for the type of global dynamics for output that is possible in such oligopoly models we considered via numerical simulations a fairly general reaction function that also allows for externalities in the cost function. For this example at least, and treating the degree of demand function uncertainty as a bifurcation parameter, the result of a limit cycle for output implied by the local analysis seems to be a global result.

We focused on differences with respect to information time lags, degree of demand function uncertainty and speeds of output adjustment. Broadly speaking we have found via numerical simulations that differences in information lags and speeds of ouptut adjustment are most likely to result in complex dynamics. Differences in the degree of demand function uncertainity lead at most to low order cyclical motion. The cost function externality also has an influence in bringing about complexity. The results of Puu (2000) for the case of discrete-time duopolies and triopolies would suggest that firm differences should lead to far more complex output dynamics than we have been able to generate. This may be due to the fact that we have only allowed firms to differ in certain ways. We leave for future research the task of allowing the firms to differ in other ways.

Rational firms would soon realize that outputs evolve cyclically, and therefore they would revise their beliefs with some appropriate learning process. We leave this task to future research.

Here we have considered a very particular way of modelling the firms' uncertainty about the inverse demand function, by using the multiplicative parameter ε. It would be of interest to consider a proper stochastic modelling using stochastic differential equations. This difficult task we also leave to future research.

References

Chiarella, C. and A. Khomin, 1996, "An Analysis of the Complex Dynamic Behaviour of Nonlinear Oligopoly Models with Time Lags", *Chaos, Solitons of Fractals,* Vol. 7, No. 2, 2049-2065

Chiarella, C. and F. Szidarovszky, 2001, "The Birth of Limit Cycles in Non-linear Oligopolies with Continuously Distributed Information Lags", In *Modeling Uncertainty* (M. Dror editor), Kluwer Academic Publishers, Dordrecht

Cournot, A., 1938, *Recherches sur les Principes Mathèmatiques de la Thèorie de Richesses,* Hachette, Paris (English translation (1960), *Researches into the Mathematical Principles of the Theory of Wealth,* Kelley, New York)

Cyert, R. M. and M. H. DeGroot, 1971, "Interfirm Learning and the Kinked Demand Curve", *Journal of Economic Theory,* Vol. 3

Cyert, R. M. and M. H. DeGroot, 1973, "An Analysis of Cooperation and Learning in a Duopoly Context", *American Economic Review,* Vol. 63

Gates, D. J., J. A. Rickard and M. Westcott, 1982, "Exact Cooperative Solutions of a Duopoly Model without Cooperation", *Journal of Mathematical Economics,* Vol. 9, 27-35

Guckenheimer, J. and P. Holmes, 1983, *Nonlinear Oscillations, Dynamical Systems and Bifurcations of Vector Fields,* Springer - Verlag, New York

Invernizzi S. and A. Medio, 1991, "On Lags and Chaos in Economic Models", *Journal Mathematical Economics,* Vol. 20, 521-550

Jin, J. Y., 2001, "Monopolistic Competition and Bounded Rationality", *Journal of Economic Behaviour and Organization,* Vol. 45, 175-184

Kirman, A., 1975, "Learning by Firms about Demand Conditions", In: *Adaptive Economic Models* (R. H. Day and T. Groves, eds.), Academic Press, New York

Kirman, A., 1983, "On Mistaken Beliefs and Resultant Equilibria", In: *Individual Forecasting and Aggregate Outcomes* (R. Frydman and E. S. Phelps, eds.), Cambridge University Press, New York

Kopel, M., 1996, "Simple and Complex Adjustment Dynamics in Cournot Duopoly Models", *Chaos, Solitions and Fractals,* Vol. 7, No. 12, 2031-2048

Léonard D., and K. Nishimura, 1999, "Nonlinear Dynamics in the Cournot Model without Full Information", *Annals of Operations Research,* Vol. 89, 165-173

Okuguchi, K., 1976, *Expectations and Stability in Oligopoly Models.* Springer-Verlag, Berlin/ Heidelberg/ New York

Okuguchi, K. and F. Szidarovszky, 1999, *The Theory of Oligopoly with Multi-Product Firms (2nd edition),* Springer-Verlag, Berlin/ Heidelberg/ New York

Puu, T., 2000, *Attractors, Bifurcations, and Chaos: Nonlinear Phenomena in Economics.* Springer-Verlag, Berlin/ Heidelberg/ New York

Russel, A.M., J. Rickard and T. D. Howroyd, 1986, "The Effects of Delays on the Stability and Rate of Convergence to Equilibrium of Oligopolies", *Economic Record,* Vol. 62, 194-198

Shapiro, C., 1989, "Theories of Oligopoly Behavior", In: *Handbook of Industrial Organization* (R. Schmalensee and R. D. Willig, eds.), North Holland, Amsterdam

Szidarovszky, F. and K. Okuguchi, 1990, "Dynamic Oligopoly: Models with Incomplete Information", *Applied Mathematics and Computation,* Vol. 38, No. 2, 161-177

Appendix 1. Deriving the General Differential Equation System For the Case of *n* Different Firms

Consider the jth firm and suppose that the polynomial power in the weighting function (16) is m_j. Consider the case $m_j \geq 1$. In order to turn the integro-differential equations (19)-(20) into a set of ordinary differential equations we need to define the subsidiary variables

$$q_{\nu_j}^{(l_j,m_j)}(t)=\int_0^t \frac{1}{l_j!}\left(\frac{m_j}{T_j}\right)^{l_j+1}(t-s)^{l_j}e^{-\frac{m_j}{T_j}(t-s)}q_{\nu_j}(s)ds \quad \text{for } l_j=0,1,2,\cdots m_j, \tag{A1}$$

where $\nu_j = j$ when firm j averages over its own past output and $\nu_j = i$ when firm j averages over rivals' output.

Note that for $l_j = m_j, m_j-1, m_j-2, \cdots, 1$ we have

$$\begin{aligned}\dot{q}_{\nu_j}^{(l_j,m_j)}(t) &= \int_0^t \frac{1}{l_j!}\left(\frac{m_j}{T_j}\right)^{l_j+1}\left[l_j(t-s)^{l_j-1}-\frac{m_j}{T_j}(t-s)^{l_j}\right]e^{-\frac{m_j}{T_j}(t-s)}q_{\nu_j}(s)ds \\ &= \frac{m_j}{T_j}\frac{1}{(l_j-1)!}\left(\frac{m_j}{T_j}\right)^{l_j}\int_0^t (t-s)^{l_j-1}e^{-\frac{m_j}{T_j}(t-s)}q_{\nu_j}(s)ds \\ &\quad -\frac{m_j}{T_j}\frac{1}{l_j!}\left(\frac{m_j}{T_j}\right)^{l_j+1}\int_0^t (t-s)^{l_j}e^{-m_j\frac{(t-s)}{T_j}}q_{\nu_j}(s)ds \\ &= \frac{m_j}{T_j}\left[q_{\nu_j}^{(l_j-1,m_j)}(t)-q_{\nu_j}^{(l_j,m_j)}(t)\right]\end{aligned} \tag{A2}$$

When $l_j = 0$ we have

$$q_{\nu_j}^{(0,m_j)}(t)=\int_0^t\left(\frac{m_j}{T_j}\right)e^{-\frac{m_j}{T_j}(t-s)}q_{\nu_j}(s)ds \tag{A3}$$

from which we readily obtain

$$\dot{q}_{\nu_j}^{(0,m_j)}(t)=\left(\frac{m_j}{T_j}\right)\left[q_{\nu_j}(t)-q_{\nu_j}^{(0,m_j)}(t)\right] \tag{A4}$$

Recapitulating we see that for firm j the quantity $q_j^{(m_j,m_j)}(t)$ required to form the term $Q_j^e(t)$ in equation (20) may be generated by the set of (m_j+1) ordinary differential equations

$$\dot{q}_{v_j}^{(l_j,m_j)}(t)=\frac{m_j}{T_j}\left[q_{v_j}^{(l_j-1,m_j)}(t)-q_{v_j}^{(l_j,m_j)}(t)\right] \quad \mathit{for}\ l_j=(m_j,m_j-1,\cdots,1),$$
$$\dot{q}_{v_j}^{(0,m_j)}(t)=\frac{m_j}{T_j}\left[q_{v_j}(t)-q_{v_j}^{(0,m_j)}(t)\right]. \tag{A5}$$

Finally consider the case $m_j=0$. Now we only need to consider

$$q_{v_j}^{(0,0)}(t)=\int_0^t \frac{1}{T_j}e^{-\frac{(t-s)}{T_j}}q_{v_j}(s)\,ds. \tag{A6}$$

Differentiating (A6) we obtain

$$\dot{q}_{v_j}^{(0,0)}(t)=\frac{1}{T_j}\left[q_{v_j}(t)-q_{v_j}^{(0,0)}(t)\right].$$

We can summarise the foregoing discussion by defining

$$q_{v_j}^{(l_j,m_j)}(t)=\int_0^t \frac{1}{l_j!}\left(\frac{r_j}{T_j}\right)^{l_j+1}(t-s)^{l_j}\,e^{-\frac{r_j}{T_j}(t-s)}q_{v_j}(s)\,ds \tag{A7}$$

where

$$r_j=\begin{cases} m_j, & m_j\geq 1\\ 1, & m_j=0\end{cases}$$

and $l_j=(m_j,m_j-1,\cdots,0)$.

The quantities $q_j^{(l_j,m_j)}(t)$ are generated by

$$\dot{q}_{v_j}^{(l_j,m_j)}(t)=\frac{r_j}{T_j}\left[q_{v_j}^{(l_j-1,m_j)}(t)-q_{v_j}^{(l_j,m_j)}(t)\right], \quad \text{for } l_j=(m_j,m_j-1,\cdots 1),$$

$$\dot{q}_{v_j}^{(0,m_j)}(t)=\frac{r_j}{T_j}\left[q_{v_j}(t)-q_{v_j}^{(0,m_j)}(t)\right].$$

11 Hotelling Type Duopoly and Oligopoly

Tönu Puu[1] *and Laura Gardini*[2]

[1] Centre for Regional Science, University of Umeå, Sweden
[2] Department of Economics, University of Urbino, Italy

1 Introduction

Hotelling's seminal contribution of 1929 was one of several successful attempts to give a precise interpretation to Bertrand's sweeping criticism of Cournot's duopoly model of 1838. Chamberlin's of 1932 was another. The common point was that if the commodity were homogenous, and the competitors were quantity adjusters, as originally assumed by Cournot, then any one competitor could, by undercutting the other competitor's price, however slightly, recover the entire market as its share. This would lead to a price war ending first when marginal costs were barely covered. Both Hotelling and Chamberlin assumed price, not quantity, to be the decision variable. Chamberlin assumed the commodity to be perceived as *heterogeneous* by the consumers, so that they would prefer one brand or dealer among similar ones, and only gradually desert their favourite when price differences grew too much adverse. Hotelling, in contrast, assumed the commodity to be perceived as *perfectly homogenous* by the consumers, but incorporated space, location, and transportation costs, which provided each competitor with a local monopoly area, with competition only at the fringes.

In this way Hotelling's variant included a location problem. Hotelling assumed demand to be *completely inelastic*. This made the location choice unstable. In order to maximise market shares, both competitors would eventually crowd in the same point. So, what Hotelling showed was that his suggested solution was no solution. To judge from passing comments, Hotelling understood how the case would work out with elastic demand: The competitors would tend to gravitate closer than placing themselves in the centres of their respective markets, but would no longer cluster in the same point as with inelastic demand.

However, Hotelling did not analyse the case of elastic demand formally, and this set the standard for following re-treatments. Hotelling's theory became a theory of location, and it was almost forgotten that it primarily aimed at being a theory of duopoly pricing.

Therefore, we think it rewarding to pin down the facts of Hotelling's model with elastic demand. Lerner and Singer, in their ingenious graphic analysis of 1937, took the first step. They assumed a given reservation price. Whenever the actual price was lower, the customers would buy a fixed quantity, quite as Hotelling assumed, whenever it was higher, they would, however, buy nothing. Later, in 1941, Smithies replaced this step function by a linear decreasing demand function, and presented an insightful verbal analysis for it, though he still considered the problem "*too complex to be treated by rigorous methods*".

However, the neat formal analysis of monopoly pricing by Beckmann 1968 and 1976, can easily be extended also to duopoly. With linear demand, profits become cubic in prices and quadratic in locations even in 1D, so inner profit maxima for both location and pricing do exist for a wide range of parameter values and can even be obtained in closed form.

The case hence evades all problems pointed out by d'Aspremont *et. al.* in their note 1979 on the original Hotelling case, where profits become quadratic in price and linear in location. Linear demand is a much more interesting case to study than the artificial case of transportation costs which increase with the square of the distance, as suggested by d'Aspremont *et. al.*

2 Local Oligopoly Conditions

Suppose a firm is located along a 1D line at point x_i. To the left there is another firm at point x_{i-1}, to the right of it another at point x_{i+1}. The firms charge "mill" prices p_{i-1}, p_i, p_{i+1}, so the consumers pay for transportation and the good becomes more and more expensive farther away from the producing firm.

Unlike supply, which is concentrated to a discrete set of locations, demand is continuously distributed. At any point x, provided the good is transported from the firm located at x_i, the density of demand is given by the local demand function:

$$q = f(z) \qquad \text{with} \qquad z = p_i + k|x - x_i| \tag{1}$$

where k denotes the (constant) transportation cost per unit distance. Now take the linear specification

$$f(z) = \alpha - \beta z \tag{2}$$

defined for $z \geq 0$, and with $f(z) = 0$ if $z \geq \alpha / \beta$. Total demand for the *i:th* firm then becomes

$$Q_i = \int_{a_i}^{b_i} f(p_i + k|x - x_i|)dx \tag{3}$$

For the linear function (2) we get (3) as the closed form integral:

$$Q_i = (\alpha - \beta p_i)(b_i - a_i) - \frac{\beta k}{2}\left((a_i - x_i)^2 + (b_i - x_i)^2\right) \tag{4}$$

Here a_i, b_i denote the boundary points for the market interval. Given the firm is located between competitors, these points are not fixed, but determined by the conditions that prices from different suppliers be equal in boundary points:

$$p_i + k|a_i - x_i| = p_{i-1} + k|a_i - x_{i-1}| \tag{5}$$

$$p_i + k|b_i - x_i| = p_{i+1} + k|b_i - x_{i+1}| \tag{6}$$

As $x_{i-1} < a_i < x_i < b_i < x_{i+1}$, the conditions (5)-(6) yield:

$$a_i = \frac{x_{i-1} + x_i}{2} + \frac{p_i - p_{i-1}}{2k} \qquad b_i = \frac{x_i + x_{i+1}}{2} + \frac{p_{i+1} - p_i}{2k} \tag{7}$$

This implies that always $b_i = a_{i+1}$. The geometry of the "price landscape" of this case is illustrated in Fig. 1.

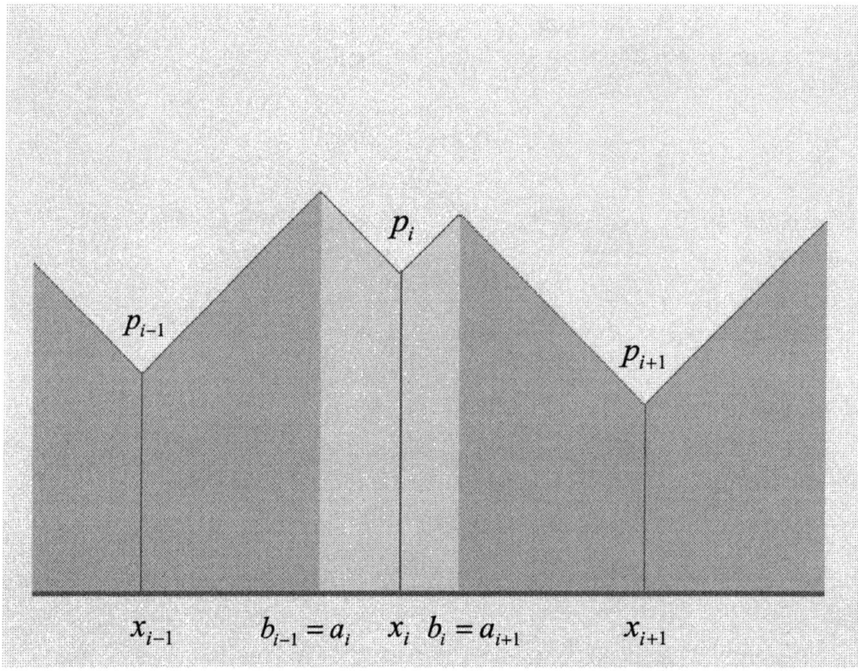

Fig. 1. Local price variation over the region around the i:th firm.

Substituting from (7) in the quantity integral (4) it now becomes:

$$Q_i = (\alpha - \beta p_i)\frac{1}{2k}\left(k\left(x_{i+1} - x_{i-1}\right) + p_{i+1} + p_{i-1} - 2p_i\right) - \frac{\beta}{8k}\left(\left(kx_{i-1} - kx_i + p_i - p_{i-1}\right)^2 + \left(kx_{i+1} - kx_i + p_{i+1} - p_i\right)^2\right) \tag{8}$$

The i:th firm will maximize its profits: $\Pi_i = p_i Q_i - C_i(Q_i)$, where we also take a linear production cost function $C_i(Q_i) = c_i Q_i$. Thus:

$$\Pi_i = (p_i - c_i)Q_i \tag{9}$$

Each firm maximizes Π_i, by choosing its price p_i *and* its location x_i.

Suppose we first want to find the optimal location. Taking $\partial\Pi_i / \partial x_i = 0$, or even $\partial Q_i / \partial x_i = 0$, as the multiplicative factor $(p_i - c_i)$ of (9) does not contain location x_i, and solving we get:

$$x_i = \frac{x_{i-1} + x_{i+1}}{2} + \frac{p_{i+1} - p_{i-1}}{2k} \tag{10}$$

This makes sense: If the competitors' prices are equal, the firm locates midway in between, whereas a higher price of the competitor left or right drags the firm in that direction. The second order condition presently becomes $\partial^2\Pi_i / \partial x_i^2 = -\frac{1}{2}\beta k(p_i - c_i) < 0$, so as price has to exceed marginal cost, i.e. $p_i > c_i$, the second order condition is negative, and the location choice indeed yields maximum profit. This is quite nice, because in the original Hotelling model, where $\beta = 0$, there were no inner solutions.

Given the firm has chosen an optimal location, we can substitute for x_i from (10) back in the expression (8) for Q_i and hence in (9) $\Pi_i = (p_i - c_i)Q_i$. In order to attain a concise formula, define a new compound variable:

$$\lambda_i = p_{i-1} + p_{i+1} + k(x_{i+1} - x_{i-1}) \tag{11}$$

We can also write $\lambda_i = p_{i-1} + k(x_i - x_{i-1}) + p_{i+1} + k(x_{i+1} - x_i)$, so λ_i denotes the *sum* of the prices of the commodity, including transportation, if transported to x_i from the left neighbour x_{i-1} *and* from the right neighbour x_{i+1}. The formula (11) is, however, more useful for stressing that λ_i only depends on x_{i-1} and x_{i+1}, but not on x_i.

With the new variable defined, the profits of the *i*:th firm become:

$$\Pi_i = \frac{\beta}{16k}(p_i - c_i)(\lambda_i - 2p_i)\left(8\frac{\alpha}{\beta} - \lambda_i - 6p_i\right) \tag{12}$$

Next, differentiating (12), a cubic in p_i, with respect to p_i and solving for the variable, we get two solutions:

$$p_i = \frac{4}{9}\frac{\alpha}{\beta} + \frac{1}{3}c_i + \frac{1}{9}\lambda_i \pm \frac{1}{18}\sqrt{36\left(\frac{\alpha}{\beta} - c_i\right)^2 - 12\left(\frac{\alpha}{\beta} - c_i\right)\left(2\frac{\alpha}{\beta} - \lambda_i\right) + 13\left(2\frac{\alpha}{\beta} - \lambda_i\right)^2} \tag{13}$$

We have the second derivatives:

$$\frac{\partial^2 \Pi_i}{\partial p_i^2} = \pm \frac{1}{4k}\sqrt{36\left(\frac{\alpha}{\beta} - c_i\right)^2 - 12\left(\frac{\alpha}{\beta} - c_i\right)\left(2\frac{\alpha}{\beta} - \lambda_i\right) + 13\left(2\frac{\alpha}{\beta} - \lambda_i\right)^2} \tag{14}$$

so the second solution is the maximum. From (13)-(14) the profit minimising price is the higher, and from (12) we find that profits increase without limit after that minimum, but never mind. Such prices do not count because they exceed the maximum admissible value α / β.

3 Disjoint Monopolies

The linear demand function has certain discontinuity problems where it cuts the axes. Above the price $p_i = \alpha / \beta$, demand drops to zero and remains zero. Of course, prices cannot be negative either. They must even be higher than the positive unit production costs $p_i > c_i$. What was said above about maximum price must hold for *price plus transportation cost*. Hence, the highest total prices for the consumers, at the boundary points a_i, b_i, must not exceed α / β, i.e.:

$$\frac{\alpha}{\beta} \geq p_i + k|a_i - x_i| \qquad \frac{\alpha}{\beta} \geq p_i + k|b_i - x_i| \tag{15}$$

Both actually boil down to the same condition once we consider how a_i, b_i and x_i were determined in the above formulas (7) and (10):

$$\frac{\alpha}{\beta} \geq \frac{1}{4}(p_{i-1} + 2p_i + p_{i+1}) - \frac{k}{4}(x_{i+1} - x_{i-1}) \tag{16}$$

If we want to avoid discontinuities due to the linear demand function, we must check that condition (16) holds all the time.

Should the condition not be satisfied, then market diameter $2R_i$ becomes less than the interval available. Demand drops to zero at $\alpha - \beta p_i - \beta k R_i = 0$, which yields

$$R_i = \frac{1}{k}\left(\frac{\alpha}{\beta} - p_i\right) \tag{17}$$

Then the quantity integral (4) becomes:

$$Q_i = \int_{x_i - R_i}^{x_i + R_i} f(p_i + k|x - x_i|)dx = 2(\alpha - \beta p_i)R_i - \beta k R_i^2 \tag{18}$$

or, with substitution for the market radius from (17),

$$Q_i = \frac{\beta}{k}\left(\frac{\alpha}{\beta} - p_i\right)^2 \tag{19}$$

Profits then become:

$$\Pi_i = \frac{\beta}{k}(p_i - c_i)\left(\frac{\alpha}{\beta} - p_i\right)^2 \tag{20}$$

Differentiating this cubic with respect to p_i, equating the derivative to zero, and solving, we get two solutions:

$$p_i = \frac{\alpha}{\beta} \qquad \text{and} \qquad p_i = \frac{1}{3}\frac{\alpha}{\beta} + \frac{2}{3}c_i \tag{21}$$

The second derivatives of (20) for the two solutions (21) are respectively

$$\frac{\partial^2 \Pi_i}{\partial p_i^2} = \pm \frac{2\beta}{k}\left(\frac{\alpha}{\beta} - c_i\right) \tag{22}$$

so, again, given marginal cost c_i does not exceed maximum price α / β, the first solution is a minimum and the second a maximum. The second solution (21) in fact is the well known solution to a spatial monopoly (mill) pricing problem in 1D with a linear demand function.

We can also easily calculate maximum monopoly profits by substituting from the second expression of (21) into (20)

$$\Pi_i = \frac{4\beta}{27k}\left(\frac{\alpha}{\beta} - c_i\right)^3 \tag{23}$$

This solution is relevant when demand drops to zero at a distance before the prices with accumulated transportation costs break even for the competitors. The result then is that the market areas of neighbouring firms no longer touch, but are isolated, possibly with intervals in between which are not served by any firm - the price would simply be too high for anybody to buy the commodity. Whether this occurs seems to be a question of how many firms crowd on a given distance, what the maximum price is, and what the marginal costs are. The present case represents a number of *non-competing monopolies*. It is first when the firms are squeezed closer together that oligopoly arises.

To see that disjoint monopolies seamlessly go over into oligopoly, take (16) as an equality, and substitute into (11). Then we obtain:

$$\lambda_i = 4\frac{\alpha}{\beta} - 2p_i \tag{24}$$

which is further substituted into equation (12) and after simplification yields (20) quite as above, obtained for the case of monopoly. Further, substituting from (24) into the solution for oligopoly price (13), rearranging, taking squares to get rid of the root sign, and factoring, we get:

$$\left(\frac{\alpha}{\beta}-p_i\right)\left(\frac{1}{3}\frac{\alpha}{\beta}+\frac{2}{3}c_i-p_i\right)=0 \tag{25}$$

which gives the monopoly solutions (21) back.

We can transform formula (17), by expressing market radius in terms of marginal cost, which is a parameter, in stead of in terms of price, which is an endogenous variable. Just substitute from the second expression (21) in (17):

$$R_i=\frac{2}{3}\frac{1}{k}\left(\frac{\alpha}{\beta}-c_i\right) \tag{26}$$

The total space $\tilde{L}$ occupied by n touching monopolies is hence:

$$\tilde{L}=\sum_{i=1}^{n}2R_i=\frac{4}{3}\frac{1}{k}\sum_{i=1}^{n}\left(\frac{\alpha}{\beta}-c_i\right) \tag{27}$$

It is clear that the total length L of the space available must be less than $\tilde{L}$ in order that oligopolistic competition should develop, i.e.

$$L<\frac{4}{3}\frac{1}{k}\sum_{i=1}^{n}\left(\frac{\alpha}{\beta}-c_i\right) \tag{28}$$

Note that a market radius can be defined not only in the case of monopoly, but also in the case of oligopoly, provided we deal with a firm having competitors to both sides. The choice rule for optimal location (10) implies that the firm always places itself in the *midpoint of its market interval* $[a_i, b_i]$. To see this, substitute from (10) into (7), and form the differences:

$$x_i-a_i=\frac{x_{i+1}-x_{i-1}}{4}+\frac{p_{i-1}-2p_i+p_{i+1}}{4k} \tag{29}$$

$$b_i-x_i=\frac{x_{i+1}-x_{i-1}}{4}+\frac{p_{i-1}-2p_i+p_{i+1}}{4k} \tag{30}$$

The expressions being equal, we can define market radius:

$$R_i = \frac{x_{i+1} - x_{i-1}}{4} + \frac{p_{i-1} - 2p_i + p_{i+1}}{4k} \tag{31}$$

The result was based on the fact that there were other firms to the right and left of each firm, so what if a_i or b_i is fixed, i.e. that the firm is the leftmost or rightmost in a fixed interval? Then things become very different, as we will see. It is no longer possible to define a market radius, because the firms do not locate in the centres of their market areas. They may even crowd in the same point, as in the original Hotelling case.

4 Space Dimensions and Boundary Conditions

The above discussion started out from the original Hotelling case, firms located on a line in 1D, though we formulated the matter more generally by considering one firm in relation to its neighbours. As we know, Hotelling considered *two* firms on a *fixed interval*, and then each firm only has a neighbour to one side, the boundary point to the other being fixed from the outset. We will consider this case at some length. It is, however, interesting to consider also different types of similar models.

First, we may note that it makes a profound difference whether we have fixed boundaries or not. We already mentioned Hotelling's conjecture about a remaining tendency to gravitate towards the midpoint even when demand is elastic. This will in fact be shown to be true in the sequel. However, consider instead three oligopolists on a circle, hence without any boundary points. Then we already know from the above discussion that each firm always locates in the midpoint between the competitors, so there is no gravitation towards the middle. The Hotelling clustering phenomena hence depend on the presence of fixed endpoints.

Second, we should take note that most interesting phenomena in geographical space occur in 2D, the linear case only being a way of making things more manageable in analysis. In 2D areas have shape in addition to size, and there hence arise many more geometrical configurations than just beads on a string.

In unbounded space the regular tessellations first come into mind, triangles, squares, and hexagons with the firms located in the midpoints. Given all mill prices are equal, things become simple enough, but, as we know from the Launhardt funnel construction, any price difference sets up fourth order curves for market boundaries, which make closed form integration for demand or even for market area almost impossible to perform.

At least this is the case when we relate transportation costs to an Euclidean distance metric. This, however, is not the only choice. Given a chessboard lattice of locations, where each firm has neighbours to four sides, we can combine it with a "Manhattan" North-South, East-West transportation cost metric of the form $|x+y|+|x-y|$, where x and y are the Euclidean space coordinates. Then the loci for constant delivered price from each dealer become squares, quite in conformity with the location pattern. Likewise we can consider a triangular lattice of locations, combined with the metric $\left|x+\sqrt{3}y\right|+\left|x-\sqrt{3}y\right|+|2x|$. This case arises when there are streets in three directions intersecting at angles of 60 degrees, quite as there are streets in two directions intersecting at angles of 90 degrees in the Manhattan case. Each firm is now surrounded by six neighbours, and the curves of constant delivered price in conformity become hexagonal. These cases seem to be the simplest candidates for treatment in 2D, though we will not attempt these at present.

As for *bounded* space, it is not even obvious which case is the natural counterpart to Hotelling's duopolists on a line segment, maybe it is three competitors in an equilateral triangle, and there seems not to be even an obvious guess about whether the crowding phenomenon carries over to 2D.

We should also note that in 2D the powers for integrals of total sales and profits are raised by one. To see this, we can just check the trivial cases of independent monopoly market areas.

Given there are no close competitors, each firm would have a circular market area of radius R_i, determined by the vanishing of demand on the boundary circle, i.e. through the condition : $\alpha-\beta(p_i+kR_i)=0$. From this we find:

$$R_i=\frac{1}{k}\left(\frac{\alpha}{\beta}-p_i\right) \tag{32}$$

Given an Euclidean transport cost metric, the total sales for the firm are

$$Q_i=\int_0^{2\pi}\int_0^{R_i}\left(\alpha-\beta(p_i+kr)\right)rdrd\theta \tag{33}$$

which simplifies to

$$Q_i = 2\pi(\alpha - \beta p_i)\frac{R_i^2}{2} + 2\pi\beta k\frac{R_i^3}{3} \tag{34}$$

or, substituting from (32), to

$$Q_i = \frac{\pi\beta}{3k^2}\left(\frac{\alpha}{\beta} - p_i\right)^3 \tag{35}$$

The corresponding profits then are:

$$\Pi_i = \frac{\pi\beta}{3k^2}(p - c_i)\left(\frac{\alpha}{\beta} - p_i\right)^3 \tag{36}$$

It is now appropriate to maximise (36) with respect to p_i, which yields the profit maximising solution:

$$p_i = \frac{1}{4}\frac{\alpha}{\beta} + \frac{3}{4}c_i \tag{37}$$

This is as well known as (21) in the 1D case, and similar, only the coefficients are different. In both cases the mill price chosen is a weighted average of maximum price and marginal production cost.

Equations (35)-(36) illustrate the raising of powers referred to. The independent monopoly case is possible only if the space available allows circular market areas to be packed together without deforming the circular boundaries. If not, we have to consider other possibilities. There are intermediate cases discussed in the literature (for instance hexagons with rounded corners), and, of course, the three regular tessellations of the plane: triangles, squares, and hexagons. No matter how interesting, we, however, now leave the 2D topics.

5 The Hotelling Case: Equilibrium

5.1 Between Monopoly and Crowding. It is now time to deal, not only with a firm in its relation to the closest competitors, but the global setup. Suppose we just have two competitors. Further, suppose $a_1 = -1$ and $b_2 = 1$ are fixed. This corresponds to the original Hotelling model. From (7):

$$a_2 = b_1 = \frac{x_1 + x_2}{2} + \frac{p_2 - p_1}{2k} \tag{38}$$

From (4) then:

$$Q_1 = (\alpha - \beta p_1)(1 + b_1) - \frac{\beta k}{2}\left((1 + x_1)^2 + (b_1 - x_1)^2\right) \tag{39}$$

$$Q_2 = (\alpha - \beta p_2)(1 - a_2) - \frac{\beta k}{2}(a_2 - x_2)^2 + \left((1 - x_2)^2\right) \tag{40}$$

The optimal locations can be found by differentiating (39)-(40) with respect to x_1, x_2 respectively, as location does not enter the multiplicative factor for profits. Thus we obtain:

$$x_1 = \frac{x_2 - 4}{5} + \frac{p_2 - 3p_1}{5k} + \frac{2\alpha}{5\beta k} \tag{41}$$

$$x_2 = \frac{x_1 + 4}{5} + \frac{3p_2 - p_1}{5k} - \frac{2\alpha}{5\beta k} \tag{42}$$

Note that (41)-(42) are different from (10), as we now have one boundary point fixed for each firm. Also note that if we solve the location choice equations (41)-(42) *as a simultaneous system*, we get $x_1 + x_2 = (p_2 - p_1)/k$, and so from (38):

$$a_2 = b_1 = \frac{p_2 - p_1}{k} \tag{43}$$

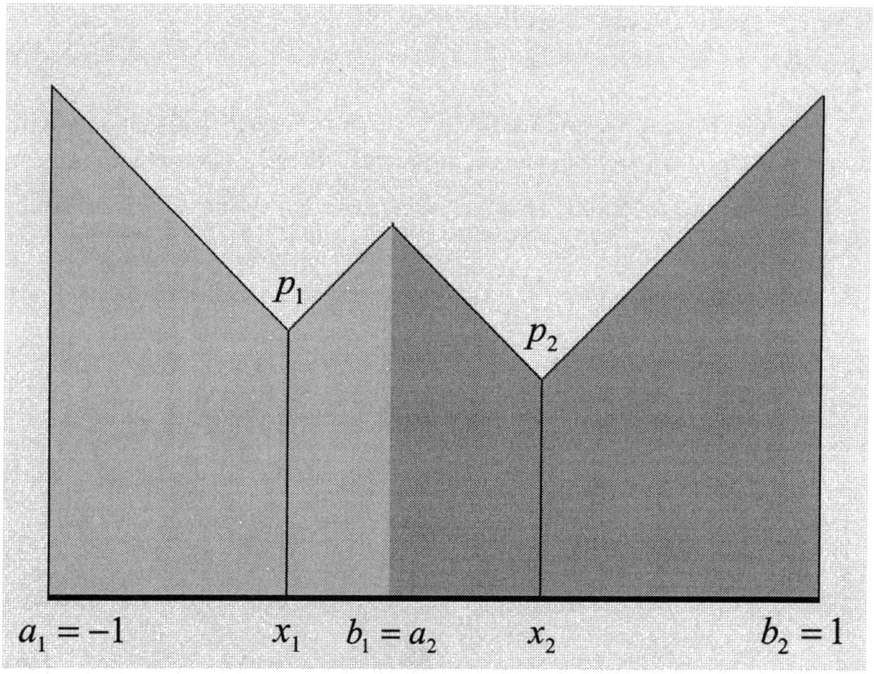

Fig. 2. Global price variation with two firms on the interval [-1, 1].

Hence *in equilibrium* the inner market boundary point only depends on the price difference divided by transportation cost. The geometry of the Hotelling case is displayed in Fig. 2.

In disequilibrium, the inner market boundary point, resulting from location choice by the first firm, taking the location of the second as given, is obtained by substituting (41) into (38):

$$b_1 = \frac{3x_2 - 2}{5} + \frac{3p_2 - 4p_1}{5k} + \frac{\alpha}{5\beta k} \tag{44}$$

Similarly, by substituting from (42) in (38), the choice of the second firm results in:

$$a_2 = \frac{3x_1 + 2}{5} + \frac{4p_2 - 3p_1}{5k} - \frac{\alpha}{5\beta k} \tag{45}$$

Expressions (41)-(42) and (44)-(45) substituted in (39)-(40), result in one single formula:

$$Q_i = \frac{\beta}{10k}\left(6\left(\frac{\alpha}{\beta}-p_i\right)^2 - 4\left(\frac{\alpha}{\beta}-p_i\right)\left(\frac{\alpha}{\beta}-\lambda_i\right) - \left(\frac{\alpha}{\beta}-\lambda_i\right)^2\right) \tag{46}$$

provided we define

$$\lambda_1 = p_2 + k(1+x_2) \quad \text{and} \quad \lambda_2 = p_1 + k(1-x_1) \tag{47}$$

Note that λ_i have the following interpretation: They are the prices accumulated by transportation costs if the commodity is transported from the firm to the right all the way to the left endpoint, or from the left firm all the way to the right endpoint.

From (46) we have profits:

$$\Pi_i = \frac{\beta}{10k}(p_i - c)\left(6\left(\frac{\alpha}{\beta}-p_i\right)^2 - 4\left(\frac{\alpha}{\beta}-p_i\right)\left(\frac{\alpha}{\beta}-\lambda_i\right) - \left(\frac{\alpha}{\beta}-\lambda_i\right)^2\right) \tag{48}$$

which looks more complex than (12). This again is due to the fixed endpoints.

Optimising (48) with respect to p_i we obtain:

$$p_i = \frac{4}{9}\frac{\alpha}{\beta} + \frac{3}{9}c_i + \frac{2}{9}\lambda_i \pm \frac{1}{18}\sqrt{36\left(\frac{\alpha}{\beta}-c_i\right)^2 - 24\left(\frac{\alpha}{\beta}-c_i\right)\left(\frac{\alpha}{\beta}-\lambda_i\right) + 34\left(\frac{\alpha}{\beta}-\lambda_i\right)^2} \tag{49}$$

Again the second derivative has the sign of the root term of (49), i.e.

$$\frac{\partial^2 \Pi_i}{\partial p_i^2} = \pm 2\sqrt{36\left(\frac{\alpha}{\beta} - c_i\right)^2 - 24\left(\frac{\alpha}{\beta} - c_i\right)\left(\frac{\alpha}{\beta} - \lambda_i\right) + 34\left(\frac{\alpha}{\beta} - \lambda_i\right)^2} \quad (50)$$

so the root with the minus sign corresponds to the maximum.

Let us check out equilibrium. Then equations (41)-(42), (47), and (49) hold as a simultaneous system. It is nonlinear and not so obvious for a closed form solution, unless the firms are identical, i.e. $c_i = c$ for i = 1, 2, so for simplicity we assume this to be the case. A good guess is that they then in equilibrium also charge the same mill prices, i.e. $p_i = p$ for i = 1,2. From (41)-(42) we obtain, substituting $p_i = p$

$$x_1 = -\frac{2}{3} + \frac{1}{3k}\left(\frac{\alpha}{\beta} - p\right) \quad \text{and} \quad x_2 = \frac{2}{3} - \frac{1}{3k}\left(\frac{\alpha}{\beta} - p\right) \quad (51)$$

As we see, $x_1 + x_2 = 0$, so *the firms locate symmetrically around the zero point*, the midpoint of the whole interval [-1, 1], though not as a rule in the midpoints, -0.5 and 0.5, of their respective market areas. From (43) we already saw that in equilibrium with equal prices the markets are separated by the zero point.

Try to substitute zero in the left hand sides of (51), and note that both firms *locate at the same point*, the Hotelling case, if

$$\frac{1}{k}\left(\frac{\alpha}{\beta} - p\right) = 2 \quad (52)$$

Also note that the firms locate exactly in the middle of their intervals $x_1 = -0.5$, $x_2 = 0.5$, *the case of disjoint monopolies*, if

$$\frac{1}{k}\left(\frac{\alpha}{\beta} - p\right) = \frac{1}{2} \quad (53)$$

Normally they tend to locate closer together. With infallible intuition Hotelling noted that with elastic demand "*the tendency ... to establish business exces-*

sively close ... will be less marked", but the competitors would "*not go as far ... as public welfare would require*", i.e. locating in the midpoints of their markets, due to the "*tempting intermediate market*". As a matter of fact, they would, but only in the extreme case of adjacent monopolies. However, monopoly pricing would not correspond to the demands of public welfare either, for a different but quite obvious reason. In Hotelling's original case it was possible that once the firms located in the same point, competition would take the form of price cutting even until the marginal cost was reached (the classical competitive solution), but then the location choice would imply loss of public welfare due to excessive transportation costs.

In order to avoid both the extreme cases, monopoly and crowding in the centre, we would require $k/2 \le \alpha/\beta - p \le 2k$, i.e. maximum price must overshoot the equilibrium price by between half and double the transportation cost rate. Once we established equilibrium price we will again be able to translate this condition in terms of production cost.

Note that Hotelling's crowding phenomenon has to do with the global setup of the model: How many competitors there are, and also with the boundary conditions. In the present case we deal with a region that is a fixed interval. But if we in stead consider three firms on a circle periphery, then, as we see from the rule (10), each firm locates halfway between its neighbours. In equilibrium the firms will be equally spaced, in the centres of their respective subintervals, and crowding would never occur.

Next, substitute (51) in (47):

$$\lambda_1 = \lambda_2 = -\frac{\alpha}{3\beta} + \frac{4}{3}p + \frac{5}{3}k \tag{54}$$

So the auxiliary variables become equal for the two identical competitors, and given also that marginal costs are equal, we get by substituting from (54) and $c_i = c$ in (49):

$$p = \frac{2}{5}\frac{\alpha}{\beta} + \frac{3}{5}c + \frac{8}{5}k - \frac{3}{10}\sqrt{4\left(\frac{\alpha}{\beta} - c\right)^2 - 8\left(\frac{\alpha}{\beta} - c\right)k + 34k^2} \tag{55}$$

Note that after substitution from (54) in (49), p is substituted under the root sign, so we have to rearrange, take squares and solve for p anew in order to get (55). Also note that we took the smaller root right away because we already know it to correspond to maximum. We can now obtain *equilibrium price p* by substituting for the three *parameters* of the model: maximum price α / β according to the demand function, the unit production cost c, and the transportation cost rate k in (55). Then, substituting the equilibrium price p, the maximum price α / β, and the transportation rate k into (51) we also obtain the equilibrium locations x_1, x_2.

Before continuing note that we can use (55) for replacing the endogenous variable p by the parameter c in equations (52)-(53). The condition for crowding and for noncompetitive monopolies thus read:

$$\frac{1}{k}\left(\frac{\alpha}{\beta}-c\right)=\frac{11}{4} \quad \text{and} \quad \frac{1}{k}\left(\frac{\alpha}{\beta}-c\right)=\frac{3}{4} \tag{56}$$

The parameter interval between monopoly and crowding is hence $0.75 < (\alpha / \beta - c) / k < 2.75$.

5.2 Price Cutting. We have disregarded one complication. As we saw, the Hotelling identical duopolists on a fixed interval may locate in the same point, if $(\alpha / \beta - c) / k = 2.75$, even when demand is elastic. They may then start a price cutting war until the level of marginal costs is reached, or they may form a collusive monopoly, if law permits.

It is notable that this competitive situation may occur even if the firms do *not* locate in the same point *but just sufficiently close*. Each firm can then, by undercutting the competitor's price *by the transportation cost over the distance between the locations* take the whole market. This, of course, is feasible only if the undercut price still exceeds marginal cost for production. But a more interesting criterion is when the maximum duopoly profits, from each firm taking half the total interval, break even with undercutting the competitor's price and taking the whole market. As we will see it is possible to establish an exact condition for this to happen.

First, calculate the profits, when each firm takes half the market:

$$\begin{aligned}\Pi_1 &= (p-c)\int_{-1}^{0}\left(\alpha-\beta\left(p+k|x-x_1|\right)\right)dx \\ &= (p-c)\left(\alpha-\beta p-\frac{\beta k}{2}\left(x_1^2+(1+x_1)^2\right)\right)\end{aligned} \tag{57}$$

$$\Pi_2 = (p-c)\int_0^1 (\alpha - \beta(p + k|x - x_2|))dx \tag{58}$$
$$= (p-c)\left(\alpha - \beta p - \frac{\beta k}{2}\left(x_2^2 + (1-x_2)^2\right)\right)$$

Note that these profits are equal because from (51) the firms locate symmetrically around the origin, i.e. $x_2 = -x_1$. Further, prices and marginal costs are equal for the firms, as indicated by dropping the indices. Substituting for price and location from (55) and (51) in (57)-(58), and using:

$$\kappa = \frac{1}{k}\left(\frac{\alpha}{\beta} - c\right) \tag{59}$$

we obtain *maximum* profits for the duopolists (equal for both firms):

$$\Pi_1 = \Pi_2 = \tag{60}$$
$$\frac{\beta k^2}{250}\left(\left(2\kappa^2 - 44\kappa + 377\right)\sqrt{4\kappa^2 - 8\kappa + 34} + 4\kappa^3 - 92\kappa^2 + 482\kappa - 2194\right)$$

Note that maximum duopoly profit, apart from the multiplicative factor βk^2, only depends on the compound parameter κ, as defined in (59). Note from (23) that, using (59), we can also put maximum *monopoly* profit in the same form: $\Pi_i = (4/27)\beta k^2 \kappa^3$. So if we want to find at which point monopoly profit breaks even with duopoly profit we would equate this expression to (60) and could find the break even point in terms of the compound parameter κ alone. Further from (56) we find that the criteria not only for monopoly, but also for clustering are in terms of this parameter alone. It so turns out that κ, the difference between maximum price and production cost divided by the transportation cost rate, is crucial for most of the topics we discuss.

Now, consider that if any of the firms undercuts the equilibrium price p, as given by (55), with the amount $k(x_2 - x_1) = -2kx_1 = 2kx_2$, which is the transportation cost between the duopoly locations of the firms, it will take the whole market as its share. However, this takes for granted that the undercutting firm does *not* relocate, as assumed by d'Aspremont *et. al.* in their discussion for the case of inelastic demand. Putting things in this way we get a *too restrictive estimate* of when price undercutting becomes profitable.

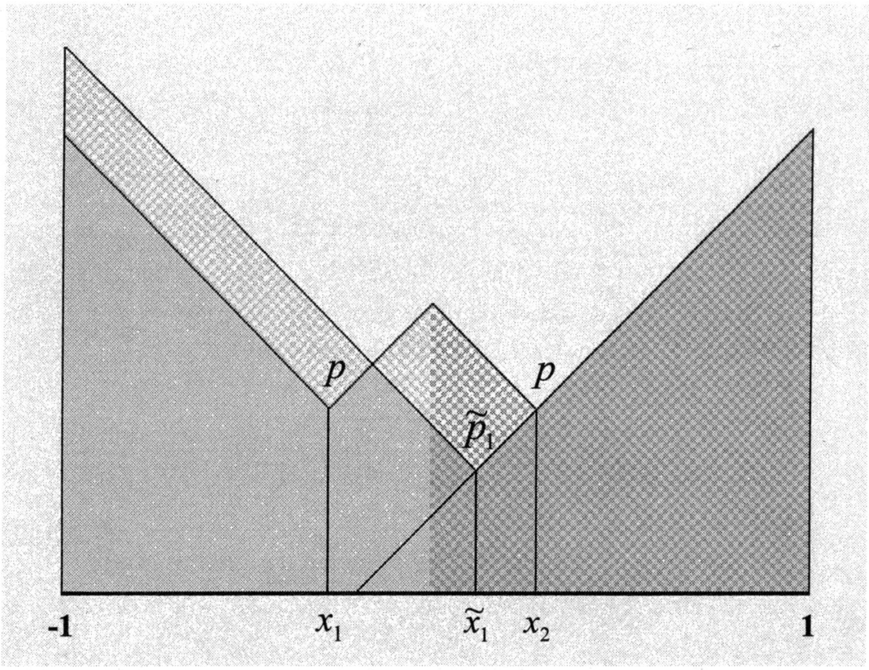

Fig. 3. Price undercutting and relocation by the first firm.

As a rule the firm trying price undercutting would *not remain where it is located.* It would relocate and choose the perfect location in combination with price, which always yields a higher profit than if it were constrained to remain where it is.

There are constraints for the choice of location and price for the firms:

$$\tilde{p}_1 = p - k(x_2 - \tilde{x}_1) \quad \text{and} \quad \tilde{p}_2 = p - k(\tilde{x}_2 - x_1) \tag{61}$$

Here *p* denotes the original duopoly price. It has no index as it is the equilibrium price for both duopolists to which the other firm is assumed to continue to adhere. So *p* is given according to (55), as before, and so are x_1, x_2, by (51). But the firms can choose $\tilde{x}_1$, $\tilde{p}_1$ and $\tilde{x}_2$, $\tilde{p}_2$ as they wish provided (61) is fulfilled. Note that the constraints result in undercutting without change of location as indicated above if we take $\tilde{x}_1 = -x_2$ or $\tilde{x}_2 = -x_1$. The undercutting situation is illustrated in Fig. 3. Note also that given the way we have drawn the picture, unchanged location x_1 is no choice at all, because the first firm could then not even charge a positive undercutting price, so there would be no profit even if production costs were zero.

Now the undercutting profits for the first firm are:

$$\Pi_1 = (\tilde{p}_1 - c)\int_{-1}^{1}\left(\alpha - \beta\left(\tilde{p}_1 + k\left|x - \tilde{x}_1\right|\right)\right)dx \tag{62}$$

$$= (\tilde{p}_1 - c)\left(2(\alpha - \beta\tilde{p}_1) - \frac{\beta k}{2}\left((1+\tilde{x}_1)^2 + (1-\tilde{x}_1)^2\right)\right)$$

and for the second firm:

$$\Pi_2 = (\tilde{p}_2 - c)\int_{-1}^{1}\left(\alpha - \beta\left(\tilde{p}_2 + k\left|x - \tilde{x}_2\right|\right)\right)dx \tag{63}$$

$$= (\tilde{p}_2 - c)\left(2(\alpha - \beta\tilde{p}_2) - \frac{\beta k}{2}\left((1-\tilde{x}_2)^2 + (1+\tilde{x}_2)^2\right)\right)$$

We now have just one free optimization variable for each firm, either price or location, the other is dependent so as to fit on the relevant linear constraint of (61). Suppose we choose location for optimising. Then we substitute for $\tilde{p}_1$ from (61) in (62) and optimise with respect to $\tilde{x}_1$. Likewise we substitute for $\tilde{p}_2$ from (61) in (63) and optimise with respect to $\tilde{x}_2$. The solutions for location become more messy than before, as we now solve quadratic equations:

$$\tilde{x}_1 = \frac{1}{3}(x_2 + 1) - \frac{1}{3}\frac{p-c}{k} - 1 + \frac{1}{3k}\sqrt{\left(c - p + k(1+x_2)\right)^2 + 6k\left(\frac{\alpha}{\beta} - p + kx_2\right)} \tag{64}$$

$$\tilde{x}_2 = \frac{1}{3}(x_1 - 1) + \frac{1}{3}\frac{p-c}{k} + 1 - \frac{1}{3k}\sqrt{\left(c - p + k(1-x_1)\right)^2 + 6k\left(\frac{\alpha}{\beta} - p - kx_1\right)} \tag{65}$$

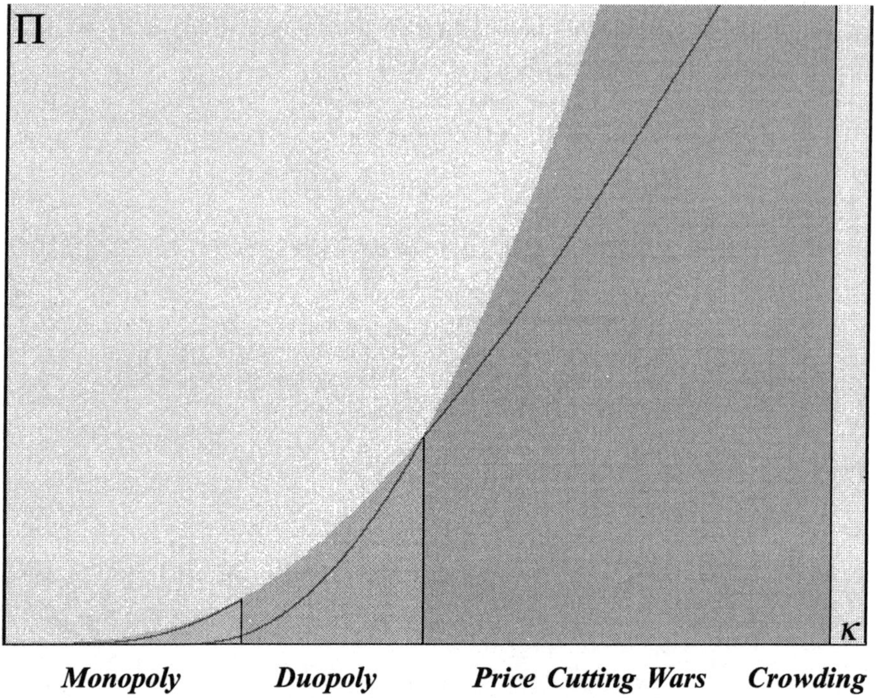

Fig. 4. Profit curves and competition regimes.

Given $x_2 = -x_1$ we hence also get $\tilde{x}_1 = -\tilde{x}_2$ which is reassuring. However, note that undercutting is a strategy under the condition that the competitor adheres to previous duopolistic behaviour. The case $x_2 = -x_1$ is a conceivable equilibrium, but $\tilde{x}_1 = -\tilde{x}_2$ is not. If both try to undercut at the same time both get disappointed. Undercutting is hence like the von Stackelberg 1938 case, rather than like the Cournot 1838 case.

But, supposing that only one firm undercuts, we can now first substitute the location according to (64)-(65) in (61) to obtain the corresponding undercutting prices, and then substitute for both in the profit expressions (62)-(63). Further, in either case, we again substitute for duopoly equilibrium price p (assumed to be retained by the other firm) from (55), and for the location of the other firm x_2 or x_1 from (51). Finally, again using the compound parameter from (59), profits become:

$$\Pi_1 = \Pi_2 = \frac{2\beta k^2}{3375}\left(13\kappa^2 + 34\kappa - 17 - 6(\kappa - 6)\sqrt{4\kappa^2 - 8\kappa + 34}\right)^{\frac{3}{2}} \tag{66}$$

$$+\frac{2\beta k^2}{3375}\left((31\kappa^2 + 28\kappa - 29)\sqrt{4\kappa^2 - 8\kappa + 34} - 63\kappa^3 + 369\kappa^2 - 819\kappa - 2187\right)$$

Note that (66) states that undercutting profits are equal for both firms, provided one firm undercuts but the other does not. As the undercutting firm takes the whole market, the profits of the other firm become zero. If both try undercutting at the same time, we have a new situation which is worth exploring in order to make explicit the dynamics of economic warfare.

It is yet interesting to establish criteria for when such warfare does *not* occur, i.e. the firms find it most profitable to adhere to duopoly behaviour. Equating (66), price cutting profit, to (60), duopoly profit, cancelling the equal factors βk^2, and rearranging to get rid of the roots, we finally obtain the nasty looking tenth order polynomial equation:

$$\begin{aligned} &3584\kappa^{10} - 216832\kappa^9 + 4696592\kappa^8 - 50448000\kappa^7 \\ &+294183456\kappa^6 - 1142701440\kappa^5 + 3361000744\kappa^4 \\ &-6631985696\kappa^3 + 9946507384\kappa^2 - 8248430736\kappa + 2165411397 = 0 \end{aligned} \tag{67}$$

It has three pairs of complex roots, and four real ones. As always, repeated squaring produces lots of nonsense roots. In fact only one real root makes sense. It is approximately equal to $\kappa \approx 1.2146$. So it seems that *we should have* $0.75 < \kappa < 1.2146$ *to ensure that true oligopolistic competition be established, without monopoly, and without price cutting competition.* The various regimes: monopoly, duopoly, and price cutting competition, are illustrated in Fig. 4 by their respective profit functions, and we see that their intersections define sharp points in which one regime becomes more profitable than another.

6 The Hotelling Case: Dynamics

6.1 Immediate Location Adjustment. We can now set up the Hotelling model as a dynamical system from (41)-(42) and (49):

$$x_1' = \frac{x_2 - 4}{5} + \frac{p_2 - 3p_1}{5k} + \frac{2\alpha}{5\beta k} \tag{68}$$

$$x_2' = \frac{x_1 + 4}{5} + \frac{3p_2 - p_1}{5k} - \frac{2\alpha}{5\beta k} \tag{69}$$

$$p_1' = \frac{4}{9}\frac{\alpha}{\beta} + \frac{3}{9}c_1 + \frac{2}{9}\lambda_1 - \frac{1}{18}\sqrt{\Delta_1} \tag{70}$$

$$p_2' = \frac{4}{9}\frac{\alpha}{\beta} + \frac{3}{9}c_2 + \frac{2}{9}\lambda_2 - \frac{1}{18}\sqrt{\Delta_2} \tag{71}$$

where

$$\lambda_1 = p_2 + k(1 + x_2) \tag{72}$$

$$\lambda_2 = p_1 + k(1 - x_1) \tag{73}$$

as stated in equation (47), and

$$\Delta_i = 36\left(\frac{\alpha}{\beta} - c_i\right)^2 - 24\left(\frac{\alpha}{\beta} - c_i\right)\left(\frac{\alpha}{\beta} - \lambda_i\right) + 34\left(\frac{\alpha}{\beta} - \lambda_i\right)^2 \tag{74}$$

for $i = 1, 2$. The dash indicates advancing the map one period, from t to $t + 1$. We are interested in the dynamic behaviour of the system (68)-(71). As we shall see the system is a contraction, so any initial condition gives rise to a trajectory converging to the attracting fixed point, which exists and is necessarily unique.

To see this, consider the Jacobian matrix of the system (68)-(71):

$$J = \frac{\partial(x_1', x_2', p_1', p_2')}{\partial(x_1, x_2, p_1, p_2)} = \begin{bmatrix} 0 & \frac{1}{5} & -\frac{3}{5k} & \frac{1}{5k} \\ \frac{1}{5} & 0 & -\frac{1}{5k} & \frac{3}{5k} \\ 0 & k\Phi_1 & 0 & \Phi_1 \\ -k\Phi_2 & 0 & \Phi_2 & 0 \end{bmatrix} \tag{75}$$

where

$$\Phi_i = \frac{\partial p_i'}{\partial x_j} = \frac{2}{9} - \frac{1}{9\sqrt{\Delta_i}}\left(6\left(\frac{\alpha}{\beta} - c_i\right) - 17\left(\frac{\alpha}{\beta} - \lambda_i\right)\right) \tag{76}$$

for $i, j = 1, 2$ and $i \neq j$. As a preliminary lemma, let us show that the values of these partial derivatives are always bounded (whatever the parameter values and the points in phase space).

Lemma. *The following inequalities hold*:

$$-\frac{1}{4} < \Phi_i < \frac{3}{4} \tag{77}$$

for i = 1, 2.

Proof: We have $\Phi_i > -1/4$ iff

$$\frac{24}{17}\left(\frac{\alpha}{\beta} - c_i\right) - 4\left(\frac{\alpha}{\beta} - \lambda_i\right) < \sqrt{\Delta_i} \tag{78}$$

The inequality is certainly true when the left hand side is negative, otherwise we solve:

$$\left(\frac{24}{17}\left(\frac{\alpha}{\beta} - c_i\right) - 4\left(\frac{\alpha}{\beta} - \lambda_i\right)\right)^2 < \Delta_i \tag{79}$$

which gives:

$$18\left(\frac{\alpha}{\beta}-\lambda_i\right)^2-12.706\left(\frac{\alpha}{\beta}-\lambda_i\right)\left(\frac{\alpha}{\beta}-c_i\right)+34.5052\left(\frac{\alpha}{\beta}-c_2\right)^2>0 \quad (80)$$

As the discriminant of the left hand side is negative, this inequality is always satisfied.

On the other hand, $\Phi_i < 3/4$ iff

$$-\frac{24}{19}\left(\frac{\alpha}{\beta}-c_i\right)+4\frac{17}{19}\left(\frac{\alpha}{\beta}-\lambda_i\right)<\sqrt{\Delta_i} \quad (81)$$

Again, this inequality is certainly true if the left hand side is negative. If not we solve for

$$\left(-\frac{24}{19}\left(\frac{\alpha}{\beta}-c_i\right)+4\frac{17}{19}\left(\frac{\alpha}{\beta}-\lambda_i\right)\right)^2<\Delta_i \quad (82)$$

which gives

$$24.19\left(\frac{\alpha}{\beta}-\lambda_i\right)^2-14.958\left(\frac{\alpha}{\beta}-\lambda_i\right)\left(\frac{\alpha}{\beta}-c_i\right)+34.4\left(\frac{\alpha}{\beta}-c_2\right)^2>0 \quad (83)$$

Again, the discriminant of the left hand side is negative, so the inequality is always satisfied. ■

We hence prove the following:

Theorem. *The map defined in* (68)-(71) *is a contraction.*

Proof: In order to show that the map is a contraction it is enough to prove that all the eigenvalues, say z, of the matrix J as stated in (75) are less than 1 in absolute value at all the points in the phase space. To this end we solve the characteristic polynomial equation:

$$\mathrm{Det}(J)=z^4+C_1z^3+C_2z^2+C_3z+C_4=0 \quad (84)$$

where from the Jacobian (75)

$$C_1 = 0 \tag{85}$$

$$C_2 = -\left(\frac{1}{5} - \Phi_1\right)\left(\frac{1}{5} - \Phi_2\right) \tag{86}$$

$$C_3 = \frac{3}{25}(\Phi_1 + \Phi_2 - 10\Phi_1\Phi_2) \tag{87}$$

$$C_4 = -\frac{9}{25}\Phi_1\Phi_2 \tag{88}$$

It is known from Fairbrother (1973) that the necessary and sufficient conditions for (84) to have all roots less than 1 in absolute value are:

$$1 - C_4 > 0 \tag{89}$$

$$3(1 + C_4) - C_2 > 0 \tag{90}$$

$$1 + C_2 + C_4 - |C_1 + C_3| > 0 \tag{91}$$

$$(1 - C_4)^2(1 + C_4 - C_2) + (C_1 - C_3)(C_3 - C_1C_4) > 0 \tag{92}$$

which upon substitution from (85)-(88) become:

$$1 + \frac{9}{25}\Phi_1\Phi_2 > 0 \tag{93}$$

$$3\left(1 - \frac{9}{25}\Phi_1\Phi_2\right) + \left(\frac{1}{5} - \Phi_1\right)\left(\frac{1}{5} - \Phi_2\right) > 0 \tag{94}$$

$$1 - \left(\frac{1}{5} - \Phi_1\right)\left(\frac{1}{5} - \Phi_2\right) - \frac{9}{25}\Phi_1\Phi_2 - \left|\frac{3}{25}(\Phi_1 + \Phi_2 - 10\Phi_1\Phi_2)\right| > 0 \tag{95}$$

$$\left(1+\frac{9}{25}\Phi_1\Phi_2\right)^2\left(1-\frac{9}{25}\Phi_1\Phi_2+\left(\frac{1}{5}-\Phi_1\right)\left(\frac{1}{5}-\Phi_2\right)\right) \tag{96}$$
$$-\left(\frac{3}{25}(\Phi_1+\Phi_2-10\Phi_1\Phi_2)\right)^2>0$$

Given the result of the above Lemma, stating that $-\frac{1}{4}<\Phi_i<\frac{3}{4}$ for $i=1,2$, it is easy to see that conditions (93)-(96) are always satisfied.

(*i*) The inferior limit value for the expression on the left in (93) is $\left(1+\frac{9}{25}\left(-\frac{1}{4}\right)\frac{3}{4}\right)$ which is positive.

(*ii*) Next, rearranging the condition stated on the left in (94) we get $\frac{1}{25}\left(76-2\Phi_1\Phi_2-5(\Phi_1+\Phi_2)\right)>\frac{1}{25}\left(76-2\left(\frac{3}{4}\right)^2-5\left(\frac{3}{2}\right)\right)>0$.

(*iii*) As for (95) we consider two cases:

(*a*) Suppose $(\Phi_1+\Phi_2-10\Phi_1\Phi_2)>0$. Rearranging expression (95), we get $\frac{2}{25}(12-2\Phi_1\Phi_2+\Phi_1+\Phi_2)=\frac{2}{25}(12+8\Phi_1\Phi_2+\Phi_1+\Phi_2-10\Phi_1\Phi_2)$ $>\frac{2}{25}(12+8\Phi_1\Phi_2)>\frac{2}{25}\left(12+8\left(-\frac{1}{4}\right)\frac{3}{4}\right)>0$.

(*b*) Next, suppose $(\Phi_1+\Phi_2-10\Phi_1\Phi_2)<0$. Then, rearranging (95), its left hand side equals $\frac{8}{25}(3-8\Phi_1\Phi_2+\Phi_1+\Phi_2)$, the inferior limit value of which is zero (obtained for $\Phi_1=\Phi_2=\frac{3}{4}$).

(*iv*) As for (96) we use the maximum value of $|\Phi_1+\Phi_2-10\Phi_1\Phi_2|$, which is 33/8 (obtained for $\Phi_1=\Phi_2=\frac{3}{4}$). Rearranging the left hand side of (96): $\frac{1}{25}\left(1+\frac{9}{25}\Phi_1\Phi_2\right)^2\left(26+16\Phi_1\Phi_2-5(\Phi_1+\Phi_2)\right)-\left(\frac{3}{25}\right)^2(\Phi_1+\Phi_2-10\Phi_1\Phi_2)^2>$ $\frac{1}{25}\left(1+\frac{9}{25}\left(-\frac{1}{4}\right)\frac{3}{4}\right)^2\left(26+16\left(-\frac{1}{4}\right)\frac{3}{4}-5\left(\frac{3}{2}\right)\right)-\left(\frac{3}{25}\right)^2\left(\frac{33}{8}\right)^2>0$. ■

6.2 Delayed or Adaptive Location Adjustment. Let us now generalize the model (68)-(71) as follows:

$$x_1' = (1-\sigma_1)x_1 + \sigma_1\left(\frac{x_2-4}{5} + \frac{p_2-3p_1}{5k} + \frac{2\alpha}{5\beta k}\right) \tag{97}$$

$$x_2' = (1-\sigma_2)x_2 + \sigma_2\left(\frac{x_1+4}{5} + \frac{3p_2-p_1}{5k} - \frac{2\alpha}{5\beta k}\right) \tag{98}$$

$$p_1' = \frac{4}{9}\frac{\alpha}{\beta} + \frac{3}{9}c_1 + \frac{2}{9}\lambda_1 - \frac{1}{18}\sqrt{\Delta_1} \tag{99}$$

$$p_2' = \frac{4}{9}\frac{\alpha}{\beta} + \frac{3}{9}c_2 + \frac{2}{9}\lambda_2 - \frac{1}{18}\sqrt{\Delta_2} \tag{100}$$

where $\sigma_i \in [0,1]$ while the λ_i and Δ_i are as given in (72)-(73) and (74) respectively.

The difference to the previous model is that we now allow for a slower location adjustment than price adjustment by assuming an adaptive process. The firms need not immediately jump to the new optimum locations but may experience a certain inertia due to the considerable cost associated with relocation. In reality, the firms would relocate only now and then, whenever the potential gains from relocation offset the substantial relocation costs, but then take full relocation steps.

In a model it makes no harm to represent this conservatism to relocation in terms of smaller steps in stead. Over a longer period the outcomes are equivalent. The coefficients σ_i can represent any speed of reaction, from total inertia when $\sigma_i = 0$, to extreme agility when $\sigma_i = 1$.

We note that the system (97)-(100) in fact becomes identical to (68)-(71) for $\sigma_i = 1$, and that the equilibrium of the system (97)-(100) is the same as for (68)-(71), which we already know to exist and to be unique. The study of its stability, however, is now more complex. In order to investigate the properties of (97)-(100), we make use of the Jacobian matrix which now is:

$$J = \frac{\partial(x_1', x_2', p_1', p_2')}{\partial(x_1, x_2, p_1, p_2)} = \begin{bmatrix} (1-\sigma_1) & \frac{\sigma_1}{5} & -\frac{3\sigma_1}{5k} & \frac{\sigma_1}{5k} \\ \frac{\sigma_2}{5} & (1-\sigma_2) & -\frac{\sigma_2}{5k} & \frac{3\sigma_2}{5k} \\ 0 & k\Phi_1 & 0 & \Phi_1 \\ -k\Phi_2 & 0 & \Phi_2 & 0 \end{bmatrix} \tag{101}$$

where the Φ_i are the derivatives defined in (76), with upper and lower bounds still as stated in the above Lemma. The characteristic polynomial of the Jacobian matrix (101) is

$$\text{Det}(J) = P(z) = z^4 + C_1 z^3 + C_2 z^2 + C_3 z + C_4 \tag{102}$$

where now

$$C_1 = -(1-\sigma_1) - (1-\sigma_2) \tag{103}$$

$$C_2 = (1-\sigma_1)(1-\sigma_2) - \left(\frac{\sigma_1}{5} - \Phi_1\right)\left(\frac{\sigma_2}{5} - \Phi_2\right) \tag{104}$$

$$\begin{aligned} C_3 = {} & \left(2 - 8\frac{\sigma_1}{5} - 8\frac{\sigma_2}{5}\right)\Phi_1\Phi_2 \\ & + \left(3\frac{\sigma_1}{5}\frac{\sigma_2}{5} - \frac{\sigma_2}{5}(1-\sigma_1)\right)\Phi_1 + \left(3\frac{\sigma_1}{5}\frac{\sigma_2}{5} - \frac{\sigma_1}{5}(1-\sigma_2)\right)\Phi_2 \end{aligned} \tag{105}$$

$$C_4 = -\left(9\frac{\sigma_1}{5}\frac{\sigma_2}{5} - 3\frac{\sigma_1}{5}(1-\sigma_2) - 3\frac{\sigma_2}{5}(1-\sigma_1) + (1-\sigma_1)(1-\sigma_2)\right)\Phi_1\Phi_2 \tag{106}$$

It is clear that in order to have a contraction map the conditions can be obtained by introducing the above coefficients in (89)-(92), but the analysis is now more complex, and even if numerical experiments suggest that the system is still always stable, we are not able to prove this rigorously, except for the limiting cases $\sigma_i = 1$, dealt with above, and $\sigma_i = 0$, dealt with below.

However, we can prove that the equilibrium is always stable in the symmetric case. Considering identical firms ($c_1 = c_2 = c$) we already stated the explicit expressions for Nash equilibrium, whence $p_1 = p_2 = p$, as given in (55), and $\Phi_1 = \Phi_2 = \Phi$. Further assume $\sigma_1 = \sigma_2 = \sigma$ for the sake of complete symmetry. Evaluated in the equilibrium point, (103)-(106) become:

$$C_1 = -2(1-\sigma) \tag{107}$$

$$C_2 = (1-\sigma)^2 - \left(\frac{\sigma}{5} - \Phi\right)^2 \tag{108}$$

$$C_3 = \left(4\frac{\sigma}{5}\Phi - 2\Phi^2\right)\left(8\frac{\sigma}{5} - 1\right) \tag{109}$$

$$C_4 = -\left(8\frac{\sigma}{5} - 1\right)^2 \Phi^2 \tag{110}$$

With these coefficients the characteristic polynomial factorizes:

$$P(z) = P_1(z)P_2(z) \tag{111}$$

where

$$P_1(z) = z^2 + \left(4\frac{\sigma}{5} + \Phi - 1\right)z + \left(8\frac{\sigma}{5} - 1\right)\Phi \tag{112}$$

$$P_2(z) = z^2 + \left(6\frac{\sigma}{5} - \Phi - 1\right)z - \left(8\frac{\sigma}{5} - 1\right)\Phi \tag{113}$$

so that it is possible to get the expressions for the four eigenvalues explicitly. To prove stability we can employ the usual necessary and sufficient conditions to have the two eigenvalues of a quadratic polynomial in modulus less than 1:

$$P_1(1) > 0, \quad P_1(-1) > 0, \quad P_1(0) < 1 \tag{114}$$

$$P_2(1) > 0, \quad P_2(-1) > 0, \quad P_2(0) < 1 \tag{115}$$

From (107)-(110) and (112)-(113) we have:

$$P_1(1) = 8\frac{\sigma}{5}\Phi + 4\frac{\sigma}{5} > 0 \quad \text{iff} \quad \Phi > -\frac{1}{2} \tag{116}$$

$$P_1(-1) = \frac{10-4\sigma}{5} - \frac{10-8\sigma}{5}\Phi > 0 \quad \text{iff} \quad \Phi < \frac{5-2\sigma}{5-4\sigma} \tag{117}$$

$$P_1(0) = \left(8\frac{\sigma}{5} - 1\right)\Phi < 1 \quad \text{iff} \quad \Phi < (>)\frac{5}{8\sigma-5} \quad \text{for} \quad \sigma > (<)\frac{5}{8} \tag{118}$$

$$P_2(1) = -8\frac{\sigma}{5}\Phi + 6\frac{\sigma}{5} > 0 \quad \text{iff} \quad \Phi < \frac{3}{4} \tag{119}$$

$$P_2(-1) = \frac{10-6\sigma}{5} + \frac{10-8\sigma}{5}\Phi > 0 \quad \text{iff} \quad \Phi > -\frac{5-3\sigma}{5-4\sigma} \tag{120}$$

$$P_2(0) = -\left(8\frac{\sigma}{5} - 1\right)\Phi < 1 \quad \text{iff} \quad \Phi > (<) -\frac{5}{8\sigma-5} \quad \text{for} \quad \sigma > (<)\frac{5}{8} \tag{121}$$

As $-\frac{1}{4} < \Phi < \frac{3}{4}$ it is easy to see that the six conditions (116)-(121) are always satisfied for any $\sigma \in (0,1]$. Thus the Nash equilibrium is locally stable, whatever the values of the parameters. Hence a bifurcation can never occur.

The case $\sigma = 0$ is degenerate. Given any choice of locations x_1, x_2 these are never changed, and there is a pure price dynamics given by the two equations:

$$p_1' = \frac{4}{9}\frac{\alpha}{\beta} + \frac{3}{9}c_1 + \frac{2}{9}\lambda_1 - \frac{1}{18}\sqrt{\Delta_1} \tag{122}$$

$$p_2' = \frac{4}{9}\frac{\alpha}{\beta} + \frac{3}{9}c_2 + \frac{2}{9}\lambda_2 - \frac{1}{18}\sqrt{\Delta_2} \tag{123}$$

This two-dimensional system is a contraction. The Jacobian matrix is:

$$J = \begin{bmatrix} 0 & \Phi_1 \\ \Phi_2 & 0 \end{bmatrix} \tag{124}$$

with $-\frac{1}{4} < \Phi_i < \frac{3}{4}$ for $i = 1, 2$. The two eigenvalues (real or complex conjugates), always less than 1 in absolute value, are $z = \pm\sqrt{\Phi_1 \Phi_2}$.

Throughout the discussion we kept the coefficients α and β from the demand function, though nowhere in the reaction functions did these parameters enter except in terms of their ratio α/β, so we could just have used one symbol for their ratio. Further, inspecting the reaction formulas it is obvious that if we redefine all value variables (prices, production costs, transportation costs) as ratios to this maximum price α/β, then nothing at all is changed. Accordingly, we could even put $\alpha/\beta = 1$ without any loss of generality. We will profit from this possibility in Section 8 below.

7 Triopoly with an Intermediate Firm

Despite the seeming stability inherent in the model of two firms on a fixed interval, the introduction of a third intermediate firm introduces some manifest instability. The intermediate firm will, as we have seen, locate between the competitors, in the centre of its market area. On the other hand the latter will still tend to gravitate towards the centre. At a certain point it may hence occur that the intermediate firm becomes so squeezed by its neighbours that it will find larger market areas by moving out to the left or right on one or the other side of the competitor there. As a rule, in equilibrium, the intermediate firm will also have to charge a lower price than the competitors, even when all three are identical in terms of equal production costs. This enhances the tendency for the intermediate firm to move out, and so introduces a locational instability in the process.

Let us now reconsider the Hotelling case, but suppose that there is an intermediate firm, i.e. three firms on a fixed interval. As we will see the case has some unexpected surprises. Again assume the interval is [-1,1]. Now there are three firms located at $x_1, x_2, x_3 \in [-1, 1]$. The demand quantities can be recovered immediately from (39)-(40) for the extremal firms, and from (4) for the intermediate one:

$$Q_1 = (\alpha - \beta p_1)(1 + b_1) - \frac{\beta k}{2}\left((1 + x_1)^2 + (b_1 - x_1)^2\right) \tag{125}$$

$$Q_2 = (\alpha - \beta p_2)(b_2 - a_2) - \frac{\beta k}{2}\left((a_2 - x_2)^2 + (b_2 - x_2)^2\right) \tag{126}$$

$$Q_3 = (\alpha - \beta p_3)(1 - a_3) - \frac{\beta k}{2}\left((a_3 - x_3)^2 + (1 - x_3)^2\right) \tag{127}$$

where we used $a_1 = -1$ and $b_3 = 1$. We can recover the two remaining market boundary points $a_2 = b_1$ and $a_3 = b_2$ directly from (7) above.

So, substitute for these, and maximise sales according to (125)-(127) with respect to the choice of location. As we recall, the multiplicative factor (price minus unit cost) in profits does not influence location. Hence, differentiating (125)-(127) and solving for location, we obtain:

$$x_1 = \frac{x_2 - 4}{5} + \frac{p_2 - 3p_1}{5k} + \frac{2\alpha}{5\beta k} \tag{128}$$

$$x_2 = \frac{x_1 + x_3}{2} + \frac{p_3 - p_1}{2k} \tag{129}$$

$$x_3 = \frac{x_2 + 4}{5} + \frac{3p_3 - p_2}{5k} - \frac{2\alpha}{5\beta k} \tag{130}$$

As we see, (128) and (130) are quite similar to (41)-(42), whereas (129) resembles (10) above.

Adding (128) and (130) and substituting from (129) we obtain the handy expression:

$$x_2 = x_1 + x_3 = \frac{p_3 - p_1}{k} \tag{131}$$

Considering equilibrium with three identical firms (equal marginal production costs), we could conjecture that, (125) and (127) being similar, as well as (128) and (130), we would have equal prices at least for the extremal firms. So, substituting $p_1 = p_3$ in (131) we find that: i) *the extremal firms locate symmetrically around the origin*; ii) *the intermediate firm locates midway between the competitors*. The conjecture about prices of the extremally located firms in fact turns out to be true, though the intermediate firm will as a rule have to charge a lower price.

The intermediate firm always locates in the middle of its market area $[a_2, b_2]$ - whether in equilibrium or not! From (7) and (129) we find that for $i = 2$, equations (29)-(30) always hold. This is, however not true for the extremal firms, which have a location bias towards the centre.

Consider for a moment the case of disjoint identical monopoly firms (i.e. having equal production costs). Then all three firms charge equal prices. Another case of price equality is if all three cluster in the centre. Substituting $p_1 = p_2 = p_3 = p$ in (128)-(130), we obtain:

$$x_1 = -\frac{4}{5} + \frac{2}{5}\frac{1}{k}\left(\frac{\alpha}{\beta} - p\right) \tag{132}$$

$$x_2 = 0 \tag{133}$$

$$x_3 = \frac{4}{5} - \frac{2}{5}\frac{1}{k}\left(\frac{\alpha}{\beta} - p\right) \tag{134}$$

First try the case of clustering by substituting zeros in all left hand sides. Then we easily obtain: $\alpha / \beta - p = 2k$ quite as in the case of two competitors, so if maximum price overshoots equilibrium price by twice the transportation cost, then all firms crowd in the centre, not only two, but even three.

Next try the case when the firms locate in the midpoints of their respective markets. The central one always is at the origin according to (133), but the outer ones would have to locate in the points $\mp 2/3$. Substituting in (132) and (134) we then have $\alpha/\beta - p = k/3$. It is easy to check that this corresponds to the case of disjoint monopolies. The denominator in the condition is different from the duopoly case, but the firms are now three in stead of two, so a shorter segment is available for each.

For the general case of duopoly equilibrium we have to consider one special problem. As the extremal firms tend to locate close it may happen that the intermediate firm finds that it will obtain a larger market interval by leaving its central position and moving left or right of their competitors. (We suggested that the extremal firms may also charge higher prices, so a larger market with a higher potential price might result in higher profit.) If so there will be set up a location instability, and we have to check the exact point at which this happens.

We now have to settle the issue of pricing. Assume the firms are identical, i.e. that all the marginal costs are equal to c. From (125)-(127) we then get output, which we multiply by $(p_i - c)$ to get profits and then substitute for locations x_i from (128)-(130). Again define auxiliary variables to simplify the formulas. Let:

$$\lambda_1 = p_2 + k(1 + x_2) \quad \lambda_2 = p_1 + p_3 + k(x_3 - x_1) \quad \lambda_3 = p_2 + k(1 - x_2) \qquad (135)$$

Note that the first and last resemble (47), whereas the middle one resembles (11). Using (135) we get profits:

$$\Pi_1 = \frac{\beta}{10k}(p_1 - c)\left(6\left(\frac{\alpha}{\beta} - p_1\right)^2 - 4\left(\frac{\alpha}{\beta} - p_1\right)\left(\frac{\alpha}{\beta} - \lambda_1\right) - \left(\frac{\alpha}{\beta} - \lambda_1\right)^2\right) \qquad (136)$$

$$\Pi_2 = \frac{\beta}{16k}(p_2 - c)(\lambda_2 - 2p_2)\left(8\frac{\alpha}{\beta} - \lambda_2 - 6p_2\right) \qquad (137)$$

$$\Pi_3 = \frac{\beta}{10k}(p_3 - c)\left(6\left(\frac{\alpha}{\beta} - p_3\right)^2 - 4\left(\frac{\alpha}{\beta} - p_3\right)\left(\frac{\alpha}{\beta} - \lambda_3\right) - \left(\frac{\alpha}{\beta} - \lambda_3\right)^2\right) \qquad (138)$$

Note that (136) and (138) look similar, and indeed become identical in equilibrium when $x_2 = 0$ according to (131), given equal prices for the extremal firms and hence $\lambda_1 = \lambda_3$ from (135). Note also the similarity of (136) and (138) to equation (48), as well as of (137) to (12).

Next differentiate with respect to p_i, equate to zero, and solve. In order to make the formulas more concise let us define the two expressions:

$$A_i = 36\left(\frac{\alpha}{\beta} - c\right)^2 - 24\left(\frac{\alpha}{\beta} - c\right)\left(\frac{\alpha}{\beta} - \lambda_i\right) + 34\left(\frac{\alpha}{\beta} - \lambda_i\right)^2 \tag{139}$$

$$B_i = 36\left(\frac{\alpha}{\beta} - c\right)^2 - 12\left(\frac{\alpha}{\beta} - c\right)\left(2\frac{\alpha}{\beta} - \lambda_i\right) + 13\left(2\frac{\alpha}{\beta} - \lambda_i\right)^2 \tag{140}$$

Using (139)-(140) we can write:

$$p_1 = \frac{4}{9}\frac{\alpha}{\beta} + \frac{3}{9}c + \frac{2}{9}\lambda_1 - \frac{1}{18}\sqrt{A_1} \tag{141}$$

$$p_2 = \frac{4}{9}\frac{\alpha}{\beta} + \frac{3}{9}c + \frac{1}{9}\lambda_2 - \frac{1}{18}\sqrt{B_2} \tag{142}$$

$$p_3 = \frac{4}{9}\frac{\alpha}{\beta} + \frac{3}{9}c + \frac{2}{9}\lambda_3 - \frac{1}{18}\sqrt{A_3} \tag{143}$$

Again we note the similarity of (141) and (143) to (49), and of (142) to (13). In equilibrium with identical firms we have $p_1 = p_3$ but, as a rule, $p_2 < p_1 = p_3$, except in the obvious case of disjoint monopolies.

We can now easily state the maximum profits of the various firms when a profit maximising price policy is chosen according to (141)-(143). Substituting from (141)-(143) in (136)-(138) we obtain:

$$\Pi_1 = \frac{\beta}{2430k}\left(\frac{1}{2}(A_1)^{\frac{3}{2}} + \left(2\frac{\alpha}{\beta} - 3c + \lambda_1\right)\left(A_1 - 120\left(\frac{\alpha}{\beta} - \lambda_1\right)^2\right)\right) \tag{144}$$

$$\Pi_2 = \frac{\beta}{3888k}\left((B_2)^{\frac{3}{2}} - \left(4\frac{\alpha}{\beta} - 6c + \lambda_2\right)\left(16\frac{\alpha}{\beta} - 6c - 5\lambda_2\right)\left(8\frac{\alpha}{\beta} + 6c - 7\lambda_2\right)\right) \tag{145}$$

$$\Pi_3 = \frac{\beta}{2430k}\left(\frac{1}{2}(A_3)^{\frac{3}{2}} + \left(2\frac{\alpha}{\beta} - 3c + \lambda_3\right)\left(A_3 - 120\left(\frac{\alpha}{\beta} - \lambda_3\right)^2\right)\right) \tag{146}$$

As we consider identical firms in equilibrium (135) apply for the auxiliary variables, further with $p_1 = p_3$.

Consider now the case when the middle firm moves out from its central position either right or left of one of the extremal firms. Suppose it moves right, to the position $\vec{x}_2$. Note that all the symbols associated with a right move are identified by the arrow pointing right. The market boundary points then become $\vec{a}_2$, to be obtained from (7), and $\vec{b}_2 = 1$, as prescribed by the fixed interval. From (7) we immediately get:

$$\vec{a}_2 = \frac{x_3 + \vec{x}_2}{2} + \frac{\vec{p}_2 - p_3}{2k} \tag{147}$$

where $\vec{p}_2$ denotes the yet undetermined price of the previously middle firm in its new rightmost position. Similarly, for a left move to $\overleftarrow{x}_2$ (indicated by arrows pointing left) we get $\overleftarrow{a}_2 = -1$ in the new position and $\overleftarrow{b}_2$, likewise obtained from (7):

$$\overleftarrow{b}_2 = \frac{x_1 + \overleftarrow{x}_2}{2} + \frac{p_1 - \overleftarrow{p}_2}{2k} \tag{148}$$

Next, from (4) we get demand for each case:

$$\vec{Q}_2 = (\alpha - \beta\,\vec{p}_2)(1-\vec{a}_2) - \frac{\beta k}{2}\left((\vec{a}_2 - \vec{x}_2)^2 + (1-\vec{x}_2)^2\right) \tag{149}$$

$$\overleftarrow{Q}_2 = (\alpha - \beta\,\overleftarrow{p}_2)(1+\overleftarrow{b}_2) - \frac{\beta k}{2}\left((1+\overleftarrow{x}_2)^2 + (\overleftarrow{b}_2 - \overleftarrow{x}_2)^2\right) \tag{150}$$

We can substitute for $\vec{a}_2$ from (147) in (149) and then as usual differentiate with respect to $\vec{x}_2$, put equal to zero, and solve for the optimum location in the new interval. In the same way substitute for $\overleftarrow{b}_2$ from (148) in (150) and do the same with respect to $\overleftarrow{x}_2$. We get:

$$\vec{x}_2 = \frac{x_3+4}{5} + \frac{3\vec{p}_2 - p_3}{5k} - \frac{2\alpha}{5\beta k} \tag{151}$$

$$\overleftarrow{x}_2 = \frac{x_1-4}{5} - \frac{3\overleftarrow{p}_2 - p_1}{5k} + \frac{2\alpha}{5\beta k} \tag{152}$$

where $\partial^2\vec{Q}_2 / \partial\vec{x}_2{}^2 = \partial^2\overleftarrow{Q}_2 / \partial\overleftarrow{x}_2{}^2 = -5\beta k / 4 < 0$ guarantee the fulfilment of the second order conditions for profit maxima in both cases.

Substituting for $\vec{x}_2$ from (151) in (149) and for $\overleftarrow{x}_2$ from (152) in (150) we obtain total sales in the new optimal locations:

$$\vec{Q}_2 = \frac{\beta}{10k}\left(6\left(\frac{\alpha}{\beta} - \vec{p}_2\right)^2 - 4\left(\frac{\alpha}{\beta} - \vec{p}_2\right)\left(\frac{\alpha}{\beta} - \vec{\lambda}_2\right) - \left(\frac{\alpha}{\beta} - \vec{\lambda}_2\right)^2\right) \tag{153}$$

$$\overleftarrow{Q}_2 = \frac{\beta}{10k}\left(6\left(\frac{\alpha}{\beta} - \overleftarrow{p}_2\right)^2 - 4\left(\frac{\alpha}{\beta} - \overleftarrow{p}_2\right)\left(\frac{\alpha}{\beta} - \overleftarrow{\lambda}_2\right) - \left(\frac{\alpha}{\beta} - \overleftarrow{\lambda}_2\right)^2\right) \tag{154}$$

respectively, where we have defined new auxiliary variables

$$\vec{\lambda}_2 = p_3 + k(1-x_3) \qquad \overleftarrow{\lambda}_2 = p_1 + k(1+x_1) \tag{155}$$

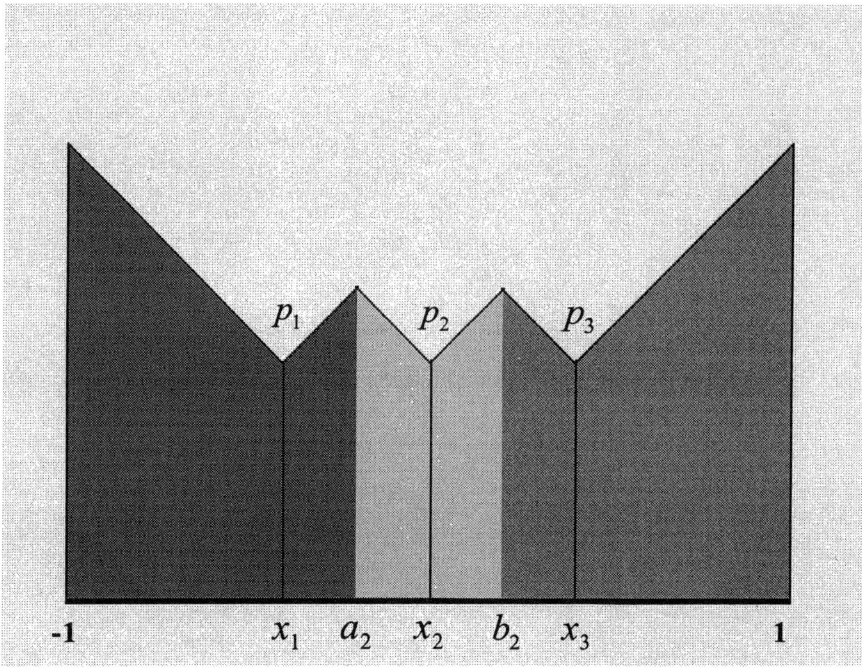

Fig. 5. Price landscape for three competitors on a line segment.

different from the middle formula (135) as the firm is now in an extremal position.

From (153)-(154) we get profits:

$$\vec{\Pi}_2 = \frac{\beta}{10k}(\vec{p}_2 - c)\left(6\left(\frac{\alpha}{\beta} - \vec{p}_2\right)^2 - 4\left(\frac{\alpha}{\beta} - \vec{p}_2\right)\left(\frac{\alpha}{\beta} - \vec{\lambda}_2\right) - \left(\frac{\alpha}{\beta} - \vec{\lambda}_2\right)^2\right) \quad (156)$$

$$\overleftarrow{\Pi}_2 = \frac{\beta}{10k}(\overleftarrow{p}_2 - c)\left(6\left(\frac{\alpha}{\beta} - \overleftarrow{p}_2\right)^2 - 4\left(\frac{\alpha}{\beta} - \overleftarrow{p}_2\right)\left(\frac{\alpha}{\beta} - \overleftarrow{\lambda}_2\right) - \left(\frac{\alpha}{\beta} - \overleftarrow{\lambda}_2\right)^2\right) \quad (157)$$

which we maximise with respect to $\vec{p}_2$ or $\overleftarrow{p}_2$, depending on the direction of jump, always using the corresponding $\vec{\lambda}_2$ or $\overleftarrow{\lambda}_2$ from (155), and solve for the variable. In order to obtain a concise expression we again define some abbreviations:

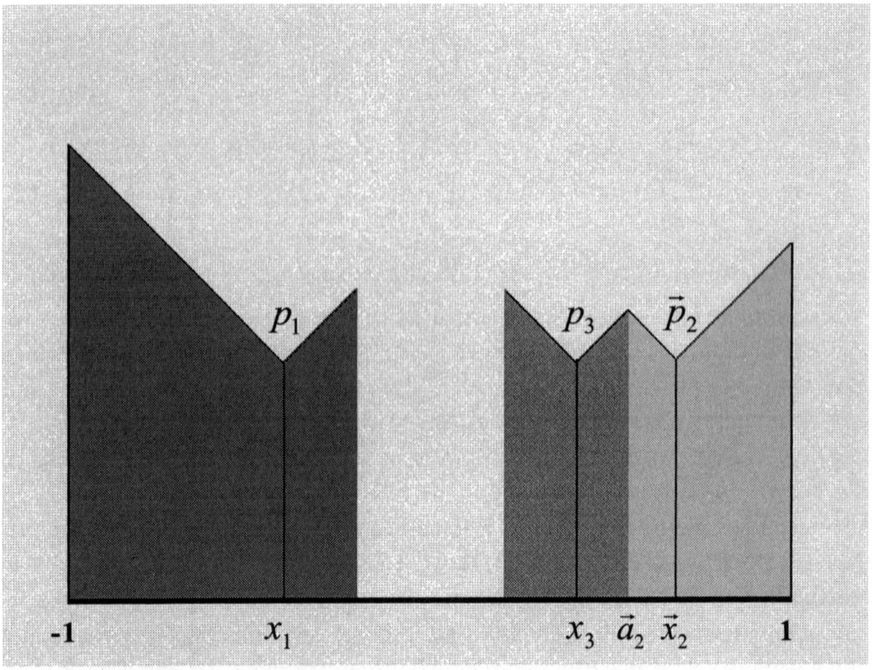

Fig. 6. Price landscape with the squeezed middle firm jumping right.

$$\vec{A}_2 = 36\left(\frac{\alpha}{\beta}-c\right)^2 - 24\left(\frac{\alpha}{\beta}-c\right)\left(\frac{\alpha}{\beta}-\vec{\lambda}_2\right) + 34\left(\frac{\alpha}{\beta}-\vec{\lambda}_2\right)^2 \tag{158}$$

$$\overleftarrow{A}_2 = 36\left(\frac{\alpha}{\beta}-c\right)^2 - 24\left(\frac{\alpha}{\beta}-c\right)\left(\frac{\alpha}{\beta}-\overleftarrow{\lambda}_2\right) + 34\left(\frac{\alpha}{\beta}-\overleftarrow{\lambda}_2\right)^2 \tag{159}$$

Using this we obtain relocation prices:

$$\vec{p}_2 = \frac{4}{9}\frac{\alpha}{\beta} + \frac{3}{9}c + \frac{2}{9}\vec{\lambda}_2 - \frac{1}{18}\sqrt{\vec{A}_2} \qquad \overleftarrow{p}_2 = \frac{4}{9}\frac{\alpha}{\beta} + \frac{3}{9}c + \frac{2}{9}\overleftarrow{\lambda}_2 - \frac{1}{18}\sqrt{\overleftarrow{A}_2} \tag{160}$$

Finally, substitute from (160) in (156)-(157) to obtain maximum profits in this new location:

$$\vec{\Pi}_2 = \frac{\beta}{2430k}\left(\frac{1}{2}\left(\vec{A}_2\right)^{\frac{3}{2}} + \left(2\frac{\alpha}{\beta} - 3c + \vec{\lambda}_2\right)\left(\vec{A}_2 - 120\left(\frac{\alpha}{\beta} - \vec{\lambda}_2\right)^2\right)\right) \tag{161}$$

$$\overleftarrow{\Pi}_2 = \frac{\beta}{2430k}\left(\frac{1}{2}\left(\overleftarrow{A}_2\right)^{\frac{3}{2}} + \left(2\frac{\alpha}{\beta} - 3c + \overleftarrow{\lambda}_2\right)\left(\overleftarrow{A}_1 - 120\left(\frac{\alpha}{\beta} - \overleftarrow{\lambda}_2\right)^2\right)\right) \tag{162}$$

These expressions (161)-(162) for profits in case of moving out, are what we have to compare to (145) in order to establish if jumping out is profitable or not. Such comparisons do not result in manageable closed form criteria, but are better carried out by the computer for numerical cases.

The geometry of the original middle location and a jump is illustrated in Figs. 5-6.

Even though *only the middle firm* may have reason to jump *from an equilibrium position* due to the fact that it is squeezed by the outer firms, it is perfectly possible that, in any disequilibrium state in the dynamic process, an outer firm may in stead consider jumping in between the competitors. So, if we want to set up an explicit dynamical process we have to consider jumping possibilities for all three firms, even the extremal ones which may want to jump in into the middle or even to the other end. Only computer programs are conceivable for checking the dynamics of such processes. Preliminary simulations indicate the possibility of never ending reshuffling of sequences of firms.

8 Different Forms of the Demand Function

If one considers the absence of complex dynamics a problem, a linear demand function may seem to be one, at least in 1D models, because, as we have seen in Section 6 above, the derivatives of the reaction functions are too low for the Nash equilibrium to be ever destabilised.

However, linear demand is not the only choice. As mentioned, Lerner and Singer first tried a step function, before Smithies tried the linear. We might want to check the whole class of demand functions $p^\rho + q^\rho = 1$, with the parameter ρ between unity (the linear), and infinity (the step function). This function is of the Minkowski metric type, or as economists might prefer to say, CES type, even if the curvature would be in the wrong direction compared to normal isoquants. Note that for the linear case we now put $\alpha / \beta = 1$, as we noted that we may without loss of generality. The problem with such a family is that we have to integrate expressions such as (3) over space, and easily end up with expressions of a monstrous complexity right at the outset.

However, we can easily find these things out with a broken line demand function in the unit square with a kink point along the positive diagonal:

$$q = \begin{cases} 1 - \dfrac{1-\mu}{\mu}\left(p_i + k\left|x - x_i\right|\right) & p_i + k\left|x - x_i\right| \le \mu \\ \mu - \dfrac{\mu}{1-\mu}\left(p_i + k\left|x - x_i\right| - \mu\right) & p_i + k\left|x - x_i\right| > \mu \end{cases} \tag{163}$$

If $\mu = 1/2$ we have the linear case back, if $\mu \to 1$, we approach the Lerner and Singer case.

From (163) we see that the kink point is located at $p_i + k\left|x - x_i\right| = \mu$. For the integration of total demand we have to locate the points in space where this kink occurs, i.e. $p_i + k\left(x_i - \xi_i\right) = p_i + k\left(\eta_i - x_i\right) = \mu$. Solving we get:

$$\xi_i = x_i - \frac{p_i - \mu}{k} \quad \text{and} \quad \eta_i = x_i + \frac{p_i - \mu}{k} \tag{164}$$

so we can easily calculate the total demand for the general firm:

$$\begin{aligned} Q_i = & \int_{a_i}^{\xi_i} \left(1 - \frac{1-\mu}{\mu}\left(p_i + k\left|x - x_i\right|\right)\right) dx \\ & + \int_{\xi_i}^{\eta_i} \left(\mu - \frac{\mu}{1-\mu}\left(p_i + k\left|x - x_i\right| - \mu\right)\right) dx \\ & + \int_{\eta_i}^{b_i} \left(1 - \frac{1-\mu}{\mu}\left(p_i + k\left|x - x_i\right|\right)\right) dx \end{aligned} \tag{165}$$

Quite as above, a_i, b_i denote the boundary points for the market interval, whereas ξ_i, η_i are defined in (164) above. Breaking up the integral in three terms makes integration easy. We get the closed form expression:

$$Q_i = \frac{\mu}{1-\mu}\left((1-p_i)(b_i - a_i) - \frac{k}{2}\left((a_i - x_i)^2 + (b_i - x_i)^2\right)\right) - \frac{\mu(2\mu-1)(\mu-p_i)^2}{k(1-\mu)\mu^2} \tag{166}$$

This is quite similar to (4), only note that we have simplified by putting α and β equal to unity. Further, recall that the linear demand function case we dealt with has $\mu = 1/2$, so then the coefficient of the first term becomes unitary whereas the second term vanishes altogether. This is indeed reassuring. We note that the effect of the kink in the demand function is to scale up the first term and subtract an additional term.

We can now easily obtain all the relevant closed form expressions for the Hotelling duopolists on a line. Just recall that we have $a_1 = -1$, $b_2 = 1$, and quite as in (10) $a_2 = b_1 = (x_1 + x_2)/2 + (p_2 - p_1)/2k$. Optimising (166) with respect to location we note that *i*) location does not enter the additional term and *ii*) the multiplicative factor makes no difference for the optimum location condition. Accordingly, (41)-(42) still hold, as restated here:

$$x_1 = \frac{x_2 - 4}{5} + \frac{p_2 - 3p_1 + 2}{5k} \qquad x_2 = \frac{x_1 + 4}{5} + \frac{3p_2 - p_1 - 2}{5k} \tag{167}$$

Again there is a slight simplification due to our choice of α and β equal to unity.

The rest of the procedure follows as above. Define the auxiliary variables:

$$\lambda_1 = p_2 + k(1 + x_2) \qquad \lambda_2 = p_1 + k(1 - x_1) \tag{168}$$

and we can write the total demand for the Hotelling duopolists:

$$Q_i = \frac{\mu}{1-\mu}\frac{1}{10k}\left(6(1-p_i)^2 - 4(1-p_i)(1-\lambda_i) - (1-\lambda_i)^2\right) - \frac{\mu(2\mu-1)(\mu-p_i)^2}{k(1-\mu)\mu^2} \tag{169}$$

Profits are next obtained by multiplying the quantity expressions through by the factor $(p_i - c_i)$, which still makes profits cubic in prices. Optimization with respect to prices is straightforward, though the occurrence of the kink makes the expressions slightly more complicated.

The important thing, however, is that the derivatives of the reaction functions still seem to retain the property of being bounded in a range which makes the Nash equilibrium stable.

By conclusion we should stress that we left many loose ends worth further study, such as phenomena in 2D, processes where a third competitor in a fixed interval may feel squeezed and set up a series of location order changes, or phenomena in unbounded space, such as three firms on a circle in 1D, or a sphere in 2D.

9 Selected References

D'Aspremont, C., Gabszewicz, J. J., and Thisse, J. F., 1979, "On Hotelling's 'stability in competition'", *Econometrica* 47:1145-1150

Beckmann, M. J., 1968, *Location Theory* (Random House, New York)

Beckmann, M. J., 1976, "Spatial Price Policies Revisited", *Bell Journal of Economics* 7:619-630

Bertrand, J., 1883, "Théorie Mathématique de la Richesse Sociale", *Journal des Savants* 48:499-508

Chamberlin, E. H., 1932, *The Theory of Monopolistic Competition* (Harvard University Press)

Cournot, A., 1838, *Récherches sur les Principes Mathématiques de la Théorie des Richesses* (Hachette, Paris)

Edgeworth, F. Y., 1897, "La Teoria Pura del Monopolio", *Giornale degli Economisti* 15:13-31

Fairbrother, R. W., 1973, "Simplified Samuelson conditions for cubic and quartic equations", *The Manchester School of Economic and Social Studies* 41:396-400

Greenhut, M. L. and Ohta, H., 1972, "Monopoly Output under Alternative Spatial Pricing Techniques", *American Economic Review* 62:705-713

Hotelling, H., 1929, "Stability in Competition", *Economic Journal* 41: 41-57

Lerner, A. P., and Singer, H. W., 1937, "Some notes on duopoly and spatial competition, *Journal of Political Economy* 45:145-186

Palander, T. F., 1935, *Beiträge zur Standortstheorie* (Almqvist & Wiksell, Uppsala)

Ponsard, C., 1954, *Économie et Espace - essai d'integration du facteur spatial dans l'analyse économique* (Sedés, Paris)

Ponsard, C., (Ed.), 1988, *Analyse économique spatiale* (Presses Universitaires de France)

Puu, T., 2001, "Bertrand oligopoly revisited", *Discrete Dynamics in Nature and Society* 6:1-10

Puu, T., 2001, "Hotelling's 'ice cream dealers' with elastic demand", *Annals of Regional Science*

von Stackelberg, H., 1938, Probleme der unvollkommenen Konkurrenz, *Weltwirtschaftliches Archiv* 48:95-114

Smithies, A., 1941, "Monopolistic Price Policy in a Spatial Market", *Econometrica* 9:63-73

Smithies, A., 1941, "Optimum Location in Spatial Competition", *Journal of Political Economy* 49:423-439

Contributors

Professor Anna Agliari
Istituto di Econometria e Matematica
per le Applicazioni Ec. Fin. e Att.
Università Cattolica del Sacro Cuore
Largo Gemelli, 1,
20123 Milano
ITALY
e-mail: agliari@unipr.it

Professor Gian-Italo Bischi
Università di Urbino,
Istituto di Scienze Economiche
via Saffi, 42,
I-61029 Urbino
ITALY
*e-mail:*bischi@econ.uniurb.it

Professor Carl Chiarella
School of Finance and Economics
University of Technology, Sydney
P.O. Box 123,
Broadway 2007, NSW
AUSTRALIA
e-mail: carl.chiarella@uts.edu.au

Professor Laura Gardini
Università di Urbino,
Istituto di Scienze Economiche
via Saffi, 42,
I-61029 Urbino
ITALY
e-mail: gardini@econ.uniurb.it

Professor Michael Kopel
Department of Managerial Economics and Industrial Organization
University of Technology
Theresianumgasse 27,
A-1040 Vienna
AUSTRIA
e-mail: kopel@ebwnov.tuwien.ac.at

Professor Fabio Lamantia
Department OAAP, University of Calabria.
Via Pietro Bucci Cubi 3B -3 C
I- 87036 Arcavacata di Rende - Cosenza
ITALY
e-mail: lamantia@mbox.unict.it

Professor Tönu Puu
Centre for Regional Science
Umeå University
SE-90187 Umeå
SWEDEN
e-mail: tonu.puu@econ.umu.se

Professor Barkley Rosser
Program in Economics, MSC 0204
James Madison University
Harrisonburg, VA 22907
USA
e-mail: rosserjb@jmu.edu

Dr. Irina Sushko
Institute of Mathematics
National Academy of Sciences of Ukraine
3 Tereshchenkivska st ,
01601 Kiev-4
UKRAINE
e-mail: sushko@imath.kiev.ua

Professor Ferenc Szidarovszky
Systems and Industrial Engineering Department
University of Arizona
PO Box 210020, Tucson
AZ 85721-1120
USA
e-mail: szidar@sie.arizona.edu

Dr. Shahriar Yousefi
Department of Statistics and Demography
Odense University
Campusvej 55,
5230 Odense M
DENMARK
e-mail: syo@sam.sdu.dk

Dr. Peiyuan Zhu
School of Finance and Economics
University of Technology, Sydney
P.O. Box 123,
Broadway 2007, NSW
AUSTRALIA
e-mail: peiyuan.zhu@uts.edu.au

Druck: Strauss Offsetdruck, Mörlenbach
Verarbeitung: Schäffer, Grünstadt